PHYSICS FOR THE HEALTH PROFESSIONS

PHYSICS FOR THE HEALTH PROFESSIONS

THIRD EDITION

J. Trygve Jensen, Ed.D.

Professor of Chemistry
Director, Nuclear Medicine Technology Program
Wagner College
Staten Island, New York
Formerly Lecturer in Physics for Nurses
Teachers College
Columbia University
New York, New York

A WILEY MEDICAL PUBLICATION
JOHN WILEY & SONS
New York • Chichester • Brisbane • Toronto • Singapore

Cover design by Wanda Lubelska

The first two editions of this book were published by J. B. Lippincott, Publishers, 1960, 1976.

LC 82-70068
ISBN 0-471-08696-7

Printed in the United States of America

10 9 8 7 6 5 4 3 2 1

PREFACE TO THE THIRD EDITION

We mentioned in the Preface to the Second Edition that there has been a tremendous growth in biomedical knowledge, combined with increasing reliance on physical methods of diagnosis. That was certainly true in 1976; but in the years since that time, the growth in instrumentation has been far greater. The explosive growth in nuclear medicine procedures, using scanning and Anger cameras; the use of computers in processing and storing data; the rapid growth of radioimmunoassay procedures; the development of various types of computed axial tomography and ultrasonography; all have revolutionized medical diagnoses.

The chart opposite page 1 of the text shows some applications of physics to health care in the operating room, the patient's room, the clinical laboratory, the intensive care unit, the radiologic department, and special areas. It demonstrates how widely principles of physics are involved in the hospital.

Additional quantitative material includes two types of problems—both worked out, step by step in text, and in the form of questions at the end of each chapter. The problems make use of the International System of Units (SI), as well as the more familiar British system.

As in the previous edition, topics are presented fully with an aim to make the material as easy to understand as possible by "talking out" the explanations. It is hoped that the large number of medical applications discussed will be helpful to students and practitioners in both nursing and allied health, as well as others in the medical field.

J. T. J.

v

PREFACE TO THE SECOND EDITION

Practitioners in the health profession daily use physical principles in their work. To function effectively, they need to understand the basis of those principles. The tremendous growth of biomedical knowledge, combined with increased reliance on physical methods of treatment, places greater demands on all members of the health care team than might have been imagined only a few short years ago.

The incredibly sensitive and versatile electronic sensors now available have made it possible to measure changes in pressure, temperature, and volume not only on the body surface of the patient, but internally as well. Hospitals are equipped with numerous sophisticated instruments that see routine use. In the United States alone, more than 1,200 companies design and manufacture the instrumentation required for patient care, in research, and by the clinical laboratory. In all of these activities, both Newtonian and modern physical principles are called into play. *Physics for the Health Professions* is designed to show how some of these principles operate, by demonstrating their immediate relevance to various procedures. It is hoped that the practitioner will gain a clear understanding of the "how" and the "why" of the data obtained and will thus be prepared to apply such principles to new or unfamiliar situations.

The hallmark of professional education is an appreciation of fundamental principles. As Dunlap Smith, former Provost of Carnegie Institute of Technology, has put it, "He had learned something still more valuable—to have confidence in his capacity to think things out for himself, and to know the joy of doing so."

J. T. J.

ACKNOWLEDGMENTS

John Donne's famous line, "No man is an island unto himself," is true for us all. I am deeply aware of all the help and support I have received through the years from so many colleagues, family, and friends. To my many students, both at Columbia University and Wagner College, goes my appreciation for their inquisitive exploration of much of the material covered in this book and for the large part they have played in educating the author.

I appreciate very much the professional efforts of the staff of John Wiley & Sons. I wish especially to thank Andrea Stingelin, Health Sciences Editor, for her enthusiastic guidance and encouragement based on her extensive background in this area, Janet Walsh for her careful, patient editing and positive support, and David Feren, my Production Editor, for his excellent management of this book through production. I am also indebted to David T. Miller, Managing Editor of the Nursing Department of the J. B. Lippincott Company, for his help in supplying illustrations. I would also like to thank Orlando Manfredi, M.D., Director of Radiology and Nuclear Medicine, and John Schaumburg and Maria Boccardo, Chief Nuclear Medicine Technologists at St. Vincent's Medical Center of Richmond, N.Y., for supplying photographs of various nuclear and ultrasonic scans.

I am also indebted to many at Wagner College, including Eleanore Sweatman and Rocco Delassandro, for their willing help, and Dr. John Satterfield, former President of the College, for encouraging publications by the faculty.

Finally, but most important of all, I am thankful for the support and encouragement of my wife, Marie, and our children, James and Linda, during the writing of the various editions of the book.

J. T. J.

CONTENTS

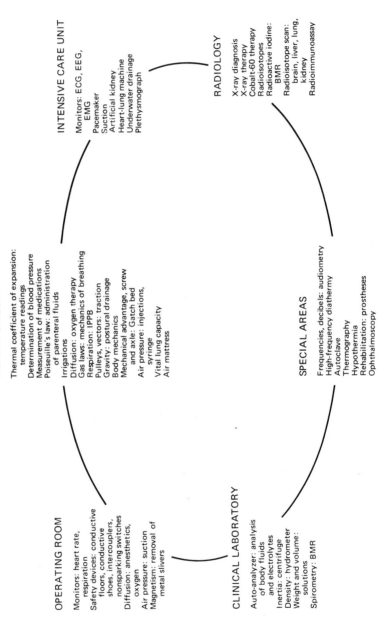

SOME APPLICATIONS OF PHYSICS TO HEALTH CARE

PATIENT'S ROOM

Thermal coefficient of expansion:
    temperature readings
Determination of blood pressure
Measurement of medications
Poiseuille's law: administration
    of parenteral fluids
Irrigations
Diffusion: oxygen therapy
Gas laws: mechanics of breathing
Respiration: IPPB
Pulleys, vectors: traction
Gravity: postural drainage
Body mechanics
Mechanical advantage, screw
    and axle: Gatch bed
Air pressure: injections,
    syringe
Vital lung capacity
Air mattress

INTENSIVE CARE UNIT

Monitors: ECG, EEG,
    EMG
Pacemaker
Suction
Artificial kidney
Heart-lung machine
Underwater drainage
Plethysmograph

RADIOLOGY

X-ray diagnosis
X-ray therapy
Cobalt-60 therapy
Radioisotopes
Radioactive iodine:
    BMR
Radioisotope scan:
    brain, liver, lung,
    kidney
Radioimmunoassay

SPECIAL AREAS

Frequencies, decibels: audiometry
High-frequency diathermy
Autoclave
Thermography
Hypothermia
Rehabilitation: prostheses
Ophthalmoscopy

CLINICAL LABORATORY

Auto-analyzer: analysis
    of body fluids
    and electrolytes
Inertia: centrifuge
Density: hydrometer
Weight and volume:
    solutions
Spirometry: BMR

OPERATING ROOM

Monitors: heart rate,
    respiration
Safety devices: conductive
    floors, conductive
    shoes, intercouplers,
    nonsparking switches
Diffusion: anesthetics,
    oxygen
Air pressure: suction
Magnetism: removal of
    metal slivers

CHAPTER 1

THE SCIENTIFIC METHOD

The spirit of free inquiry and open-mindedness to new facts characterizes the scientific method. Scientists use controlled experimentation and inductive and deductive reasoning to develop scientific theories and laws. "Flashes of insight" also account for a great deal of progress in science.

Physics deals with the study of matter and energy. It is one of the sciences, and as such is subject to the discipline of that fundamental system of thought and action known as the scientific method. The scientific method is basically a careful, commonsense method of seeking truth. If we have a problem, we first of all search for some information—that is, facts. Based upon the information, a tentative hypothesis is postulated—an "educated guess" is made and proposed. We do not stop here, but test the hypothesis by further experimentation or collection of data. Now, if each new piece of information agrees with our tentative hypothesis, the hypothesis finally becomes accepted as a theory or a law. No scientific law is completely proved until *all* cases have been tried, and this is seldom possible. However, only one negative fact is sufficient to require a revision or discarding of the hypothesis.

Two things characterize the scientific method. One is a "humility before the facts," an open-mindedness to new information, even though it means changing the hypothesis. Second, in securing these data, controlled experimentation is used, if possible.

For example, if we were to feed your class a certain breakfast food every morning for a month, and then give the whole class a reputable IQ examination, we might find the class average to be 115. Keeping in mind that the average of the population is about 100, we might conclude that eating "Crunchy" breakfast food every morning for a month had raised the IQ by 15 points.

This, of course, does not seem very logical. What is wrong with the experiment? The experiment was not set up with a *control*. We should have done one of two things: (1) test the group before and after the month's culinary adventure; or (2) divide the group into two subgroups matched with respect to intelligence, and give the breakfast food to one group but not to the other. Then, if a significant difference in IQ were found, we might possibly attribute the increase in the IQ to the food.

Actually, this would not be an absolute proof. If the other group received nothing, this could cause psychological factors to operate. To guard against this event, it is usual to give both groups similar treatment except for the difference in the item being tested. For example, in administering medication, the control group is also given a pill that looks like the test medicine; the control medication is called a *placebo*. Neither group knows whether it is receiving the medication or the placebo. To avoid unintentionally providing clues to the subject, often the person administering the test does not know which samples are the placebos. This is called a *double-blind study*. The word *placebo* means "I will please" in Latin, quite an appropriate word.

An interesting exercise, called the Hawthorne experiment, was done some years ago to test the effect of illumination on the productivity of factory workers. In one section of the factory the lamps then in use were exchanged for brighter ones. Productivity increased immediately. Then larger bulbs were installed, and output in that section further increased. You can imagine how excited the people running the experiment were. However, someone suggested checking to see what would happen if the light intensity were reduced. Productivity still increased! What had happened? It was not the level of illumination that was the critical factor, but simply that the workers felt they were being singled out for attention. The importance of the psychological factors that can affect performance should never be forgotten. This is also true in patient care; TLC and technical competence are probably of equal significance.

An unusual set of experiments was reported in the June 1980 issue of *Science*. Two groups of rabbits were fed a high-cholesterol diet. The only difference in the treatment was that one group was petted for some time several times daily. When they were analyzed, both groups had high-cholesterol blood levels. However, when the arteries were stained, the petted group showed a statistically significant lower level of lesions in the blood vessels. In other words, even though the level of cholesterol in the blood was high, the damage to the blood vessels was reduced. These results were found in three experiments over a period of 3 years. Again, we see the close psychosomatic connection and the importance of placebos.

Recent studies seem to indicate that the placebo effect is not only psychological, but that it has a physiological component as well. As a result of receiving a placebo the brain produces its own pain-killing substances, the endorphins, a family of chemicals that resemble morphine. In a very recent experiment, patients who had a tooth pulled were given a placebo, and many reported relief of pain. They were then given the compound maloxone, a drug known to be antagonistic to the action of the endorphins, and the pain returned. In other words, this seems to prove that the brain actually produced the pain-killer compounds, the endorphins, and when the maloxone inhibited the action of the endorphins, the pain returned.

It is essential in all experiments that, so far as possible, adequate controls be present. If there are too many variables, it is difficult, if not impossible, to attribute a certain result to a particular cause.

In this connection we may also mention that a statistical correlation does not necessarily mean a causal relationship. For example, just because lung cancer occurs more frequently among heavy smokers, it does not *necessarily* mean that one is caused by the other. However, the case against smoking does seem to become stronger every year, and there is little doubt that there is a *causal* relationship as well as a *statistical correlation* between smoking and lung cancer.

Science uses both the *inductive* and the *deductive* methods of reasoning. By inductive is meant going from specific cases to a general principle. This is the method used in developing scientific laws. The deductive method, on the other hand, works from a principle to a prediction of what might happen in a specific situation. The value of understanding scientific principles is that this enables a person to predict in a new situation what will happen.

Not all progress in science is attributable to the application of the scientific method. In fact, many great advances have come as flashes of insight, or hunches. W. Furness Thompson, a former research director of a large company, suggests that we tend to overemphasize the value of the "scientific method."

> The scientist you know as a genial, sensible, rather humble man, often becomes pompous and formal when he puts pen to paper. You'll probably never read a paper that begins, "For no good reason at all I had a hunch that . . ." or, "I was just fooling around one day when . . ." No sir! All trace of anything haphazard, anything human, is removed from published reports. The paper will more likely begin, "In view of recent evidence concerning the validity of the Glockenspiel theory, it seemed advisable to conduct. . . ." And the report goes on to describe a carefully thought out experiment that inevitably followed not only a logical but also a chronological order. This was done, then this resulted; therefore, these conclusions were suggested. . . .

An example of the importance of serendipity or a fortuitous accident is demonstrated in the discovery of penicillin by Alexander Fleming. He discovered that the mold *Penicillium notatum* was able to induce lysis in *Staphylococcus aureus,* and also to inhibit the growth of many pathogenic bacteria. When the petri dish was incubated at 37°C overnight, no effect was found, but when he looked several weeks later, he discovered the effect.

The interesting thing is that, according to Ernst Chain, a co-worker of Fleming in England, Fleming did not run a neat, tidy laboratory. Most laboratories wash the petri dishes a few hours after they are inspected. Fleming left his dishes around for weeks at a time. In September 1928, he looked at a dish that had been standing on a shelf, and found an inhibition in the growth of colonies of *Staphylococcus.* By leaving the dish at room temperature for several weeks, he had given the faster-growing penicillin an opportunity to grow, while the staph was growing very slowly. Thus because Fleming ran an untidy laboratory and left his dishes lying around, penicillin was discovered.

The scientific method is not infallible. It is a valuable method in searching for truth, but it is not the *only* method. Science, as Dr. Robert Oppenheimer has stated,

is activity, a becoming, rather than a body of doctrine. The scientific process demands qualities of optimism, tolerance, of selfless curiosity that civilization needs for the preservation of its future. Let me put it in ridiculously oversimplified terms: there's a point in anybody's train of reasoning when he looks back and says, "I didn't see this straight."

Men in other walks of life need the ability to say without shame, "I was wrong about that." Science is a method of having this happen all the time. You notice a conflict or an oddity about a number of things you've been thinking for a long time. It's the shock that may cause you to think another thought. That is the opposite of worldly men's endless web of rationalization to vindicate an initial error. For the scientist the compensation is the marvelous revelation of a connection you had never suspected before and the—perhaps occasional—reassurance, to use Jefferson's words, that there are no barriers to human knowledge.

The spirit of inquiry is not limited to science, but is important in all areas of life. Children seem to be naturally inquisitive, but quite often as they become older they tend to lose this ability to wonder. From the microcosm of the atom to the macrocosm of the solar system, from the workings of a machine to the wonders of the human body, there is much to learn. Lucretius stated it well in the first century BC,

no fact is so simple that it is not harder to believe than to doubt at the first presentation. Equally, there is nothing so mighty or marvelous that the wonder it evokes does not tend to diminish in time. Take first the pure and undimmed lustre of the sky and all that it enshrines: the stars that roam across its surface, the moon and the surpassing splendor of the sunlight. If all these sights were now displayed to mortal view for the first time by a swift unforeseen revelation, what miracle could be recounted greater than this? What would men before the revelation have been less prone to conceive as possible? Nothing, surely. So marvelous would have been that sight—a sight which no one now, you will admit, thinks worthy of an upward glance into the luminous regions of the sky. So has satiety blunted the appetite of our eyes.

CHAPTER 2

MEASUREMENTS

The process of obtaining quantitative information has come a long way since the twelfth century, when King Henry I decreed that the yard should be the distance from his nose to his thumb. Today, science and medicine use the metric system. As always, errors—their cause and prevention—are of great interest and import, especially in health care.

The importance of quantitative data in science and medicine is already well appreciated by anyone caring for a patient. The correct amount of medication, oxygen concentration, electric stimulation, or x-ray radiation may make the difference between life and death. *A lost decimal point can mean the loss of a patient.*

General acceptance of some system of measurement is a necessity of civilization. Imagine the confusion at a marketplace where rope or cloth was measured according to the span of the seller's arms. A salesman with short arms would certainly have been an asset to the business in those days! Therefore, general acceptance of a definite unit for a yard was quite progressive.

The two systems of measurement in common use are the British system and the metric system. As every schoolchild knows, the conversion from one unit to another in the British system involves peculiar factors such as 12 inches in a foot and 5,280 feet in a mile, not to mention rods, acres, and such. The archaic measurement systems of the United States and Great Britain are only approximately similar. For instance, a British yard is only 0.999997 as long as a U.S. yard. The British bushel equals 36.37 metric liters; the U.S. bushel only 35.2383. The British gallon equals 4.5459631 liters; the U.S. gallon 3.785332 liters.

That the British gallon is about 20 percent larger than the U.S. gallon once was forcefully brought to my remembrance on a vacation trip. I was checking the number of miles per gallon of gasoline for our new car and discovered that in Canada I was getting tremendous mileage. I finally realized that the 20 percent extra mileage was because of the larger British gallon.

In addition, one has to remember such things as the fact that the British hundredweight is not 100 pounds, but 112, and that tons can be long or short, there being a 240-pound difference between them.

King Henry I of England (reigned AD 1100 to 1135) attempted to set a standard when he decreed that the yard should be the distance from the end of his nose to the end of the thumb of his outstretched hand.

More than a century later, King Edward I of England took a big stride in the science of measurement. To reduce the confusion resulting from the arm's length as a standard, he ordered a permanent standard to be made of iron. This master yardstick was to be the standard for the entire kingdom. At the same time, the king also decreed that the foot should be one-third of this length of iron, and that the inch should be 1/36th of the length.

THE METRIC SYSTEM

The metric system was originally developed by a committee of the French Academy in 1791. In 1875 the Treaty of the Meter, establishing an International Bureau of Weights and Measures near Paris, was signed by the United States and 16 other nations. Since then, most countries of the world have adopted the metric system, whereas the United States, except for the scientific and medical fields, still uses the British system.

The *meter* was originally defined as one ten-millionth of the north polar quadrant of the earth on the meridian through Paris. This was then changed to the distance between two lines on a platinum–iridium bar kept in Paris. From this original, a measured meter and a kilogram were sent to Washington in 1890 and placed in the keeping of the National Bureau of Standards. In 1960, the Eleventh General Conference of Weights and Measures redefined the meter to be equal to 1,650,763.73 wavelengths, in vacuum, of the red radiation from krypton-86. Measurements now available are accurate to about one part in a billion, or about 1 centimeter compared with the distance across the United States.

The *kilogram* was legalized in 1889 by the First General Conference on Weights and Measures on the International System of Units (CGPM). It is the mass of the platinum–iridium prototype kept at the International Bureau of Weights and Measures.

The *second* was for centuries defined as 1/86,400th of a mean solar day. But because of some irregularities in the rotation of the earth, a new, more precise definition was adopted in 1967 by the CGPM: "The second is the duration of 9,192,631,770 periods of the radiation corresponding to the transition between the two hyperfine levels of the ground state of the cesium-133 atom."

The metric system is a logical, sensible, functional system, in which conversion from one unit to another involves only factors of 10, so that only a change in the decimal point is required. Then why is it that we have not adopted the metric system in place of the outmoded British system? Besides human inertia (the tendency of people to resist change), there are economic reasons: the cost of converting all rulers, scales, tools, and instructions would probably run into billions of dollars. However, there is a strong movement in this country to make the conversion to the metric system; soon the shopper may ask for 3 liters of milk and half a kilo of butter.

Below are listed some units of the metric system:

Length	*Mass*	*Volume*
kilometer—km	kilogram—kg	liter—l
meter—m	gram—g	milliliter—ml
millimeter—mm	milligram—mg	

Note that *milli* means one-thousandth; thus 1 mg means 1 milligram, or one one-thousandth of a gram, 1 mm means 1 millimeter (one one-thousandth of a meter), and 1 ml means 1 milliliter (one one-thousandth of a liter). The other unit of volume, which is practically the same as the milliliter, is the cubic centimeter (cc, or cm^3). This is the volume occupied by a cube the edge of which measures 1 centimeter.

METRIC—BRITISH EQUIVALENTS

1 inch = 2.54 cm = 25.4 mm   1 pound = 453.6 g

1 meter = 39.4 inches   1 liter = 1.06 quarts

1 kilogram = 2.2 pounds

The metric system has two basic subdivisions, mks and cgs units. As seen below, the mks stands for the meter–kilogram–second system, whereas cgs refers to the centimeter–gram–second system.

	British	*MKS*	*CGS*
Length	foot—ft	meter—m	centimeter—cm
Mass	slug	kilogram—kg	gram—g
Time	second—sec	second—sec	second—sec
Velocity	ft/sec	m/sec	cm/sec
Acceleration	ft/sec^2	m/sec^2	cm/sec^2

The *liter* was defined in 1879 as the unit of volume. It was supposed to equal 1 cubic decimeter, but it was found by careful work that the liter differed from the cubic decimeter by 28 parts per million (ppm), that is, 1 l = 1.000028 cubic decimeters. This means that the milliliter differed from the cubic centimeter as well by 1.000028. The International System of Units, or Système Internationale (SI) in 1964 defined the unit of volume as the cubic meter (m³), hence 1.000000 ml = 1.000000 cc. Except for very precise work, this distinction is not important, and we will use milliliter (ml) and cubic centimeter (cc) interchangeably.

Another derived unit of measurement is the density of a substance. *Density* is defined as the mass per unit volume.

$$D = \frac{M}{V}$$

In the cgs metric system, M is expressed in grams and V in milliliters; thus, D is g/ml, or, in the case of gases, g/l. Density is a measurement of how closely a substance is packed. For example, 1 kg of lead and 1 kg of cork will weigh the same, but 1 ml of cork will weigh much less than 1 ml of lead. We will discuss the difference between mass and weight in the next chapter.

The buoyancy of an object was found by Archimedes to be equal to the weight of the volume of the liquid displaced by the object. Therefore, if an object has a density of less than 1 g/ml, it will float in water. If it has a density of greater than 1 g/ml, it will sink. The human body does float (despite what some nonswimmers maintain); therefore, the density of the human body must be a little less than the density of water.

Let us illustrate this with a problem. Will a solid weighing 25.0 g and occupying 20.0 ml sink or float in water?

$$D = \frac{M}{V} = \frac{25.0 \text{ g}}{20.0 \text{ ml}} = 1.25 \text{ g/ml}$$

Because the density of water is approximately 1 g/ml, this object will sink.

A very similar unit is *specific gravity,* the ratio between the density of a liquid compared with the density of water at the same temperature. As the density of water is approximately 1 g/ml in the metric system, the specific gravity is numerically equal to the density. That is, a solution having a density of 2 g/ml will also have a specific gravity of 2. Note the absence of units for specific gravity. You will find several applications of specific gravity in Chapter 8.

CALIBRATION

Calibration is the marking of a scale on an instrument, such as the lines on a ruler or a thermometer or the count-per-minute scale on a Geiger counter. Generally, instruments commercially available are fairly accurately calibrated. However, you should always keep in mind that there is nothing absolute about such a scale, and that it might be inaccurate. Occasionally, thermometers or rulers are incorrectly calibrated.

PARALLAX

When a scale is read, if there is any space between the scale and the pointer, there is some possibility of error, unless the observer is directly in front of and at eye level with the scale. This deviation in reading caused by the space between scale and pointer is called *parallax*. This is one reason for holding the thermometer at eye level when you are reading it. Otherwise, the mercury will appear to be in front of the wrong line on the scale. To ensure accurate reading of the liquid level in a graduate cylinder, the surface of the liquid also should be held at eye level (Fig. 2-1).

EXPONENTIAL NUMBERS

We often encounter very large and very small numbers in science. Instead of spending a lot of time counting zeros, we write the numbers in exponential form.

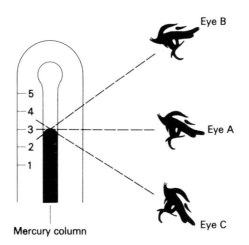

Figure 2-1.
Error of parallax. An observer at A, at eye level with the mercury, will read the correct value of 3. An observer at B will make an error of parallax and read 2, whereas the observer at C will read 4.

You probably remember that 10^2 means 10×10, or 100. Similarly:

$10^3 = 1000$ $10^{-3} = 0.001$

$10^4 = 10{,}000$ $10^{-4} = 0.0001$

$10^5 = 100{,}000$ $10^{-5} = 0.00001$

$10^6 = 1{,}000{,}000$ $10^{-6} = 0.000001$

$10^9 = 1{,}000{,}000{,}000$ $10^{-9} = 0.000000001$

For a number with a positive exponent, start with the first digit, and count off to the *right* the same number of places as the exponent. For a negative number, again start with the decimal point after the first digit, and count off to the *left*. Thus, 200,000,000,000 equals 2×10^{11} and 0.00000004 equals 4×10^{-8}. The value of expressing numbers in this manner can be shown by considering a number called Avogadro's number: 602,300,000,000,000,000,000,000; more simply, this is written: 6.023×10^{23}.

SIGNIFICANT FIGURES

It is important to be aware of significant figures. The significance of a figure depends on the experimental precision. An answer obtained as the result of an experiment is only as precise as the least precise measurement made. In contrast to numbers obtained by measurement, we might consider the so-called *pure* numbers of mathematics, which have no particular significance regarding actual measurements.

As an example of pure numbers:

$$10.0642 \times 1.5 = 15.09630$$

However, when *measurements* are made, the answer is only as precise as the least-known figure.

If it is given that a table top has the dimensions of 5.42 ft $\times$ 6.5 ft, the answer would be 35.230 ft^2. However, because the least well-known figure involved, 6.5, contains only two significant figures, the result should be reported as 35 ft^2, or because the number is known to 0.1 ft, the result might be reported as 35.2 ft^2. With the use of pocket calculators, students often report results to 8 to 10 decimal places when the data are valid only to one decimal place.

ERRORS

No measurement is ever completely accurate. No matter how carefully we measure something, there is usually some doubt as to the exact value. Errors in measurement may be determinate or indeterminate. *Determinate* (constant) errors are caused by factors such as incorrect calibration of the measuring instrument; *indeterminate* (random or accidental) errors are usually caused by errors in interpretation.

Indeterminate errors can be minimized by taking the mean or average of a number of readings, but determinate errors cannot be reduced by taking an average. For example: If you were to shoot at a target with a gun having a bent barrel, you might have all your shots close together but far from the bull's-eye. Taking more shots would not improve the situation because of the determinate error caused by the faulty gun barrel. In other words, there is a difference between consistency of measurement and accuracy.

Why will taking a number of readings help reduce indeterminate errors? If the errors are truly random or accidental, that is, just as likely to be high or low, the mean or average result should be closer to the correct answer compared with a single result. Random events tend to follow the familiar normal distribution, or Gaussian curve (Fig. 2-2).

The mean value (n) of a series of results is the best estimate of the "true" value. The deviation of a series of results is often calculated as the *standard deviation,* indicated by sigma (σ). Defined as the square root of the mean of the squares of the deviations from the mean, the standard deviation is a measure of the *scatter* or variance of the data:

$$\sigma = \sqrt{\frac{\Sigma\ d^2}{n\ -\ 1}}$$

This means the sum of the deviations from the mean squared, divided by the number of results minus 1. It gives the mean of the squares of the deviations from the mean. The square root is then taken to give σ.

For large numbers of results, if the readings are truly random, that is, indeterminate, we would expect to find 68% within $\pm 1\sigma$, 95% within $\pm 2\sigma$, and 98% within $\pm 3\sigma$.

Figure 2-3*b* has a much smaller deviation, that is, a smaller standard deviation. This would mean better precision. We have to remember, however, that we do not

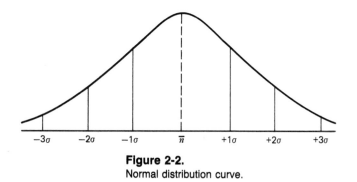

Figure 2-2.
Normal distribution curve.

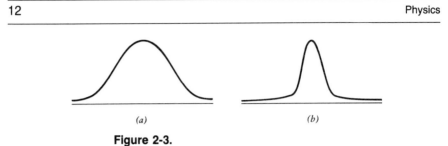

Figure 2-3.
Distribution curves showing difference in variance.

know what the accuracy is, as accuracy means how close the result is to the correct answer. Because we do not know what the correct answer is, the degree of accuracy is not known. It is possible to have both good precision and a constant, determinate error; therefore, the accuracy is very poor.

To prove to yourself that errors can exist both in your measuring tools and in your subjective interpretation of measurements, try the following experiments:

1. Compare the calibration of three meter sticks against a fourth stick. Place the zero mark of each stick to be checked directly over the zero mark on the standard stick, and check whether the 80-cm lines coincide. Repeat, using the inch scale, and check on the lines for 36 inches.

2. Measure the length of the laboratory table as accurately as possible. Have your partners do the same without knowing each other's results. Then list the results obtained. Make a list of factors that could account for any discrepancies that occurred. How can you determine the "correct" answer?

3. If you have four or five sphygmomanometers available, have the same person take the blood pressure for one person. What type of errors could be involved? Using the same sphygmomanometer, have a number of people take the blood pressure on the same person. (Allow some time between readings for the subject to recuperate.) Explain the results.

CELSIUS–FAHRENHEIT TEMPERATURE SCALES

There are three temperature scales in common use: the Celsius, the Fahrenheit, and, in some scientific work, the Kelvin. The Celsius temperature scale was worked out by Anders Celsius (1701–1744), a Swedish astronomer, who proposed in 1742 that the freezing point of water be called zero degrees and the boiling point of water at 1 atmosphere pressure be called 100 degrees. (Atmospheric pressure is the pressure exerted by the air on any object.) As this gives 100 degrees between freezing and boiling, the scale was for many years referred to as centigrade. Today the scale is known as Celsius. The Fahrenheit scale is named after the German physicist, Gabriel Fahrenheit (1686–1736), who set 32 degrees for the freezing point and 212 degrees for the boiling point of water. The Kelvin scale, named for the British mathematician and physicist, William Thomson, Lord Kelvin (1824–1901), incorporates units of the same size as the Celsius, but adds 273 to the degrees C. For example, 0°C is equal to 273°K, and 100°C is equal to 373°K.

In a previous course, you have probably studied how to make the conversion from Celsius to Fahrenheit, and vice versa. If you simply memorized a formula, you probably have a vague notion that you multiply by $^9/_5$ or $^5/_9$, before or after subtracting or adding 32. If this is the case, such knowledge is not functional. Figure 2-4 will help us attempt to describe the operation involved.

If we use the symbol Δ, or delta, for the difference between certain levels, on the Celsius scale the delta between the freezing and boiling point is 100 degrees, whereas on the Fahrenheit scale the delta is 180 (212 minus 32). In other words, there are 100 steps on the Celsius scale for every 180 on the Fahrenheit scale. One hundred and one hundred-eighty divided by 20 (the highest common factor) reduce to 5 and 9, respectively. That is, for every five units on the Celsius scale, there are nine units on the Fahrenheit scale. Now, let us take an example. Let us convert 40°C to Fahrenheit. The delta from 0 to 40 is, of course, 40. On the Fahrenheit scale there are more degrees for an equivalent interval, so 40 must be multiplied by $^9/_5$ (rather than by $^5/_9$). This equals 72 units. However, at the freezing point the Fahrenheit is already 32; the 72 plus the 32 equals 104 degrees Fahrenheit (Fig. 2-4).

Let us work this in reverse, and change 104°F to Celsius. First of all, the delta is not 104, but 104 minus 32 or 72. Now, on the Celsius scale there are fewer units, so the 72 is multiplied by $^5/_9$. $72 \times ^5/_9 = 40$. This on the Celsius scale is both the delta and the actual reading. What we are trying to point out is that memory work can be kept to a minimum if the procedure is discussed and understood. In practice, conversion tables are available.

Make sure that you understand Celsius–Fahrenheit conversions by working out the following problems:

1. Change 98.6°F to Celsius.

2. Change 41°F to Celsius.

3. Change 60°C to Fahrenheit.

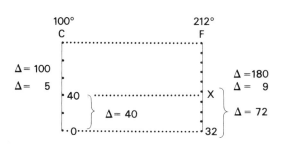

Figure 2-4.
Celcius–Fahrenheit conversions.

QUESTIONS

1. A baby weighed 3.2 kg at birth. What was his weight in pounds?
2. Convert 10 inches to centimeters (cm). To millimeters (mm).
3. How many milliliters (ml) are there in 1.2 liters? In 10 cc?
4. In reading a thermometer, suppose you hold it to the right of center. What type of error will this produce? Will it tend to increase or decrease the reading?
5. The master meter stick in France is housed in a room in which the temperature is kept constant. Why is this necessary?
6. An object has a mass of 106 g and a volume of 100 ml. Find the density of the object. Will it sink or float?
7. Explain why it would be incorrect to say that New York City's population is, for instance, 7,263,925?
8. A building lot is surveyed and found to measure 60.2 × 90.6 ft. Calculate the area, and report it with the proper significant figures.
9. One patient has the following blood pressure readings 88, 82, 90, 91, 86, 89, 84, 87, 90, 86. Calculate the mean, the individual deviations from the mean, and the standard deviation.
10. In the above problem, discuss what factors might contribute to the variations in the readings and indicate whether they are determinate or indeterminate.
11. Ethyl alcohol has a density of 0.79 g/ml. What will be the weight of 25 ml?
12. A patient has a temperature of 40.0°C. What is it in degrees Fahrenheit?

CHAPTER 3

FORCE AND MOTION

Newton's laws of motion and gravitation are operative in all areas of modern life, from driving, flying, and traveling in space to carrying out common patient-care activities. Knowledge of these laws, as well as of vectors and forces, helps us understand many familiar phenomena.

NEWTON'S THREE LAWS

The three laws of motion developed by Isaac Newton marked one of the great achievements in science. Newton was born on Christmas Day in 1642 in England. He was a premature, weak baby and was not expected to live. Newton had one of the greatest minds of all time, but his gifts were slow in appearing, and he was at first one of the poorest students in his class. As a youth he was absent-minded, but by the time he was 24 years of age, he had worked out the laws of motion and gravitation as well as the system of calculus.

Newton was one of the greatest scientists of all time. The revolution in man's way of looking at nature, both terrestrial and celestial, beginning with Copernicus, Kepler, and Galileo, was systematized by Newton in his three laws of motion. His book, *Philosophiae Naturalis Principiae Mathematica,* commonly called *Principia,* published in 1687, truly revolutionized the thinking about force, motion, and gravity.

In practice, you are already aware of Newton's three laws, but it is of value to state them formally in order to make them more useful in your thinking.

FIRST LAW: *ALL MATTER HAS INERTIA*

This is a brief definition but may not explain a great deal. A fuller definition is: *A body at rest will remain at rest, and a body in motion will continue in motion, traveling in a straight line unless a force is applied.* You have experienced this when a car starts or stops suddenly. The many accidents in which people are thrown against the dashboard when a car stops suddenly indicate the operation of this law.

To demonstrate Newton's First Law to yourself, place a fairly heavy object on a piece of paper and give a quick pull sideways on the paper. Did the object move with the paper or remain where it was? Explain. A more interesting demonstration of this law is to pull a tablecloth out from underneath a set of dishes—if you have the nerve to try it.

If there is no gravitational pull on an object, and no friction, an object will continue moving in a straight line with the same speed. If we were to blast off in a rocket ship for the moon and miss the target, once we were out of the gravitational field of any planet, we could shut off the rocket engine and would continue traveling indefinitely in space at a constant speed.

SECOND LAW: *FORCE EQUALS MASS TIMES ACCELERATION (F = M × a)*

When a force, such as a push or a pull, is exerted on an object, it will produce an acceleration of that object. What is acceleration? It is not synonymous with velocity (speed); rather it is a *change* in velocity. A car changing its speed from 0 to 10 miles per hour in 1 minute is undergoing acceleration. So is a car going from 10 miles per hour to 0 miles per hour. The latter can also be called deceleration or negative acceleration, but it is still defined as acceleration, which is *any* change in velocity. However, a car moving steadily at 30 miles per hour is undergoing zero acceleration.

This formula means that for an object of a certain mass or weight, the greater the force exerted on it, the greater its acceleration. Similarly, it takes a larger force on a heavy object to produce the same acceleration as a smaller force on a lighter object.

If you are pushing a patient in a wheelchair, you will find that the more force you exert, the greater the *change* in the speed of the wheelchair—that is, the greater the acceleration. Similarly, a larger force is required to achieve the same acceleration with a heavy patient than with a light one. (It goes without saying that patients will not appreciate being the subjects of Newtonian experiments. Accelerate your patient as gently as possible and remember—he could slide off a stretcher if you stop suddenly.)

Another way of stating this law is: *The acceleration of an object is directly proportional to the applied force, and inversely proportional to its mass.*

Let us work a couple of problems to illustrate this law. We shall omit frictional forces to avoid having to subtract these from the applied force in order to get the net force. The units of force depend on the units used for mass and acceleration. In the British system mass is expressed in *slugs* and acceleration is in *feet per second squared*, so force is in slugs · ft/sec^2. This unit of force is called the *pound*.

In the mks system (meter–kilogram–seconds), mass is expressed in *kilograms* and acceleration in *meters/sec²*, so force is in kg · m/sec². This unit of force is called the *newton*. A force of 1 newton applied to a mass of 1 kg will change its velocity at the rate of 1 m/sec/sec. That is, it will have an acceleration of 1 m/sec².

If we exert a force of 20 newtons on an object having a mass of 10 kg, what will be the acceleration?

$$F = Ma \qquad a = \frac{F}{M} = \frac{20 \text{ newtons}}{10 \text{ kg}} = 2 \text{ m/sec}^2$$

What will the same force of 20 newtons do to a 2-kg object?

$$a = \frac{F}{M} = \frac{20 \text{ newtons}}{2 \text{ kg}} = 10 \text{ m/sec}^2$$

You will notice that the same force of 20 newtons will give a greater acceleration to the object having less mass. This is something we have known intuitively all our lives. If you get stuck in a car and have to push it, you are better off pushing a Volkswagon than a Lincoln Continental.

Using units in the British system, what acceleration could be expected of an object having a mass of 2 slugs if a 100-lb force is applied?

$$F = Ma$$

$$lb = slugs \times ft/sec^2$$

$$a = \frac{F}{M} = \frac{100 \text{ lb}}{2 \text{ slugs}} = 50 \text{ ft/sec}^2$$

What acceleration would a force of 100 lb produce on a 64-lb object? Note that here the *weight* of the object is given in pounds, so we have to convert to slugs to get the mass.

$$a = \frac{F}{M} = \frac{100 \text{ lb}}{64 \text{ lb/32 lb/slug}} = \frac{100 \text{ lb}}{2 \text{ slugs}} = 50 \text{ ft/sec}^2$$

THIRD LAW: FOR EVERY ACTION THERE IS AN EQUAL AND OPPOSITE REACTION

This law can also be called the *recoil principle* or the *jet principle*. It means that if one object exerts a force on another object, the other object will exert the same amount of force in the opposite direction on the first object.

A few examples will make this clear. In jumping from a canoe to a dock, a

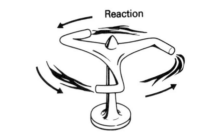

Figure 3-1.
Water sprinkler illustrating action–reaction principle.

person will push on the boat to get momentum to reach shore, but the boat also receives momentum away from the shore. (And sometimes the jumper gets wet.) The firing of a gun produces a force on the bullet, but there is also a reaction on the gun, generally called the recoil or "kick." The rotating water sprinkler (Fig. 3-1) also works on this principle, and so do rockets and jets. The exhaust gas does not push against the outside air, because a rocket will also work in a vacuum; instead, the thrust is attributable to the reaction between the exhaust gas and the rocket.

Some time ago, I read a newspaper article entitled, "Experts Explain Fireboat Enigma." The fire department was asked to explain why some of the hoses of the fireboats are sometimes turned away from the blaze, simply watering the ocean. At first glance it appears ridiculous. However, unless a fireboat can be moored to a pier, the reaction of the water will cause about 1,000 pounds backpressure per nozzle and will force the boat away from the fire. Thus, to keep the boat in place, some nozzles will have to point away from the fire.

Newton's Third Law can be explained by the concept of *momentum*. Momentum is defined as mass times velocity, and in every collision momentum must be conserved. If we assume that a gun has a mass 50 times that of the bullet, and the bullet has a muzzle velocity of 100 miles per hour, how fast will the gun move in the opposite direction?

$$\text{Momentum} = \text{momentum}$$

$$\text{(gun)} \quad \text{(bullet)}$$

$$\text{Mass}_1 \times \text{velocity}_1 = \text{Mass}_2 \times \text{velocity}_2$$

$$50 \times X = 1 \times 100$$

$$X = 2 \text{ miles per hour}$$

In other words, the bullet and the gun will have the same momentum in opposite directions, but since the gun is 50 times heavier, it will have a velocity 1/50th as much as the bullet.

CIRCULAR MOTION

As discussed earlier, according to Newton's First Law, an object will continue in motion, traveling in a straight line, unless a force is applied to it. If a force is applied, there will be a *change* in velocity—either an increase or decrease. A change in velocity with time is called acceleration. You will recall that velocity has both *magnitude* and *direction*. An object traveling in a circle at a constant speed will have a continual change in direction, or to put it another way, it will have a constant acceleration toward the center. For this change in direction, or this acceleration to occur, an inward force is necessary. This can be felt by the tension on a string when you twirl an object on a string. Similarly, for a car to negotiate a curve, there has to be a frictional force of the road on the tires. This inward force is called the *centripetal* force, a center-seeking force.

If there is no force, such as in cutting the string in the above example, the object will go in a straight line, tangent to the circle. It has been found that for *angular* accelerations, $F = Ma$ can be rewritten

$$F = \frac{mV^2}{r}$$

where M is the mass, V is the velocity, r is the radius of the curve, and F is the centripetal force. Next, let us see the effect of changing the *velocity*. We shall assume in two cases a radius of 2 ft and a weight of 128 lb, but in Figure 3-2a we will have a speed of 50 ft/sec, as in 1 (*b*) above, and in the second case (Fig. 3-2b) we shall assume a speed of 100 ft/sec.

The centripetal force in Figure 3-2a will be

$$F = \frac{mV^2}{r} = \frac{4 \text{ slugs } (50 \text{ ft/sec})^2}{2 \text{ ft}} = 5,000 \text{ lb}$$

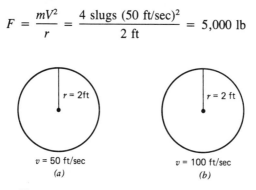

Figure 3-2.
Effect of change in velocity on centripetal force.

In Figure 3-2b, the centripetal force will be

$$F = \frac{mV^2}{r} = \frac{4 \text{ slugs } (100 \text{ ft/sec})^2}{2 \text{ ft}} = 20{,}000 \text{ lb}$$

The centripetal force in Figure 3-2b is four times larger than in Figure 3-2a when the velocity has been doubled. This checks with the equation

$$F = \frac{mV^2}{r}$$

where we see that the force is proportional to the *square* of the velocity. Thus, if we double the velocity, the force will be 2^2, or four times larger.

A common device employing the principle of rotary motion is the centrifuge. The centrifuge is a motor-driven container that rotates at a high speed. Solutions placed in the centrifuge are subjected to a force much greater than that of gravity, and any solid will precipitate out much faster than it normally would. One common use of the centrifuge is to separate the various components of blood, proteins, and microorganisms.

The fact that an object will tend to go off tangentially from the curve is often called the *centrifugal* force. This is not a real force, but is often a useful concept for describing the fact that a spinning object will throw off loose particles. For example, if you whip some cream with an electric beater, remove the blades from the dish with some cream on them, and then turn on the motor, the cream will be thrown all around the kitchen.

Let us take another look at the above equation

$$F = \frac{mV^2}{r}$$

You will notice we have a constant speed of 50 ft/sec in both diagrams (Figs. 3-3a and 3-3b). If the radius of Figure 3-3a is 1 foot and Figure 3-3b is 2 ft, what will be the centripetal force on an object having a weight of 128 pounds?

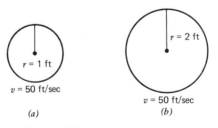

Figure 3-3.
Effect of radius on centripetal force.

$$F = \frac{mV^2}{r} = m \cdot \frac{(50 \text{ ft/sec})^2}{1 \text{ ft}}$$

The mass (m) in slugs is the weight divided by g, the acceleration caused by gravity, which is 32 ft/sec². Therefore,

$$m = \frac{128 \text{ lb}}{32 \text{ ft/sec}^2} = 4 \text{ slugs}$$

Problem a

$$F = 4 \text{ slugs} \frac{(50 \text{ ft/sec})^2}{1 \text{ ft}} = 10,000 \text{ lb centripetal force}$$

Problem b

$$F = 4 \text{ slugs} \frac{(50 \text{ ft/sec})^2}{2 \text{ ft}} = 5,000 \text{ lb centripetal force}$$

You will note that as the radius of the circle is doubled, the centripetal force is cut in half. This follows from the fact that $F \sim 1/r$, that is, the force varies inversely with the radius of the circle.

When an aviator pulls up out of a fast dive, he or she may black out because the sudden change in direction will cause the blood to travel toward the feet, causing a temporary decrease in blood to the brain.

Common household applications of centrifugal force are the spin dryer in a washing machine and the blender.

GRAVITY

THE LAW OF UNIVERSAL GRAVITATION

This law, stated by Newton, is: *Any two objects in the universe are attracted to each other with a force which is directly proportional to the product of their masses and inversely proportional to the square of the distance between them.* This statement can be summarized more easily with the formula:

$$F = \frac{K \times M_1 \times M_2}{D^2}$$

The constant K is used to make the equation correct regardless of the system of units involved. Let us see what this means. The larger the mass M_1, which could be the mass of the earth or some other planet, the greater its gravitational attraction. The weight of the earth is constant, but our weight is not. If M_2 represents a person's weight, and the substance or material of the person increases (say through overeating), the mass increases, and there will be a corresponding increase in the gravitational attraction. In other words, the person would weigh more. This is what is meant by the statement that the attraction varies directly with the mass; the more mass, the more attraction. The second part of the definition, "inversely proportional to the square of the distance between them," means that if we were to double the distance between the object and the center of the earth, the attraction or weight would be $(\frac{1}{2})^2$, or $\frac{1}{4}$ as much.

Similarly, if the distance were made three times as great, the weight would be $(\frac{1}{3})^2$ or $\frac{1}{9}$ as much. If the distance becomes very great, such as in the case of a flight into space, the weight would then approximate zero. Recent developments in space flights have given the problems of space medicine practical significance, and the solution of these problems was the object of intensive research for more than 10 years before the first space flight made by Col. John Glenn.

What is this mysterious force of attraction? We can measure how this force affects objects, but we cannot really *explain* how this attraction is exerted. We can call it a *gravitational field*, or say that it is natural for things to fall, but that does not explain it. We often name phenomena in life, and think putting a name on it explains it, but that is not the case. But we can determine how gravity affects an object.

In Aristotle's time, it was "explained" that all terrestrial objects that are *not* in their natural places will move by their own nature to these correct places.

It was found that an object in free fall, neglecting air friction, will fall faster and faster. In other words, the object is undergoing acceleration, a change in velocity. For centuries, people have been dropping objects to measure this acceleration. The question can be asked, "Is this acceleration constant or variable?" It turns out that it is constant, and equal to 32 ft per second per second (32 ft/sec^2) or 980 cm/sec^2. Because the earth is not completely spherical, the value is actually 978 cm/sec^2 at the equator and 983 cm/sec^2 at the poles.

Let us assume a stunt man decides to jump off one of the two World Trade Center buildings in New York City. We shall again conveniently ignore the effect of air

Table 3-1

Time (sec)	Velocity (ft/sec)	Distance (ft)
0	0	0
1	32	16
2	64	64
3	96	144
4	128	256
5	160	400

resistance. At time zero, when you jump, the velocity will be zero. After 1 second, the velocity will be 32 ft/sec, after 2 seconds the velocity will be 64 ft/sec, and so forth.

Table 3-1 indicates the velocities and distances after various times.

You will notice that the velocity in the table has increased 32 ft/sec each second. We can therefore say the acceleration is constant at 32 ft/sec^2.

Velocity is equal to the distance covered per unit time, $V = S/t$, such as miles per hour or feet per second.

Acceleration, on the other hand, is the *change* in velocity with time, or rate of change of velocity, such as miles per hour^2 or feet per second^2.

The velocity or distance traveled can be calculated at any time by the following equations.

$$\bar{a} = \frac{V_t - V_0}{t}$$

where V_t is the velocity at time t, and V_0 is the velocity at time 0. $\bar{a}$ is the average acceleration.

$$\bar{a}t = V_t - V_0$$
$$V_t = \bar{a}t + V_0$$

If we assume constant acceleration and $V_0 = 0$, the equation becomes $V_t = at$. The distance traveled can be calculated from the equation $S = V_0 t + \frac{1}{2} at^2$. Again, assuming constant acceleration and $V_0 = 0$, the equation becomes

$$S = \frac{1}{2} at^2$$

Let us check the values in Table 3-1 for $t = 3$ sec.

$$V = at = 32 \text{ ft/sec}^2(3) = 96 \text{ ft/sec}$$
$$S = \frac{1}{2} at^2 = \frac{1}{2} 32 \text{ ft/sec}^2(3)^2 = 16(9) = 144 \text{ ft}$$

SPACE TRAVEL

Medical experiments have been carried out in space as well as on the ground, such as those conducted during the manned Skylab missions.

The chosen medical experiments will be used to investigate the effects of long-duration space flight on the crewmen and to evaluate the metabolic effectiveness of man in space to determine future requirements for logistics resupply, environmental control, and task planning. Major areas of investigation are nutrition, musculoskeletal function, cardiovascular function, hematology and immunology, neurophysiology, pulmonary function, and metabolism.[1]

The spaceship was, in fact, a fast-moving clinical laboratory. Data secured include

1. Daily record of body weight and food intake.
2. Daily record of fluid intake and urine secretion. Samples of urine saved for analysis later for calcium, phosphorus, magnesium, potassium, nitrogen, and chromium.
3. Feces weighed, dried and returned for later analysis for calcium, phosphorus, magnesium, potassium, nitrogen, and chromium.
4. Periodic blood samples taken, processed, and analyzed postflight for calcium, phosphorus, magnesium, alkaline phosphatase, sodium, potassium, total protein, sugar, and hydroxyproline.[2]

To measure body mass in a zero-g environment, the inertial property of mass, rather than the gravitational force, was determined. The device used a linear spring mass pendulum, and recorded the time involved in the period of oscillation. The mass is accelerated uniformly by a spring, and three periods of the pendulum are timed. The device is calibrated using food trays of known masses, and a graph is made plotting time of oscillation versus mass. The crewman then sits on the "scale," the time is measured, and the mass is found from the graph.

The foods included were selected on the basis of their mineral, protein, and caloric content. The food tray contained eight cavities, four large and four small ones. Three of the large cavities contained electric heaters for heating the food. The solid and viscous foods were packaged in easy-to-open containers that fitted into the tray openings. Liquids were packaged in bellows-type containers (Fig. 3-4).

To determine whether metabolism is altered in space, the astronauts used a bicycle ergometer (an instrument that measures the amount of work muscles can perform) for about 1 hr daily. Additional physiological data secured were the following.

1. Exercise rate
2. Oxygen consumption
3. Carbon dioxide production
4. Heart rate
5. Blood pressure
6. Vectorcardiogram

[1]*Skylab and the Life Sciences,* NASA, Feb. 1973, p. 5.
[2]*Ibid,* p. 25.

Figure 3-4.
A typical meal on a tray.

Source: Photograph courtesy of the National Aeronautics and Space
Administration, Houston, Tex. 77058.

7. Respiration rate
8. Minute blood volumes
9. Body temperature and spaceship temperature

Much of this material is described in later chapters.

Although the astronauts exercised daily, they experienced a loss of muscle strength of about 20 percent, as well as a loss of red cells. Their bodies did adapt to zero gravity, and after about five days, the men seemed well and able to remain in space for a relatively long period. Normally, gravity pulls blood to the lower extremities; but in the absence of gravity, this does not happen; hence bloodflow to the heart was increased, and gradually the blood volume decreased. Cardiovascular deconditioning was measured using the lower body negative pressure device (Fig. 3-5). This container, like an iron lung, is sealed at the waist. Negative pressure up to 50 mm Hg is applied to draw blood back into the legs. A leg-volume device measures the increase in blood volume in the legs. This device, called a *plethysmograph,* is discussed in Chapter 8.

During sleep, the men wore sleep-monitoring caps. Fitted into the close-fitting

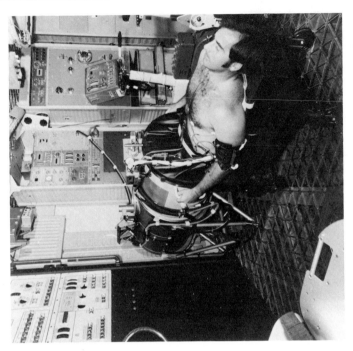

Figure 3-5.
Lower body negative pressure device.

Source: Photograph courtesy of the National Aeronautics and Space
Administration, Houston, Tex. 77058.

cap were electrodes to measure electroencephalographic activity (EEG), including
an electrooculogram (EOG). These were recorded on magnetic tape for postflight
analysis.

It had been feared that the experience of weightlessness would affect man's
ability to orient himself in space; however, the experience of the space travelers
was that they adapted quite quickly to zero gravity. They enjoyed being able to
move heavy equipment with very little effort, and even to keep it suspended in the
middle of the cabin. Upon returning to earth, the men required about 24 hr to
become acclimated. Pilot Jack Lousma said:

> It felt as though my feet had magnets on them and they wanted to stay on the
> floor. . . . Then there was the balance problem. Our legs were very strong from all
> the exercise that we did, but somehow the little sensors in the legs that want to keep
> you upright and respond to lateral motions just hadn't been exercised in a long time
> and they had gotten kind of lazy and hadn't gotten the word that we were back to earth
> yet. And so if you went into a swerve you probably stayed in a swerve and ricocheted
> off the wall or had somebody catch you.[3]

[3]Taped report of conference, Johnson Space Center, Oct. 2, 1973.

Effects of gravity and body weight distribution on equilibrium are discussed further in the next chapter and are shown in Figure 4-17. For the moment, we should note that the human body has tremendous adaptive capacity and is ceaselessly working to maintain homeostasis—the state of balance of all parts.

OTHER EFFECTS OF GRAVITY

Gravity causes liquids to flow toward a lower level. If a person is standing a great deal, the blood accumulates in the lower extremities, producing edema. Similarly, when a person feels faint, lowering his/her head will help cause the blood to flow back to the head by gravity.

The *rocking bed* is an application of gravity in which a gentle seesaw motion is used as an aid in respiration. The visceral organs press up against the diaphragm and lungs when the head is in a lower position and will move from the diaphragm as the feet are lowered, allowing the lungs to expand. Postural drainage, sometimes used in the treatment of bronchial or lung abscess, also uses gravity. The patient lies across the bed with the head and chest lower than the rest of the body; this position promotes drainage from the lungs.

CENTER OF GRAVITY

For convenience in solving problems in physics, every object is considered to possess a point—the center of gravity—at which the entire mass of the object can be considered centered. In a symmetrical, homogeneous object, this point is at the

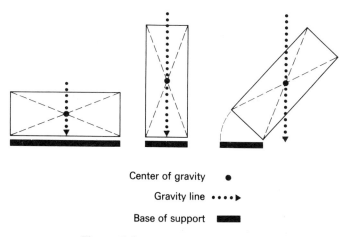

Center of gravity   ●

Gravity line ••••▶

Base of support ▬▬▬

Figure 3-6.
The effect of a broad base upon stability.

geometric center. For example, in a block of wood, the center of gravity can be found by drawing the diagonals and finding the midpoints in all directions. As long as the center of gravity falls within the base of the object, the object will be stable; but if, as in the third object shown in Figure 3-6, the center of gravity falls outside the base, the object is unstable and will topple over. In the next chapter we will see the application of this concept to body mechanics.

VECTORS AND RESOLUTION OF FORCES

A force is considered to be a pull or push. We could also define it as that which tends to cause or change the motion of matter. If we represent a force by an arrow to show the direction of the force and make the length of the arrow indicate the magnitude of the force according to an arbitrary scale, we have a *vector*.

Vectors are useful in dealing with forces. The various forces acting on an object can be represented as vectors according to a scale, and the net result of the forces, called the resultant force, can be calculated as to magnitude and direction. For example, if two people of equal strength are pulling on you at right angles, such as in Figure 3-7, would you remain in place? What would happen to you?

If we assume a 10-pound force in either direction, we can solve the problem in two ways. If the angle is a right angle, the Pythagorean theorem may be used where $a^2 + b^2 = c^2$ and c is the hypotenuse of the right triangle. $10^2 + 10^2 = c^2$, $100 + 100 = c^2$, $200 = c^2$, $c = 14.2$ pounds. If the angle is not a right angle, a

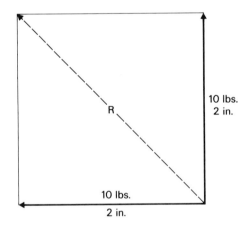

Figure 3-7.
Determining the resultant of two forces by using a parallelogram.
Let 1 inch equal 5 pounds. The measured length of the resultant is
2 27/32 inches, or 14.2 pounds.

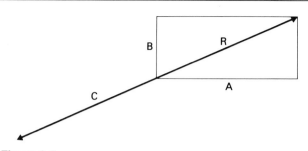

Figure 3-8.
Resultant of forces. The pull on scale C is equal to the resultant R
of forces A and B.

trigonometric formula must be used. However, most people prefer to use a *geometric* solution. By carefully making the length of the arrow represent the magnitude of the force, a parallelogram can be constructed and the resultant will be the diagonal of the parallelogram.

One point should be emphasized at this time. The resultant is the net effect of two or more forces working in different directions. Or, in other words, the forces working together could be replaced by one force, the resultant. Also, when an object is at rest, the resultant or sum of the forces must be zero. This can be shown by the use of a force board and three spring balances. No matter what tension is used, and no matter what angles are involved, the resultant of two of the forces will be equal to and in the opposite direction of the third force (Fig. 3-8).

If two forces are operating in directly opposite directions, the smaller force can be subtracted from the larger. In Figure 3-9, a force of 10 pounds to the left and a force of 5 pounds to the right are operating. The resultant will then be 5 pounds to the left.

What happens if the two forces are equal and opposite? Then the net result is zero.

If there are more than two forces operating, we can solve the problem by modifying the geometric application. Figure 3-10 illustrates four forces of known magnitude and direction. To find the resultant we may begin with any one of the forces, reproducing the angle and length. From the tip of the first vector or arrow, a second vector is drawn, again reproducing the correct angle and length for that force. From the tip of the second arrow the third vector is then drawn, and finally vector number

Figure 3-9.
Subtraction of opposite forces.

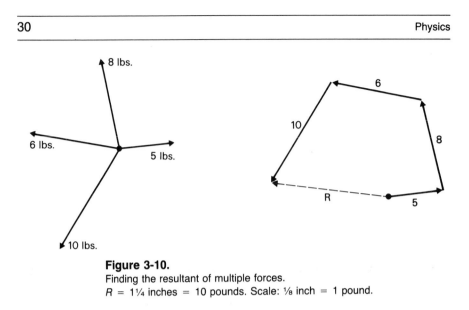

Figure 3-10.
Finding the resultant of multiple forces.
R = 1¼ inches = 10 pounds. Scale: ⅛ inch = 1 pound.

four. If the tip of the last arrow returns to the origin, the net resultant is zero, and the object is at rest. Otherwise there is a resultant. The *direction* of the resultant can be seen by the arrow going from the origin to the tip of the last arrow, and the *magnitude* of the resultant can be measured using the scale decided upon in making the vector diagram.

The Trigonometric Method of calculating resultants is more precise, but it requires the use of a table of trigonometric functions—sine, cosine, and tangent. The x and y components of each force are calculated and added. The total x and y vectors can then be resolved using the Pythagorean Theorem. This will be illustrated by several examples.

1. A force of 10 pounds is pulling North–East at a 30° angle from the horizontal. What are the x and y component vectors?

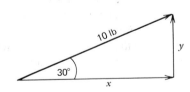

Figure 3-11.
Resolution of a force into X and Y components.

$$\text{Sine } 30° = \frac{y}{10.0} = 0.50 \qquad y = 10\,(0.50) = 5.00$$

This is the vertical vector.

$$\text{Cosine } 30° = \frac{x}{10.0} = 0.87 \qquad x = 10\,(0.87) = 8.70$$

This is the horizontal vector.

2. A force of 10.0 lb pulls North, one of 15.0 lb pulls South–East at a 45° angle, and a third force pulls 12 lb South–West at a 30° angle. Find the resultant.

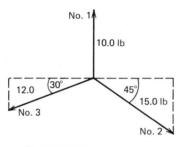

Figure 3-12.
Forces acting upon an object.

A. Determine the x and y components for each vector.

Vector 1: x component is 0 lb.

y component is 10 lb. (That one was easy.)

Vector 2: x component $-$ cosine 45° $= \dfrac{x}{15.0} = 0.707$

$x = 10.6$

y component $-$ sine 45° $= \dfrac{y}{15.0} = 0.707$

$y = 10.6$

Vector 3: x component $-$ cosine 30° $= \dfrac{-x}{12.0} = 0.866$

$x = -10.4$ (Note the negative sign.)

y component $-$ sine 30° $= \dfrac{-y}{12.0} = 0.500$

$y = -6.00$

Now we can add up all the x and y vectors.

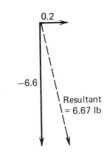

Figure 3-13.
The resultant of the sum of X and Y forces.

x vectors: $10.6 + 0.0 - 10.4 = 0.2$

y vectors: $-10.6 + 10.0 - 6.0 = -6.6$

$$a^2 + b^2 = c^2$$

$$(0.2)^2 + (-6.6)^2 = c^2$$

$$0.04 + 43.56 = c^2$$

$$43.60 = c^2 \quad c = 6.67 \text{ lb}$$

Vectors and resolution of forces will be taken up again in Chapter 4 in the discussion of traction.

THE FORCE BOARD

A force board is an apparatus used to demonstrate the resolution of forces (Fig. 3-14). If you have one available, try the following experiments:

1. On the force board take two spring balances and pull in opposite directions from the point where they are hooked together. Record the reading on one balance and then on the other. Repeat, varying the pull.

   Balance 1   Balance 2   Draw vectors
                                               representing
                                               the situation.
                                               What would the
                                               resultant be?

2. On the force board, hook three spring balances together as illustrated, with two balances at right angles and the third equidistant from the two.

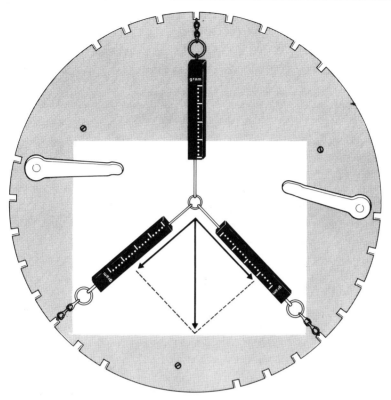

Figure 3-14.
Force board.

Source: Photograph courtesy of Sargent-Welch Scientific Company,
P.O. Box 1026, Skokie, Ill. 60077.

Record the readings on the three balances. Using either the algebraic or the geometric method of resolving the two right-angle forces, see whether the third reading agrees with the calculated resultant.

3. Hook three spring balances together and hold all equidistant from each other, that is, 120° apart. Record the pull on each balance, and calculate the resultant of two of the vectors. How does this resultant compare with the third vector?

QUESTIONS

1. Would a person in an airplane flying at 20,000 feet altitude weigh more or less or the same as on the ground? Would it make any difference if the cabin of the plane were pressurized?

2. Suppose that you were to weigh an object on the earth's surface and then reweigh it at a high altitude on (a) a spring scale, and (b) a double-beam analytical balance. Would the results be the same in all cases? Explain.

3. What is meant by "1 g"? Explain how a centrifuge can set up a greater number of g's.

4. Is velocity the same as speed? Is acceleration the same as speed? Does a car undergo acceleration when you step on the brake?

5. According to Newton's Third Law, for every action there is an equal and opposite reaction. Why does a gun not recoil just as fast as the speed of the bullet?

6. Using Newton's First Law, explain why the acceleration of a car cannot be predicted from the knowledge of the horsepower only.

7. Which usually requires less work, lifting or sliding? Discuss three patient-care situations where this question is involved.

8. Explain why the astronauts had to have a strap on their heads and chests when they were sleeping in space. Also explain how they could sleep with their heads down and their feet pointed toward the ceiling of the spaceship.

9. An object weighing 96 lb is spun in a circle with a string 2 feet in length and has a velocity of 10 ft/sec. What tension would there be on the string?

10. An object having a mass of 2 slugs has a force of 100 lb applied to it. Omitting frictional forces, what will its acceleration be? In what units?

11. If a rock is dropped from a tall building, and we omit friction with the air, what will be the acceleration after 6 seconds? What will be the velocity, and how far will it have traveled?

CHAPTER 4

MACHINES

"You can't get something for nothing." This adage applies to work expenditure as well as to other life situations. However, knowledge of the principles underlying simple machines, applied advantageously, affords efficiency and enhances general well-being.

In the preceding chapter we learned something of the nature of forces. A device used to perform work through the application of a force over a distance is a *machine*. In this chapter we shall discuss the principles of various common types of simple machines.

LEVERS

The lever is a simple machine consisting of a rigid bar that moves about a fixed point called the *fulcrum*. You are already familiar with this in the seesaw. The lever produces a rotating motion called *torque* (meaning "twist"), also called moment of force. The torque developed by a lever depends on two factors; the *force,* or the weight, and the *distance* of that force or weight from the fulcrum.

The farther out you sit on a seesaw, the more rotating tendency there will be. Torque is equal to the force times the distance from the fulcrum, or $F \times D$. If a bar is balanced, the clockwise and counterclockwise torques are equal; otherwise the bar would rotate. At equilibrium, therefore, the torque on one side of the fulcrum (torque 1) equals the torque on the other side of the fulcrum (torque 2). Expressed mathematically,

$$F_1 \times D_1 = F_2 \times D_2$$

Figure 4-1 illustrates this.

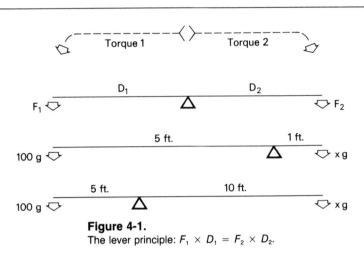

Figure 4-1.
The lever principle: $F_1 \times D_1 = F_2 \times D_2$.

In the first problem in Figure 4-1, what must the weight (*x*) be on the right end to balance the lever? 100 g times 5 feet equals 1 foot times *x*. *x* is 500 g. What is *x* in the second problem?

There are three classes of levers, differing only in the location of the fulcrum, the effort (force), and the resistance (weight).

FIRST-CLASS LEVERS

In first-class levers the fulcrum is situated between the resistance and the effort. Examples include scissors, hemostats, and hand balances.

SECOND-CLASS LEVERS

In second-class levers the resistance is between the fulcrum and the effort.

The wheelbarrow and the oxygen tank carrier illustrate levers of the second class, as does the turning of a mattress or the lifting of one end of a bed.

THIRD-CLASS LEVERS

In third-class levers the effort is between the resistance and the fulcrum.

Common examples of this class are a pair of forceps and also your arm as it holds an object.

MECHANICAL ADVANTAGE

Any machine has a *mechanical advantage,* which is the resistance overcome divided by the effort needed to overcome it. In Figure 4-2, the resistance is 5 feet from the fulcrum and the effort is 25 feet from the fulcrum; it would take 20 pounds of effort to lift a 100-pound load. ($100 \times 5 = x \times 25$. $x = 20$ lb.) Thus the mechanical advantage is 100/20, or 5 to 1. If the mechanical advantage is greater than 1, less effort is required than the resistance. The mechanical advantage can be determined also by comparing the two lever arms. In the above case, 25 feet and 5 feet equal 5 to 1, or a mechanical advantage of 5.

We can paraphrase the law, "Energy cannot be created or destroyed (except in nuclear reactions)," to read instead, "One does not get anything for nothing." In the above example we can lift 100 pounds by exerting an effort of 20 pounds. It appears that our law is breaking down. However, when we compare the distance traveled, we find that the effort will move 5 feet to lift the resistance 1 foot. In other words, we apply less force over a long distance, so the total work is still the same. This is the well-known method of buying on the installment plan, and just as in installment buying, one has to pay the full price. In addition, "interest" or friction is involved; this is discussed below. Simple machines such as levers enable us to use less force and spread it over a longer distance.

Some years ago, I had an experience illustrating the use of the mechanical advantage of a lever. As I was driving with a young friend from one Montana town

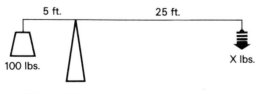

Figure 4-2.
Lever apparatus with a mechanical advantage.

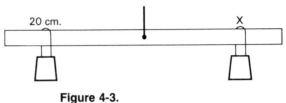

Figure 4-3.
Simple meter stick torque apparatus.

to another, the rear spring on the car broke. The body settled down on the wheel, and we were stranded at dusk some 40 miles from town. The young man, who had studied physics in high school, calmly went off in search of two fence rails. Using one rail as a lever, resting it on the tire, he easily secured a high mechanical advantage to lift up the end of the car. The second piece of wood was pushed under the body, to keep the body off the wheel. Thus we literally rode to town on a rail.

The following is a simple experiment to test the action of a lever. A torque or moment of force is equal to the product of the force times the perpendicular distance from the fulcrum

$$F_1 \times D_1 = F_2 \times D_2$$

Use a torque apparatus to check the above equation. Place a 50-g weight 20 cm from the fulcrum, and determine experimentally where a 100-g weight produces balance. Repeat for two other distances (Fig. 4-3).

If no torque apparatus is available, a very effective setup can be made by drilling a hole at the 50-cm mark on a meter stick. Tie a string through the hole, and suspend the stick from a ring stand. Suspend one clothespin at the 20-cm mark, and find where the second pin must be placed to attain equilibrium.

Repeat, hanging two pins at the 20-cm mark, and determine how many pins must be placed at the 65-cm mark. Check by calculation.

Let us work a few more problems illustrating these concepts. A wheelbarrow has a load of 80 pounds. The distance from the fulcrum, right over the wheel, to

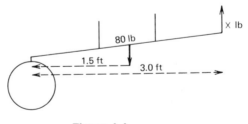

Figure 4-4.
Wheel barrow torques.

the center of the load (the resistance) is 1.5 feet. The distance from the fulcrum to the handles is 3.0 feet.

What effort will have to be exerted to hold the load? We can work the problem two ways.

$$F_1 \times D_1 = F_2 \times D_2$$

$$80 \text{ lb} \times 1.5 \text{ ft} = x \text{ lb} \times 3.0 \text{ ft}$$

$$120 = 3\, x$$

$$x = 40 \text{ lb}$$

We could also look at the relative length of the moment arms. The load is 1.5 ft and the effort 3.0 ft from the fulcrum. Since we are lifting further from the fulcrum, we have a better situation. The mechanical advantage is 3.0 ft/1.5 ft = 2/1, or 2:1. This means that we only have to lift ½ as hard, but again we pay for it by moving twice as far a distance, so the total work performed will be the same, but spread out over a longer distance. With a 2:1 mechanical advantage, therefore, to lift 80 lb we only have to use a force of 40 lb. This is the same answer we found using the $F \times D = F \times D$ equation.

If we have a crowbar 4.5 ft long, and lift a 160-lb object by resting the bar on a rock 0.5 ft from the object, how much force must be applied?

160 lb 0.5 ft 4.0 ft

Figure 4-5.
Crow bar with 8:1 mechanical advantage.

Since the total length of the bar is 4.5 ft, and the load is 0.5 ft from the fulcrum, there will be 4.0 ft left as the lifting moment arm. This will give a 4.0/0.5 = 8/1, or 8:1 mechanical advantage, and the effort will be only ⅛ the load, or ⅛ × 160 = 20 lb.

We can also calculate this by

$$F_1 \times D_1 = F_2 \times D_2$$

$$160 \times 0.5 = x \times 4.0$$

$$80 = 4\, x$$

$$x = 20 \text{ lb}$$

OTHER SIMPLE MACHINES

The wheel and axle, the wedge, and the screw are examples of other simple machines. In each case, effort is expended over a longer distance but produces a greater force. Just imagine that the steering wheel fell off your car while you were driving and you were trying to steer by holding on to the shaft. The force that could be applied would be much less. The larger the ratio of the wheel radius in comparison to the radius of the axle or shaft, the greater the force that can be applied, hence the greater the mechanical advantage. A doorknob is another example of the wheel and axle.

Examples of the wedge (also called the inclined plane) are ramps, which are sometimes used in place of stairways, needles (hypodermic, suture, etc.), and chisels.

THE SCREW

The screw is a simple machine which gives considerable mechanical advantage. Furthermore, the friction of a screw is so great that it can hold its load indefinitely without the need for locking it.

The Gatch bed is an example of a device used in nursing that uses the screw principle. The right crank regulates the position of the knees and the left crank regulates the position of the head.

How is the mechanical advantage of a screw computed? Remember that the mechanical advantage of any machine is equal to the distance moved by the force divided by the distance moved by the resistance. The knee-adjusting screw of the Gatch bed will advance 3¼ inches when the crank rotates 18 revolutions. First, we must determine the pitch of the screw. The pitch of the screw is equal to the screw advance divided by the revolutions, or

$$\frac{3.25 \text{ inch}}{18 \text{ rev}} = 0.18 \text{ inch per rev}$$

$$\text{Mechanical advantage} = \frac{\text{circumference of crank circle}}{\text{pitch}} = \frac{2\pi r}{p}$$

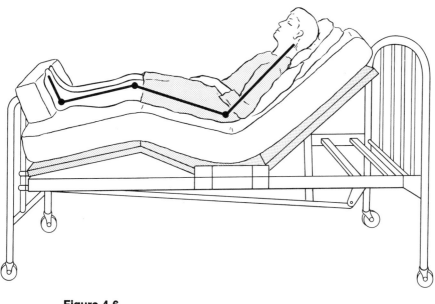

Figure 4-6.
Gatch bed.

Source: C. P. Hoffman, G. B. Lipkin, and E. M. Thompson, *Simplified Nursing,* 8th ed. J.B. Lippincott Company, Philadelphia, 1968, p. 193.

Assuming a 7-inch crank,

$$\text{M.A.} = \frac{2\pi(7)}{0.18} = \frac{2(3.14)(7)}{0.18} = 244$$

This means that the handle will travel 244 inches to move the screw one inch. Thus the force is magnified 244 times (disregarding friction) (Fig. 4-7).

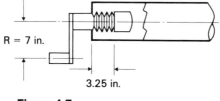

R = 7 in.

3.25 in.

Figure 4-7.
Mechanical advantage of the bed crank.

At the head of the Gatch bed, the screw will advance 8 inches when the crank rotates 44 revolutions. Therefore, the pitch of the screw is 0.18 inch.

$$\frac{8 \text{ inches}}{44r} = 0.18 \text{ per inch rev}$$

Because the crank is the same size as the other crank, the mechanical advantage is equal to 244, as before.

PULLEYS

Pulleys serve two functions: (1) to multiply a force, and (2) to change the direction of a force. The latter function makes for versatility; the pulley enables us to use the force of gravity in any direction we choose. For example, the downward pull of a weight can be changed by means of a pulley to a horizontal pull on a leg.

The use of *multiple* pulleys sets up a mechanical advantage. In Figure 4-8 the mechanical advantage is determined by counting the number of strands or ropes *excluding* the pull rope. (This rule will work in most cases, but in the case of a single *movable* pulley, the mechanical advantage is 2:1.) In the diagram there are four ropes, not counting the pull rope; thus the mechanical advantage is 4 to 1.

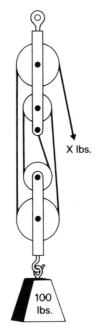

Figure 4-8.
Determining mechanical advantage of a multiple pulley system.

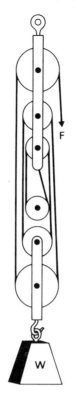

Figure 4-9.
Two-triple pulley experiment.

This means that in order to lift 100 pounds, only ¼ as much work or 25 pounds force is required.

In this case it also appears at first that we are getting something for nothing, lifting 100 pounds with a force of 25 pounds. However, in order to lift the load 1 inch, the pull rope has to travel 4 inches, so the total amount of energy expended is still the same.

EXPERIMENTS WITH PULLEYS

1. Hook up a single fixed pulley and suspend a 1-pound load. Determine the force required to lift it. How far does the force have to travel to move the load 3 inches?

2. Hook up two triple pulleys and suspend a 1-pound load (Fig. 4-9). Determine the force required to lift the weight. What is the distance that the force must travel in order to lift the weight to a height of 3 inches?

TRACTION

Now, with an understanding of pulleys and vectors, we are ready to consider the use of traction. As is well known, traction is used in cases of fracture to overcome the muscle contraction which would produce overriding and misalignment. Therefore, the object is to exert sufficient force to keep the two sections of the bone in alignment and just touching.

Skin traction is usually accomplished by applying adhesive tape or other bandage adherents to the part and attaching the ends of the tape to a "spreader" on which the pull is exerted. *Skeletal traction* exerts the pull on the bone itself by means of a steel pin (such as the Steinmann rustless steel pin) or a wire (such as the Kirschner piano wire of chromic steel) drilled through the bone and held by a metal yoke.

Provision for *countertraction* must be made if the friction of the patient's body with the bed is not sufficient to prevent him from sliding down in bed. Counter-traction may be obtained by exerting a pull against a fixed point (such as the pelvis when a Thomas splint is used) or by elevating the bed under the part being placed in traction, thus using gravity to counteract the pull on the limb in extension.

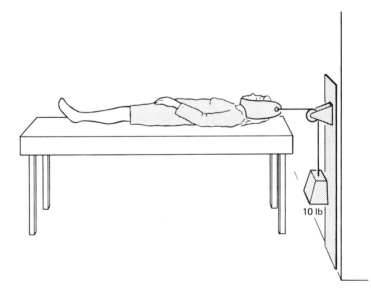

Figure 4-10.
Single pulley traction.

SIMPLE TRACTION

In simple traction, the single pulley is generally used to change direction of the force without multiplying it. Figure 4-10 shows a 10-lb weight on the rope. What will be the direction and magnitude of the force on the head?

As it is a single pulley, the mechanical advantage will be 1 : 1, and a 10-lb force downward will be changed to 10 lb horizontally. In other words, only the direction has been changed.

RUSSELL TRACTION

In Figure 4-11, an illustration of the treatment of a fracture of a femur by Russell traction, it appears that there are forces pulling in many directions except in the "correct" direction. However, if we solve for the resultant of the forces using the geometric method, we find that for a 1-pound weight the resultant will be approximately 2.2 pounds in the direction of the bone.

Let us consider the forces involved in Figure 4-11. The rope from the toe is used

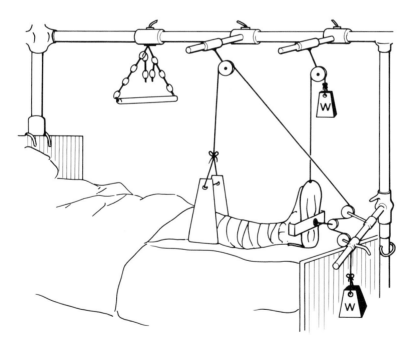

Figure 4-11.
Vectors involved in Russell traction.

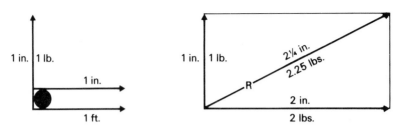

Figure 4-12.
The resultant of the vectors involved in Russell traction.

to lift the foot, and thus this force is balanced by the gravitational pull on the foot; these cancel each other. The rope going diagonally from the knee pulley to the foot pulley is just a continuation of the rope, and does not pull directly on the patient. This leaves three forces: the vertical pull at the knee and two ropes pulling on the footplate.

If we assume that a 1-pound weight has been used, the tension throughout the rope will be 1 pound. Let 1 pound equal 1 inch. Using vectors, we have the situation pictured in Figure 4-12, and the resultant is approximately 2¼ inches, or 2.25 pounds. Note that the two 1-pound forces at the foot are parallel and are directly additive; they can be treated as a 2-pound force.

If the two ropes at the foot are not parallel, as in Figure 4-13, then the three forces are treated individually.

If we use the scale in which 1 pound equals 1 inch, the resultant again is about 2.25 inches, or 2.25 pounds.

What effect would sliding down in bed have upon the traction? The vertical rope shown in Figure 4-11 would pull North-West instead of North. The 2-pound force at the foot pulling East would not be changed, even though the rope is shortened. The net effect is seen in Figure 4-14. The resultant would be changed in direction and magnitude, that is, *there would be a lesser pull in a more northerly direction.* This means that misalignment may result.

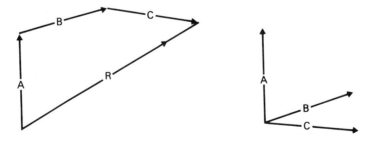

Figure 4-13.
The resultant of the vectors in Russell traction when the two ropes at the foot are not parallel.

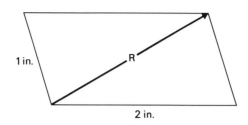

Figure 4-14.
The change in the resultant of the forces due to sliding down in bed.

Another consequence of sliding down in bed is the possibility that the weight might reach the floor; thus there will be no tension on the rope, and overriding might occur.

FRICTION

In all our calculations involving machines, we have seen that the effort times its distance equals the resistance times its distance. In other words, our hypothetical machines have been 100 percent efficient. Actually, this could never happen in practice because some power is lost in friction.

Friction is the resistance that is present whenever two surfaces slide on each other. This is because there are no perfectly smooth surfaces, and even a seemingly smooth surface has ridges and valleys. Consequently, when one surface slides on another, these ridges catch and hinder the motion.

Friction has both advantages and disadvantages. It would be impossible for us to walk without friction of our feet on the floor. The brakes on our cars employ the friction of the brake lining with the metal brake drum. The lack of friction on an icy road has caused many an accident. If there were no friction, screws and nails would fall out and buildings would collapse.

Friction costs us billions of dollars a year. It causes tools and equipment to wear out, and decreases the efficiency of machines so that they operate at a level far below the theoretical limit. A lubricant such as an oil, a grease, or graphite is used to reduce friction. This produces a film of lubricant between the two sliding surfaces, and the surfaces do not touch and wear. Instead, the sliding takes place within the layers of the lubricant. In the human body, synovial fluid keeps the joints lubricated.

A little water on a rubber stopper to be inserted into a glass tube is amazingly effective in reducing friction. When tubes or instruments are inserted into various passages of the body—the urethra, the rectum, the nasal passages, or the esophagus, for example—friction can cause not only pain but also possible damage to the membranes that line these passages unless some water-soluble lubricant is used.

Friction can be either sliding or rolling. In the former, two flat surfaces sliding against one another produce considerable friction. But if one or both surfaces are curved and either rock or roll against each other, the friction is much less than the sliding kind. It is much easier to roll a barrel across the floor than to drag it. Civilization took a long stride forward when the wheel was invented.

As a final comment on friction, a little politeness and tact in social life go a long way to reduce the "friction" between personalities, just as only a few drops of oil will do the trick in a large machine.

BODY MECHANICS

Practice of good body mechanics is very important not only in carrying out clinical procedures, but also in the everyday maintenance of posture, alignment, and motion. Good use of the body means getting the best results with the least effort. There is a right way to do everything, and the right way is the efficient and energy-saving way. The wrong way causes strain, fatigue, and frustration, and is wasteful of time and energy. Many common ailments may be caused by poor body mechanics. Among these may be mentioned low back pain and strained muscles. Maintaining the various segments of the body in correct relation to one another directly affects the working capacity of the vital organs which are supported by the skeletal structure. There is a constant interplay or state of dynamic equilibrium between the flexor and the extensor muscles of the body which produces an involuntary body sway. In the practice of good body mechanics, the less disturbance to this equilibrium the better.

There are three rules of good body mechanics: (1) use large muscles whenever possible, (2) keep feet apart for a broad base (Fig. 4-15), (3) in lifting, bend the knees with the back straight, instead of bending over.

Rule No. 1, using large muscles in lifting heavy objects, is self-evident. Rule

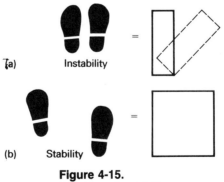

(a) Instability

(b) Stability

Figure 4-15.
Broad base for stability.

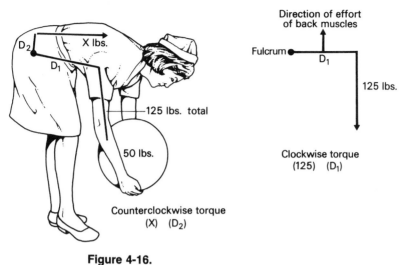

Figure 4-16.
Torques involved in lifting in a bent-over position.

No. 2, keeping the feet apart for a broad base, can be explained by the principle of gravity discussed in Chapter 3. If the feet are held close together, the base will be small and a slight tilting of the body will produce instability. If the feet are placed somewhat apart with one foot in front of the other, there will be a broader base with more stability both sideways and front to back.

Rule No. 3, about lifting objects, can be explained by the principle of levers. Torque or moment of force is equal to force times distance from the fulcrum. The farther out a weight is held, the longer the moment arm and the greater the torque.

Let us take an example. If a 50-pound object is picked up, the effort required will be close to 50 pounds if the proper procedure of bending the knees and keeping the object close to the body is used. On the other hand, if we lift in the position shown in Figure 4-16, the situation is different.

If we assume that the weight of the upper half of the body is 75 pounds, this plus the 50 pounds of the object equals 125 pounds. The clockwise torque is thus 125 times the distance from the middle of the arm socket to the hip joint, considered the fulcrum. Measure on your partner the distance described.

What keeps the person from rotating? We could see this better if we had a weight serving as a counterbalance. In the body, however, the back muscles tighten to avoid the rotation. The distance of this effort arm would be from the hip joint to the back. This short distance can be approximated if your partner stands against a wall and you measure the distance from the hip joint to the wall. The counterclockwise torque is thus equal to the force of the back muscles times the distance just measured.

Now you are ready to calculate the total force on the back muscles required to lift a 50-pound object. Your answer will probably be between 300 and 600 pounds.

This tremendous effort shows the very poor mechanical advantage of the body in this position.

EXPERIMENTS IN BODY MECHANICS

To clarify further the effect of these stresses and strains upon various parts of the body, the following experiments may be helpful.

1. Calculate the force that your back muscles must overcome when you are bending over at a right angle and holding a 25-pound weight. Estimate the weight of your upper body, and have a partner measure (in centimeters) the lengths involved.
2. Calculate the tension on the arm muscles to hold 50 pounds in the hand as shown in Figure 4-17.

PATIENT-TRANSFER TECHNIQUES

The number of health-care personnel suffering from backaches or slipped disks is eloquent testimony to the reality of torque. Imagine the effort required to lift a 200-

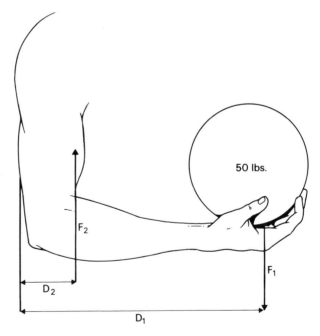

Figure 4-17.
Calculate the tension on the arm muscles.

pound patient to a position such as was described in the previous problem. The use of equipment such as a patient roller to lift the patient from the bed to a stretcher is to be encouraged. Equipment for this purpose should be available on every floor. The practice of using a pull sheet to bring the patient close to the mover's body before lifting is also desirable. Wheelchairs that can be placed in a horizontal position are also very helpful in reducing the strain of moving a patient from a bed to a wheelchair or the reverse. Use of such mechanical equipment saves staff time, and thus helps to offset its overall cost.

REDUCING BACK STRAIN

Whatever means are used to move a patient, this principle should be remembered: *keep the moment arm short*—that is, keep the weight—the patient—close to the center of gravity.

It is incongruous that industry now uses electric forklifts, enabling strong workers to lift heavy loads with a flick of a finger, while in our hospitals and nursing homes nurses and aides still strain their backs when lifting patients. It will take time (and money), but eventually mechanical help will become available in these institutions as it is now in industry.

Surveys by the National Center for Health Statistics show that more than 7 million Americans are being treated for back pain, and that about 1.5 million new cases are added each year. (In addition to poor body mechanics, other causative factors may include the person's lifestyle, situations causing tension, and so on.)

It is important to remember that it is not only the unusual, heavy lifting that can cause back trouble, but also the hundreds of daily activities that are wasteful of energy. By maintaining correct posture in all activities, whether walking, sitting, or lying down, one is able to reduce strain on the muscles. Similarly, working close to an object and keeping the moment arm short improves the mechanical advantage, so that less effort is needed and less strain experienced.

An interesting example involving torque is the so-called long scarf syndrome. Eleven cases have been reported in which a long scarf was caught in a motorcycle engine, a snowmobile engine, or, in a number of cases, a ski lift. Forty-five percent of these accidents resulted in death.[1] Most ski lift operators are aware of this danger and are on the alert for long, dangling scarves.

BODY STRESSES IN PREGNANCY

In pregnancy, the weight increase is estimated at about 22 pounds. This weight is distributed in several structures: the baby, 7 pounds; placenta, 1 pound; amniotic fluid, 1½ pounds; increased breast weight, about 1½ pounds; increased weight of

[1]*JAMA* Sept. 11, 1972.

the uterus, 1 pound; increased blood in mother's tissues, 10 pounds. By this it can readily be seen that the major portion of the weight gain is protruding anteriorly of the center of gravity.

With such a rapidly increasing weight gain and uneven distribution, body changes must occur to balance or counteract this force to maintain equilibrium and body balance. The center of gravity, normally in the pelvis, slightly anterior to the upper part of the sacrum, must shift to prevent the body from falling forward. To provide an equal and opposite reacting force in order to compensate, the upper trunk is tilted backward, and the muscles of the back, thigh, and abdomen are placed under stress (Fig. 4-18).

It should be evident that high heels, although esthetically pleasing, violate the rules of good body mechanics. They will increase the forward tendency, which will be compensated for by still greater backward tilt and an increase in the lordotic curve. In pregnancy, this is especially to be avoided. Placing one foot on a stool will release tension on the back muscles, and will reduce the amount of swayback. This can be seen in Figure 4-19.

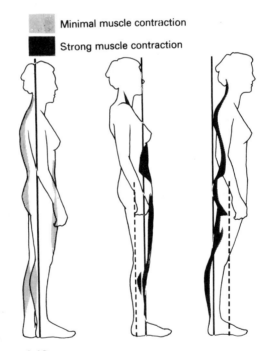

Figure 4-18.
Source: Adapted from M.C. Winters. *Protective Body Mechanics in Daily Life and Nursing.* Philadelphia: W.B. Saunders Company, 1952.

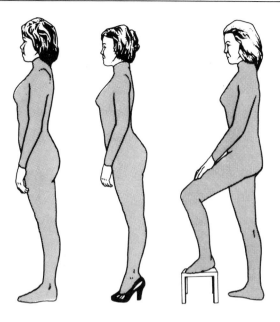

Figure 4-19.
Tension on the back in various positions.

ENERGY, WORK, AND POWER

ENERGY

Energy is the capacity to do work (work being the exertion of a force through a distance). Since energy can be neither created nor destroyed, the same law was said to hold true for matter. However, in 1905 Einstein postulated the interconvertibility of energy and matter. His energy-mass formula led directly to the development of the atomic bomb, in which some matter actually is destroyed and an enormous amount of energy created in its place. However, this does not change the fact that energy is convertible from one form to another.

Energy is considered as either potential or kinetic. *Potential* energy is energy that is available, but not actually being used at the moment, such as the chemical energy that is capable of being released in combustion, or energy due to position. Water at the top of a waterful possesses potential energy due to its position. Potential energy is static. *Kinetic* energy is energy due to motion. As the water falls, it loses its potential energy and gains kinetic energy. This, in turn, may drive a waterwheel which will produce kinetic (mechanical) energy which may drive a generator that produces electricity. Electrical energy then may be used to produce heat or light, or to perform mechanical work. It is not too farfetched, therefore, to say that we toasted our bread with the sunshine this morning. The sunlight vaporized the water, which generated hydroelectricity, which was then used in the toaster. If the electricity was generated by the burning of coal, again we can trace the source of energy to the sun, which brought about photosynthesis of the plants. The plants eventually died, and were buried. With solar energy "locked" in them, they turned to coal. Energy, therefore, can be converted from one form to another, according to our particular purpose.

WORK

If a machine is supplied with a source of energy, it can perform work. Work is defined as the product of a force (i.e., a push or a pull) times the distance over which the force is exerted ($W = F \times D$). It is important to note that the direction of the force must be the same as the movement of the object. Otherwise, the component of the force in the direction of the movement must be calculated.

The amount of work can be measured in foot-pounds. For instance, if you lift a 100-pound person a distance of one foot, you have done 100×1 or 100 foot-pounds of work. If you lift a box weighing 25 pounds to the top of a 4-foot-high table, you have once again done 100 foot-pounds of work.

Suppose that you must make a trip to the second floor of a building, and that to get there you have your choice of a ladder, a stairway, and a ramp. Assuming that your weight is 120 pounds and the vertical distance to be traveled is 10 feet, which method would require the most work? (See Fig. 4-20.)

In each instance the amount of work would be the same, 1,200 foot-pounds.

You may have seen reproductions of the statue of Atlas, who is holding the world on his back. Is Atlas doing any work? Technically he is not, because he is simply holding the world not lifting it. According to our formula, the weight of the world

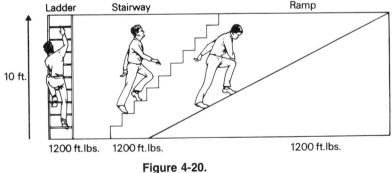

Figure 4-20.
Work done in climbing.

times zero feet equals zero foot-pounds of work (although Atlas would probably not agree).

In the metric system, if the force is measured in *dynes* and distance in *centimeters,* the work is measured in *ergs*. If the force is in *newtons* and the distance in *meters,* the work done is in *joules*. The erg is a very small amount of work and is equal to 1/10,000,000 joule.

POWER

Note that in our discussion of work, no mention was made of the time required to do the work. Power is the *rate* at which work is done, and in the English system it can be measured in foot-pounds per second (in the metric system, ergs per second). In any case, power = work ÷ time. More power is required to do a given job in less time.

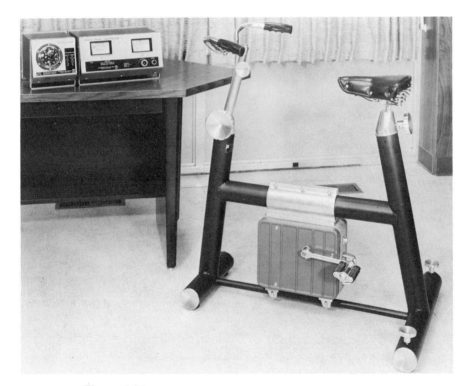

Figure 4-21.
Collins Ergometric System.
Source: Photograph courtesy of Warren E. Collins, Inc., 220 Wood Road, Braintree, Mass. 02184.

A common unit of power measurement in the British system is the horsepower. A horsepower is equal to 550 foot-pounds of work per second, or 33,000 foot-pounds of work per minute.

Horsepower also can be measured in *watts,* the unit of electrical power. One watt equals one joule per second. A horsepower is equal to 746 watts. Thus, a 1-horsepower electric motor uses approximately the same amount of electricity as fifteen 50-watt bulbs.

As noted above, the unit of work in the metric system is the erg—hence ergometer, the instrument that measures physiological work capacity. Figure 4-21 shows the Collins Ergometric System, a bike-pedaling system. The system is widely used to test patients with cardiac disorders. The machine can be preset to maintain a selected heart rate; with the addition of a cardiotachometer it is possible to limit the patient's heart rate, work load, and work duration to a safe level; a strip chart recorder provides a record of activity. Thus it is possible to measure the activity a cardiac patient can undertake.

QUESTIONS

1. What effect would the length of the crank handle of a Gatch bed have on the ease with which the patient's weight can be raised?

2. Why is it usually easier to move a patient in bed by pulling the draw sheet rather than the patient himself?

3. Why does a litter move more easily on large wheels than on casters?

4. What determines the angle of elevation of the foot of the bed when traction is used?

5. Refer back to Figure 4-20. Suppose that you were to climb the ladder in 8 seconds. How much horsepower would you have developed?

6. What effect will platform shoes with very thick soles and high heels have on the wearer's center of gravity? What effect, if any, will these have on stability?

7. If a patient in Russell traction slides way down in bed, what are two problems that may arise?

8. When a person bends forward, the upper legs and buttocks move backward. Explain why this is so.

9. A 50-lb child is seated on a seesaw 6 ft from the fulcrum. Omitting the weight of the seesaw, how far away from the fulcrum would an 80-lb child have to sit to establish equilibrium?

10. A mechanic is lifting a 600-lb engine block out of a car using two triple pulleys. How hard will he have to pull on the pull rope? What is the mechanical advantage? How far will he have to pull the rope to lift the engine 2.6 ft?

11. A crank has a radius of 8 inches. The screw moves in 0.25 inches per revolution. What is the mechanical advantage?

12. A patient is in Russell traction with a fractured femur. The rope from the knee makes a 60° angle with the horizontal, and both ropes at the foot are parallel and horizontal. Assuming a 5-lb weight, what will be the direction and magnitude of the resultant? Calculate both using the geometric and trigonometric methods.

CHAPTER 5

GASES

*Some easily conducted experiments illustrate
several of the laws governing air pressure, which
has continuous but varying effects on us and on our
environment.*

THE KINETIC THEORY

In order to understand something of the properties of gases (and, of course, of other states of matter), a brief review of the kinetic theory of matter is necessary.

The kinetic theory was an important step forward in explaining many physical phenomena. This theory states that all molecules are in continuous motion. All substances, whether solid, liquid, or gaseous, are composed of molecules that are in motion. This motion can be appreciated in Brownian movement.[1] In the solid state, the molecules are held quite firmly, vibrating in a lattice structure. As heat is applied to the substance, the molecules move faster. With more heat the bonds holding the molecules together are broken, and the material flows in the liquid form. With the addition of still more heat, the molecules leave the liquid and escape into the vapor phase, where there is the greatest movement and freedom of the molecules.

In a gas, the molecules are considered to be moving about in random fashion, continuously colliding with each other. We assume perfect elasticity of collision. This means that the sum of the kinetic energy of the particles before and after collision will be the same.

$$\text{KE (or kinetic energy)} = \frac{1}{2}m\bar{v}^2$$

[1]So-called for Robert Brown (1773–1858), who originally described this movement.

58

The symbol $\bar{v^2}$ represents the average squared *velocity* of the molecules. Just as cars on the highway have a variety of speeds, gas molecules have a whole continuum of velocities. The average kinetic energy per molecule is the same for all gases at the same temperature. This means that the heavier gases have lower velocities. The average kinetic energy $= \frac{3}{2}kT$, where k is Boltzmann's constant having a value of 1.38×10^{-16} erg per degree, and T is temperature in degrees Kelvin. This means that as the temperature increases, the average kinetic energy, or average velocity, will increase.

The velocity of gas molecules is high. For hydrogen molecules, the average velocity is over a mile per second, or over 3,600 miles per hour. For heavier molecules, it might be about 1,000 miles per hour. The gas molecules collide approximately one billion times per second. The mean free path is the distance traveled by a molecule before colliding with another molecule. This is in the range of 10^{-5} to 10^{-6} cm for a gas at standard conditions. Thus, a molecule may travel, on an average, several hundred times its diameter before colliding with another molecule.

AIR PRESSURE

Air is composed of approximately 20 percent oxygen and slightly less than 80 percent nitrogen, plus some water vapor, carbon dioxide, and inert gases. What causes air pressure? It is not enough to say that it is caused by gravity, although it is gravity that holds the air around the earth. Air pressure is exerted horizontally also, so another factor besides gravity must be involved. The kinetic theory supplies the answer. *Air pressure is the pressure exerted by the molecules striking a surface.* The random motion and collision of the molecules exert pressure on all surfaces in contact with the gas. Imagine that you are wearing a bulletproof vest and that I am shooting 22-caliber bullets at you. You would feel the force of each bullet hitting your vest. However, if many people were to fire machine guns at you at the same time, you could not detect the impact of the individual bullets, but would feel a burst of great, uninterrupted pressure. Thus it is with air pressure. The cumulative impact of millions of molecules hitting the surface each second is felt as a steady, continuous force.

Pressure is the force per unit area. For example, a total force of 9 pounds on 9 square inches of surface is equal to a pressure of 1 pound per square inch (Fig. 5-1).

The Mercury Barometer. How is air pressure measured? By the mercury barometer, which is simply a long tube closed at one end, filled with mercury and inverted in a container of mercury (Fig. 5-2). When the tube is inverted, the mercury column in it will begin to flow out. However, as the column falls, a partial vacuum

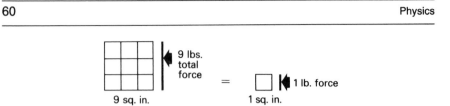

Figure 5-1.
Force and pressure diagram.

forms in the upper closed space, and atmospheric pressure will stop the flow and hold up the column. If you place a drinking straw in water and put your finger over the top, the liquid will stay in the straw. The pipette works on the same principle.

The height of the mercury column in the tube above the level of the mercury in the container will vary directly with the atmospheric pressure. The higher the air pressure, the higher the mercury column will rise. This is why barometric air pressure is reported in a unit of length, such as 760 mm or 29.9 inches of mercury. The *torr* unit,[2] a unit of pressure nearly equal to the pressure of a column of mercury 1 mm high at 0°C and standard gravity, is often used instead of mm Hg. Thus, 760mm Hg equals 760 torr. (Most writers still prefer mm Hg—perhaps because of custom and habit—so we will use the latter.)

Pressure Scales.   Air pressure is also reported in pounds per square inch, and at sea level is approximately 14.7 pounds per square inch. There are two pressure scales in use (Fig. 5-3). One is the *absolute scale* in which a complete absence of air, that is, a perfect vacuum, would read zero, and one atmosphere pressure would read approximately 15 pounds per square inch.

[2]So-called for Evangelista Torricelli (1608–1647), inventor of the barometer.

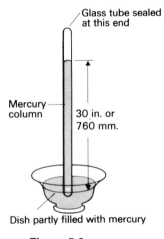

Figure 5-2.
Mercury barometer.

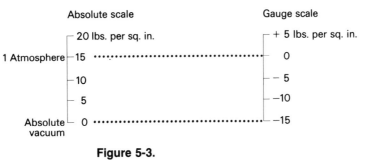

Figure 5-3.
Absolute and gauge pressure scales.

Because we live—and use most items of medical equipment—in an atmosphere of pressure, we are concerned with how much the pressure of a gas deviates from the atmospheric. The *gauge scale,* which is set at zero at atmospheric pressure, is used to calibrate practically all instruments utilized in the hospital. Thus the pressure gauge on an autoclave open to the atmosphere reads *zero;* a reading below atmospheric is called *negative pressure.* This is a misnomer in a sense, because there cannot be less than no pressure. However, it is meaningful when we mean less than *atmospheric pressure.* Any pressure less than atmospheric is a partial vacuum. An absolute vacuum can exist only in outer space.

The Aneroid Barometer.   Another instrument used for measuring pressure is the aneroid barometer. This is a round, clock-type instrument in which there is a thin-walled, hollow metal container having a partial vacuum. As the atmospheric pressure changes, the container will also change in volume since the flexible walls can move in or out. This small change is magnified by a system of levers, causing a fairly large motion of the pointer. This type of barometer must be calibrated against a mercury barometer.

The altimeter in an airplane is essentially an aneroid gauge calibrated in thousands of feet, the principle being that air pressure varies inversely with altitude. Because air pressure changes, this apparatus must be set daily for local pressures in order to ensure an accurate altitude reading.

GAS LAWS

Boyle's Law.   In 1658, a German named Otto von Guericke invented the air pump. With this device, Robert Boyle (1660) discovered the relationship between the volume of a gas and the pressure exerted on it. This relationship, known as *Boyle's law,* states that *the volume occupied by a given mass of gas is inversely proportional to the pressure exerted on it, assuming that the temperature remains constant.*

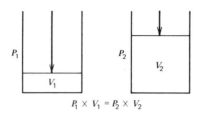

$$P_1 \times V_1 = P_2 \times V_2$$

Figure 5-4.
Pictorial representation of Boyle's Law.

The fact that the volume varies inversely with the pressure applied to the gas can be written as $V \sim 1/P$, or $V = k/P$, where k is a constant.

This should seem reasonable. If we apply pressure to a gas, the molecules will be forced closer together, resulting in less volume. Thus one liter of a gas at one atmosphere of pressure or 15 pounds per square inch will become ½ liter at a pressure of 2 atmospheres or 30 pounds per square inch.

The Snyder Hemovac demonstrates the application of Boyle's law. The apparatus is an 8-inch circular accordian-like container. It can be compressed by hand or foot, to reduce the volume. Springs inside the container cause it to expand, creating a gentle suction on the wound to remove fluid. The Hemovac will operate for about 48 hours, or until the capacity of 400 ml has been reached. The apparatus may be used to drain blood, urine, and serous fluids. The Hemovac demonstrates that as the volume increases, the pressure decreases, or, to put it another way, the pressure varies inversely with the volume.

The development of the gas laws is a good example of the "blind grouping" for scientific truth. The early Greeks had suggested that particles of matter are in constant motion. The great scientist Isaac Newton proposed a different model to explain gases. He suggested a *static* situation where the gas molecules are stationary, and held apart by repulsive forces. Boyle had already published the theory, but Newton, in 1723, showed that Boyle's Law could be explained by using Newton's static model. So great was Newton's scientific reputation that his model was accepted for quite a while—from 1723 to 1845, a period of 122 years. The Joule finally proved Newton to be wrong, and the kinetic theory of gases was accepted.

Charles's Law. In 1787, more than a century after Boyle, Jacques Charles contributed another most useful "law," which relates the *volume* of a gas to its temperature. *Charles's Law* states that *the volume occupied by a given mass of gas is directly proportional to the absolute temperature, assuming that the pressure remains constant.*

Charles's law states that the volume of a gas varies directly with the absolute temperature, provided the pressure is constant. $V \sim T$, or $V = kT$, where k is a constant. (The symbol $\sim$ means "varies with.")

If we have a balloon filled with one liter of a gas at room temperature and increase

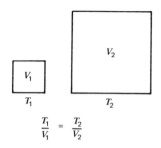

$$\frac{T_1}{V_1} = \frac{T_2}{V_2}$$

Figure 5-5.
Pictorial representation of Charles's Law.

the temperature, what effect will this have on the volume of the gas? At the higher temperature, the kinetic energy of the molecules will be greater and the molecules will move faster. This means that they will hit the walls harder, push them outward, and thereby increase the volume.

Charles's law is in operation when the cap pops off a milk bottle that has been taken from the refrigerator. As the partially empty bottle, holding cold air, warms up in the room, the increased temperature should increase the volume of the gas. Since the volume cannot change because of the rigid walls, there will be an increase in the pressure of the gas until the cap is forced off.

Another way to remember the laws of Boyle and Charles is as follows:

Boyle (inversely)   Charles (directly)

$P \uparrow V \downarrow$ $T \uparrow V \uparrow$

At this point the temperature scale must be considered. Temperature is related to the average kinetic energy of the molecules. At a higher temperature the molecules move faster. If a gas is cooled, we might expect to find that the motion of the

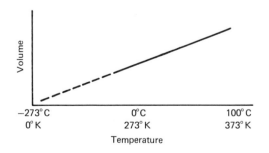

Figure 5-6.
Plot of volume as a function of temperature.

molecules is slower, and that, at a sufficiently low temperature, all molecular motion is stopped. In practice, the gases liquify before this happens. If we plot volume against the temperature of a gas, a curve as shown in Figure 5-6 is obtained.

It is found that the volume decreases 1/273th of its original volume per degree centigrade. The temperature at which the volume would theoretically become zero (if the gas did not liquefy and solidify) is $-273°C$. A new temperature scale based on this temperature, at which all molecular motion is assumed to cease, is called the absolute or Kelvin scale. (Degrees Celsius may be converted to degrees Kelvin by adding 273 to the number of Celsius degrees. Thus, $0°C = 273°K$, $10°C = 283°K$, $27°C = 300°K$, $-273°C = 0°K$, and $100°C = 373°K$.)

THE IDEAL GAS LAW

Boyle's law and Charles's law can be combined into the ideal gas law. Because $V \sim T$, and $V \sim 1/P$, $V \sim T/P$ or $PV \sim T$. In order to replace the proportionality sign with an equal sign, a constant must be used. We shall call this constant R. The equation then becomes $PV = RT$. This is valid for one mole of gas. In order to account for a different quantity of gas, however, the number of moles, n, must be used with the constant, so that

$$PV = nRT$$

The term *mole* is an abbreviated term for the *gram molecular weight*. This is found by simply adding up all the atomic weights in a molecule. For example, the gram molecular weight of hydrogen, H_2, is 2 grams, since the molecule is composed of 2 atoms of hydrogen, and the atomic weight of hydrogen is 1. Similarly, one mole of oxygen would be $2 \times 16 = 32$ grams. For a compound such as sodium hydroxide, NaOH, the molecular weight is $23 + 16 + 1 = 40$ grams.

The value of the constant R may be evaluated. If we solve the equation for R, it becomes $R = PV/nT$. A mole of an ideal gas at 760 mm pressure (one atmosphere) and $0°$ (or $273°K$) will occupy 22.4 liters. Substituting these values in the above equation, we have

$$R = \frac{1 \text{ atm} \times 22.4 \text{ liters}}{1 \text{ mole} \times 273°K} = 0.082 \frac{\text{liter-atm}}{\text{mole-}°K}$$

If other units are used (for example, mm Hg for pressure), the numerical value of R will be different.

The following examples illustrate the usefulness of this equation.

Example 5-1
What volume will 2.00 moles of oxygen occupy at a pressure of 5.00 atmospheres and 27°C?

$$PV = nRT$$

$$5.00 \, V = 2.00(0.082)(300°\text{K})$$

$$5.00 \, V = 49.2$$

$$V = 9.84 \text{ liters}$$

Example 5-2

What volume will 3.00 moles of a gas occupy at a pressure of 700 mm Hg and 57°C?

$$PV = nRT$$

$$\frac{700}{760}V = 3.00(0.082)(330)$$

$$0.921V = 8.12$$

$$V = 8.82 \text{ liters}$$

(We have to watch the units employed. The pressure must be expressed in atmospheres and the temperature in degrees Kelvin.)

Example 5-3

The equation can be used to solve for any of the factors in the equation. At what temperature would 1.50 moles of NH_3 occupy 10.0 liters if the pressure is 1520 mm Hg?

$$PV = nRT$$

$$\frac{1520}{760}(10.0) = 1.50(0.082)T$$

$$20 = 0.123T$$

$$T = 163°\text{K}$$

Another useful equation can be derived from the fact that PV/T = constant. If a certain volume of gas, V_1, is at a pressure P_1 and temperature T_1 is subjected to a different pressure P_2 and/or temperature T_2, the value of PV/T is still equal to the same constant. Therefore,

$$\frac{P_1V_1}{T_1} = \frac{P_2V_2}{T_2}$$

The following example illustrates this relationship.

Example 5-4

What volume would a gas occupy at STP (i.e., standard temperature and pressure, 0°C and 760 mm Hg) if it occupies 5.50 liters at 27°C and 600 mm Hg pressure?

$$\frac{PV}{T} = \frac{PV}{T}$$

$$\frac{600(5.50)}{300} = \frac{760(V)}{273}$$

$$11.0 = 2.78V$$

$$V = 3.96 \text{ liters at STP}$$

Another method of solving the above problem is to "talk it out." The new volume is equal to the old volume times a correction factor for the pressure and temperature.

$$V = 5.50 \times \text{(pressure fraction)} \times \text{(temperature fraction)}$$

Because the pressure increases from 600 to 760 mm Hg, the volume will *decrease,* so that the pressure fraction must be less than one, that is, 600/760 rather than 760/600. The temperature decreases from 27°C to 0°C (or 300° to 273°K). This means that the volume will also *decrease,* so that the temperature fraction will be 273/300. Then,

$$V = 5.50 \times \frac{600}{760} \times \frac{273}{300} = 3.96 \text{ liters at STP}$$

THE MOLAR VOLUME

It was mentioned earlier that *one mole of an ideal gas at STP (760 mm Hg pressure and 0°C) occupies 22.4 liters.* This value holds quite well for most gases under moderate temperatures and pressures, and it enables us to calculate the volume of gases.

Example 5-5

What volume will 16 g of oxygen occupy at STP? The molecular weight of oxygen is 32 g. Therefore,

$$\frac{16 \text{ g}}{32 \text{ g/mole}} = 0.50 \text{ mole}$$

$$0.50 \text{ mole} \times 22.4 \text{ liters/mole} = 11.2 \text{ liters at STP}$$

Example 5-6

What volume will 10.0 g of NH_3 at STP occupy? The molecular weight of NH_3 is $14 + 3 = 17$ g. Therefore,

$$\frac{10.0 \text{ g}}{17 \text{ g/mole}} = 0.588 \text{ mole}$$

0.588 moles $\times$ 22.4 liters/mole $= 13.2$ liters at STP

Example 5-7

What will be the weight of 11.2 liters of hydrogen gas at STP? The molecular weight of hydrogen is 2.00 g. Therefore,

$$\frac{11.2 \text{ liters}}{22.4 \text{ liters/mole}} = 0.500 \text{ moles}$$

0.500 moles $\times$ 2.00 g/mole $= 1.00$ g.

Example 5-8

What is the density of CO_2 gas at STP? The density is mass per unit volume. For liquids, this is usually expressed in g/ml, but for gases it is expressed in g/liter. The molecular weight of CO_2 is $12 + 2(16) = 44$ g. Therefore, one mole of CO_2 weighs

$$\frac{44 \text{ g}}{22.4 \text{ liters}} = 1.96 \text{ g/liter}$$

GRAHAM'S LAW OF DIFFUSION

The molecules of a gas are, according to the kinetic theory, in continual motion. If we release some perfume in one corner of a room, it does not take long before it can be detected all throughout the room. (The same is true, of course, for H_2S gas.) The molecules go from a region of higher to a lower concentration, and will eventually equalize throughout the room. The diffusion rates vary according to the mass of the gas. The kinetic energy is equal to $1/2 \ mV^2$, and since two gases would have the same kinetic energy at the same temperature (K.E. $= 3/2 \ kT$), the larger the mass, the smaller the velocity, and vice versa. The lighter gas will move faster than a heavier gas. This is sometimes true also in sports, where for example a lighter basketball player makes up for his lack of weight by being very fast.

Graham's law of diffusion says that *the rate of diffusion of two gases at the same temperature and pressure will vary inversely with the square roots of their masses.*

$$\frac{r_1}{r_2} = \sqrt{\frac{d_2}{d_1}}$$

Because we are using relative masses, we can use either the densities or the molecular weights, or, if both gases involved are diatomic, we can use the atomic weights. If we want to compare the relative rates of diffusion of oxygen and hydrogen, we can use either 32 g/mole for oxygen and 2 g/mole for hydrogen, or else 16 g/gram-atom for oxygen and 1 g/gram-atom for hydrogen.

$$\frac{r_{H_2}}{r_{O_2}} = \sqrt{\frac{d_{O_2}}{d_{H_2}}}$$

$$\frac{r_{H_2}}{r_{O_2}} = \sqrt{\frac{16}{1}} = 4$$

The oxygen is 16 times heavier than hydrogen, and therefore the hydrogen diffuses 4 times faster.

The fact that a lighter gas will diffuse faster is the reason for using a mixture of helium gas with oxygen rather than using nitrogen with oxygen in preparing gases for oxygen therapy for asthma. The lighter helium will diffuse faster, making it easier for the patient to breathe the gas through constricted bronchial tubes.

Another example is the separation of ^{235}U from ^{238}U by the gaseous diffusion method. This was first developed on a large scale for the Manhattan Project at Oak Ridge, Tennessee. The uranium was converted into the compound uranium hexafluoride, UF_6, which is a gas at the temperature employed. The desired isotope, the ^{235}U, makes up only 0.7 per cent of the uranium, or 1 in 140 parts. Because of the slight difference in molecular weights of ^{238}U and ^{235}U, the separation factor was only

$$\frac{r_{235}}{r_{238}} = \sqrt{\frac{238}{235}} = \frac{1.01}{1.00}$$

With such a small enrichment factor,[3] it was necessary to build a large building containing several thousand "sieves" through which the gases diffused. In addition, the gases were recycled many times using a cascade system, so that each molecule probably was diffused through a barrier over a million times. The production of sufficient ^{235}U for the atom bombs in 1945 shows the success of the method.

EXPERIMENTS WITH AIR PRESSURE

In order to demonstrate for yourself the reality of air pressure, try the following experiments:

---

[3]Actually the factor is even smaller, since the calculation should have included the weight of the fluorine also, making the ratio about 1.001 to 1.000.

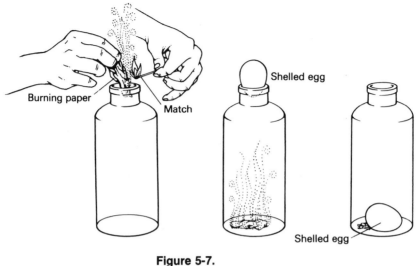

Figure 5-7.
Milk bottle experiment.

1. a. Fold a piece of paper towel until it is ½ inch wide.

   b. Light the paper at the lower end, drop it into a quart milk bottle, and quickly place a shelled hard-boiled egg on the mouth of the bottle. A balloon may be used instead of the egg.

   c. Explain the result.

   d. In order to get the egg out, wash out the burned paper, and hold the bottle upright. Blow hard into the bottle to lift the egg, and compress the air. Should you be unable to get the egg out this way, devise another method using Charles's law.

2. Repeat the first experiment, but after the lighted paper is dropped into the bottle, quickly place the bottle on its side. When the flame starts to die, press your palm over the mouth of the bottle. What happens?

3. a. Heat water in a pan until it boils. Fill a 250-ml Florence flask *half* full with *cold* water, and stop with a one-hole stopper having a glass tube drawn to a fine opening. Place the flask in the boiling water. Observe the result and explain.

   b. Prepare a bottle with a delivery tube as in Figure 5-8. Place in hot water. Observe the result and explain.

   c. Attach a balloon to the delivery tube in the bottle used in (b) above. Place in hot water. Observe the result and explain.

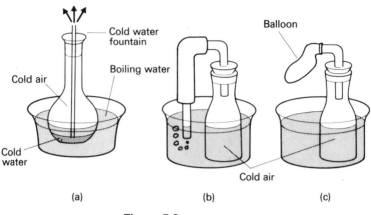

Figure 5-8.
Air pressure experiments.

STORMS AND BAROMETRIC PRESSURE

How can storms be predicted by the barometric pressure? As the air pressure in a certain area decreases rapidly, air will come rushing in from an area of higher pressure, thus causing winds. The greater the difference in pressure, the greater the winds. Rain or other precipitation occurs when warm moist air hits cooler air and releases the excess moisture. In this connection, it might be asked, "Is it possible for people suffering from corns or arthritis to predict storms?" Let us consider a corn on a toe. The possible factors that might be involved are temperature, humidity, and pressure. The first two factors do not seem to be involved. The pressure, however, may be. The pressure inside the body is the same as the atmospheric pressure. If we assume that the air pressure drops suddenly from 760 to 740 mm Hg, there will be for a brief time a difference in pressure of 20 mm Hg, and the greater internal pressure would push out. A sensitive spot such as a corn with scar tissue would be especially responsive to such a pressure differential.

AIR PRESSURE AND THE HUMAN BODY

Human curiosity and love of adventure have led us far afield of the environment into which we were born, and these ventures have given rise to many problems of adaptability, many of which deal with changes in air pressure. Only during the last three decades or so have we begun to accumulate a body of knowledge concerning the effects of reduced atmospheric pressure upon the body. Given time, the body can accommodate itself remarkably well to reduced pressures—but for a limited period only. For instance, the pilot of a nonpressurized plane usually considers it good judgment to put on an oxygen mask above 12,000 feet because that altitude is reached comparatively quickly and the pilot's body has not had time to become acclimated to the greatly reduced pressure. On the other hand, mountaineers on Everest have been able to live several days, and do heavy work besides, at altitudes above 25,000 feet without supplementary oxygen because this great altitude was reached only by slow degrees and there was time to acclimatize. The ability to adapt to low atmospheric pressures varies greatly with the person—some can do it easily, and others find it impossible. However, all experienced climbers of the high Himalayas agree that after a certain number of days above 20,000 feet or thereabouts, the lack of oxygen can produce definite mental and physical deterioration in even the most highly adaptable person. This deterioration increases rapidly with additional altitude; therefore, it appears certain that human beings simply cannot survive without a definite minimum of oxygen concentration in their bloodstream.

Most private aircraft are nonpressurized, and the rapid changes in pressure can produce distressing effects upon some individuals. For instance, if the sinuses are congested so that the opening is blocked, the volume of the trapped air will expand as the outside pressure decreases, and severe pain may result. Abdominal distention, with accompanying distress, may also occur for a similar reason. Even where cabins are pressurized, the pressure may vary somewhat and produce these same symptoms.

The December 10, 1980 issue of *The Wall Street Journal* had an interesting article by John Curley describing how J. A. C. Charles, the French physicist referred to earlier when we discussed Charles's law of gases, experienced this problem of barosinusitis, or sinus block, about 200 years ago. Charles was flying at 8,800 feet in a hydrogen balloon. "In the midst of the inexpressible rapture of this contemplative ecstasy" he reported, "I was recalled to myself by a very extraordinary pain in the interior of my right ear." Thinking the pain was caused by cold air, according to Curley, Charles donned a cap, but it was really an ear block.

It is estimated that two million air travelers suffer ear or sinus pain annually from sudden changes in cabin pressure of airplanes. It is especially a problem if one is asleep while the aircraft is landing, because there is less jaw movement or yawning. Chewing gum or yawning will help open the auditory tubes so pressure can be equalized.

Deep-sea divers have been faced with problems caused by the enormous pressures under which they must exist for a time. A very dangerous condition known as "the bends" (also known as caisson disease, and decompression sickness) is an occupational hazard of divers and tunnel workers. They work under high air pressure and consequently will have more gases dissolved in the blood (according to Henry's law, the solubility of a gas varies directly with the pressure). If the pressure is reduced too rapidly, the dissolved nitrogen may come out of solution in small bubbles (the same thing happens when you uncap a pop bottle and CO_2 bubbles fizz out of it). The bubbles of nitrogen cause excruciating pain and can cripple and even kill.

To prevent the bends, a decompression chamber is used. The diver is enclosed in the chamber and the pressure is *gradually* decreased, allowing the evolved nitrogen gas to escape slowly without the formation of bubbles.

CLINICAL APPLICATIONS OF AIR PRESSURE

Syringes.   The medicine dropper and the bulb syringe both work on the principle of reduced air pressure. When the rubber bulb of the dropper is squeezed, some air is forced out of the device. When the tip is held below the liquid surface and the bulb is released, the volume "tries" to return to normal within the instrument; but since the air pressure is lower than atmospheric, the higher outside pressure will force the liquid up into the glass barrel.

The hypodermic syringe operates the same way, except that the volume is changed by means of a piston instead of a bulb. When a drug is being removed by means of a hypodermic syringe from a container having a closed rubber cap, some precautions are in order. When the needle is pushed through the rubber cap, a good air seal is formed; and as the piston is pulled up, a partial vacuum is formed within the container in the space above the liquid. This partial vacuum will make it difficult to draw the fluid up inside the syringe. Therefore, before the needle is inserted through the cap, it is a good idea to set the piston to a point slightly above the line indicating the exact amount of the liquid to be taken. Then the needle is inserted into the container and this quantity of air is injected. Now it will be easy to pull up the proper quantity of the drug because of the increased air pressure in the vial. This procedure will work equally well with vials of any capacity.

The Alternating-Pressure Pad.   The alternating-pressure pad is used to promote local circulation in bed- or chair-ridden patients. This is a special air mattress with two series of air cells, alternately inflated and deflated every three minutes by automatic mechanical means, thereby constantly shifting the pressure points against the patient's skin. Local circulation is encouraged without the trouble and discomfort of frequent turning of the patient. This device helps prevent the formation of decubitus ulcers.

Other Devices.   Another device that promotes circulation is the water bed. It equalizes pressure against the patient's body by distributing it throughout the length of his body, thus preventing his body from touching the bed at only a few pressure points. A sheepskin placed under the patient's body also helps to provide air circulation and distribute the pressure uniformly. Synthetic pads are now available that are much cheaper than sheepskin. Recently the use of a white polyethylene foam mesh cushion was reported to be quite effective in reducing decubital dermatitis.

QUESTIONS

1. What is the meaning of "negative pressure"?
2. Why is it often helpful to chew gum while flying, especially in a nonpressurized plane?

3. Why doesn't the heavier oxygen (molecular weight = 32) settle along the floor of a room and the lighter nitrogen (molecular weight = 28) float in the upper part? (If the oxygen settled on the floor, we might have to have a snorkel tube to reach down to the oxygen in order to breathe.)

4. When you inject air into a rubber-capped container prior to filling a hypodermic syringe, why should the volume of air injected be slightly *greater* than the volume of the dosage to be withdrawn?

5. Explain how "the bends" are caused; explain the use of decompression chambers.

6. A 100-lb person has shoes with 25 inches^2 touching the floor with each shoe. What is the total force? What is the pressure? Another 100-lb young lady has high-heeled shoes with each shoe having 10 inches^2 touching the floor. Again, what is the total force? What is the pressure?

7. What will be the reading on an absolute pressure gauge if it reads 10 mm Hg on a gauge calibrated in the gauge scale?

8. Would an aneroid gauge or a mercury pressure gauge be more likely to go out of calibration? Explain.

9. What volume will 34 g of ammonia (NH_3) occupy at 0°C and 760 mm Hg pressure?

10. A balloon holds 12 liters of a gas at 700 mm Hg pressure and 27°C. What volume will it occupy at STP?

11. Explain why a mixture of oxygen and helium might be advantageous for a person with asthma?

CHAPTER 6

RESPIRATION

Knowledge of gases and air pressure is essential to understanding normal respiration; such understanding leads to the recognition and treatment of abnormal respiration, that is, respiratory disease.

In 350 BC, Aristotle conducted the first recorded experiments in respiratory physiology: he kept animals in airtight boxes until they died. He observed their panting for breath and concluded that death was due to the inability to cool themselves. Although it is true that one of the results of breathing is the elimination of heat by evaporation of water from the lungs, the primary purpose of breathing was not discovered for another 2,000 years.

Although it had been recognized that air was necessary for life, as in the Biblical story in which Elisha restored to life the son of a Shulamite woman by breathing into the child's mouth, scientists continued for many centuries to agree with Aristotle that the object of breathing was to cool the inside of the body.

We know now that respiration is the process of supplying oxygen to the body and ridding it of carbon dioxide. Respiration can be considered to take place in two phases. The first is *external respiration*. This is the inspiration of oxygen into the lungs and the removal of carbon dioxide from the lungs. The second phase is *internal respiration*. This includes the passage of oxygen through the walls of the alveoli into the bloodstream. There it is carried by the hemoglobin of the red blood cells to the body cells. The oxygen passes into the cells, where it combines with food materials to produce energy. Carbon dioxide and water are taken on by the blood. Figure 6-1 depicts these transfers. It should be noted that the passage of the gases through the walls is a case of *diffusion,* not osmosis, since the gases go from a higher to a lower concentration. (For a more detailed discussion of diffusion, see Chapter 12.)

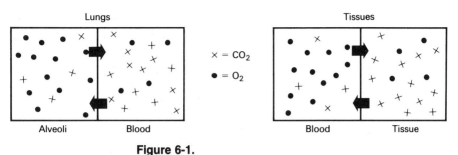

Figure 6-1.
Diffusion of oxygen and carbon dioxide.

MECHANISM OF BREATHING

How do we get air into our lungs? Air travels from a place of higher concentration to one of a lower concentration; that is, from a higher pressure to a lower pressure. Figure 6-2 shows that on inspiration the diaphragm, controlled by the phrenic nerve, moves down and the chest increases in volume. This, according to Boyle's law, should mean a decrease in pressure. When the diaphragm drops, the pressure in

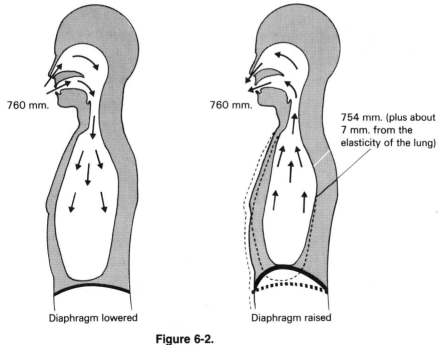

Figure 6-2.
The mechanism of breathing.

the pleural space decreases to approximately 751 mm pressure. As the external pressure is 760 mm, it seems reasonable that air will rush in to expand the lungs against the lower pressure.

On expiration the diaphragm moves up, decreasing the volume and increasing the pressure. However, the pressure attained is not greater than 760 mm as we should expect, but only 754 mm. You are probably well aware that the pleural cavity contains negative pressure, or less than atmospheric pressure. Now the question is: How can the air be squeezed out of the lungs by a pressure of 754 mm when the external pressure is 760 mm? Lung tissue is slightly elastic; and when the lung is expanded, this elasticity adds approximately 7 mm to the pressure within the lung. This is similar to a balloon that has been inflated and that tends to contract because of its elasticity. In other words, even though the pleural pressure is less than the external, addition of the 7 mm of elasticity of the expanded lung makes the total slightly greater than the external pressure, and air will be forced out.

THE SMOKING ENEMA CAN

The following simple experiment (Fig. 6-3) will demonstrate certain principles of external respiration.

1. Lower the enema can to compress the air in the can, and the can will "breathe" out.
2. Raise the can in the water, and the air inside will expand, its pressure will decrease, and air will be pulled in.

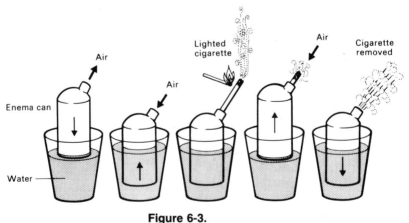

Figure 6-3.
The smoking enema can.

3. Light a cigarette without a filter, and connect it to the can. (A filter cigarette is too hard on the draw.)
4. Raise the can slowly and the smoke will be pulled in.
5. Remove the cigarette and lower the can; it will puff smoke out.

As you perform the experiment, notice the similarity to respiration. Note also that some smoke puffs out even on the third "breath." This illustrates the limited exchange of gases during respiration. This experiment demonstrates only certain aspects of the respiratory process, however.

SIMULATED BREATHING APPARATUS

A much more accurate model of the lungs and the diaphragm can be constructed with simple materials, as shown in Figure 6-4.

In this model, when the rubber diaphragm (simulating the diaphragm in the body) is pulled down, the balloons (lungs) are inflated, corresponding to inspiration. Expiration is simulated by releasing the diaphragm and thereby expelling the air.

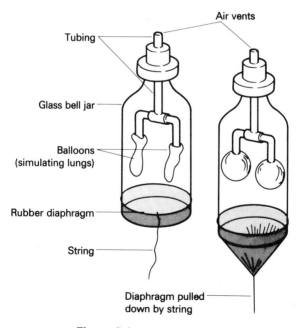

Figure 6-4.
Simulated breathing apparatus.

LUNG CAPACITY

The depth of breathing is controlled by the requirements of the body under varying conditions. In normal (quiet) breathing only a small part of the lung capacity is used: about 500 cc out of 4000 cc, or 12 percent.

The following terms are used to describe lung volumes and capacities[1]:

There are four primary, nonoverlapping lung volumes:
- *Tidal volume* (TV): volume of gas inspired or expired during each normal respiratory cycle
- *Inspiratory reserve volume* (IRV): maximal amount of gas that can be inspired in addition to tidal volume
- *Expiratory reserve volume* (ERV): maximal volume of gas that can be expired after a normal expiration
- *Residual volume* (RV): volume of gas remaining in the lungs at the end of a maximal expiration

Lung volumes are also considered in terms of four capacities; each includes two or more primary volumes:

- *Total lung capacity* (TLC): amount of gas contained in the lung at the end of a maximal inspiration
- *Vital capacity* (VC): maximal volume of gas that can be expelled by forceful effort following a maximal inspiration
- *Inspiratory capacity* (IC): maximal volume of gas that can be inspired after normal expiration
- *Functional residual capacity* (FRC): volume of gas remaining in the lungs at the end of a normal expiration

Figure 6-5 illustrates lung volumes as they appear on a spirographic tracing. The rate of gaseous exchange is controlled by the following factors:

1. Area of contact
2. Time the blood and the air are in contact
3. Volume of blood passing through the alveolar network
4. Permeability of cells forming the capillary and alveolar membranes

[1]*Medical Electronics & Equipment News,* July 1972, p. 9.

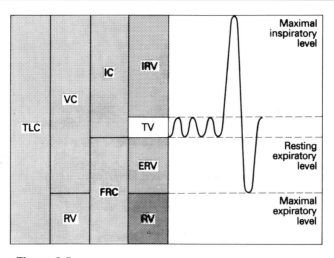

Figure 6-5.
Lung volumes and capacities.
Source: *Medical Electronics & Equipment News.* July 1972, p. 9.

5. Differences in concentration of gases in the alveolar air and the blood (also called the concentration gradient)
6. Rate of chemical reactions in the cells

The area of contact in the alveoli is 25 to 50 times the total surface area of the body, or about 50 to 75 square meters. This large area is attributable to the presence of an estimated 600,000 alveoli in the lungs. To appreciate the size of this area, visualize a floor 24 feet by 24 feet.

Respiratory rate and depth are controlled by the nervous system in response to cellular needs for oxygen intake and carbon dioxide elimination. There are three respiratory centers in the brain. The *medullary* center receives signals from all parts of the body; it is believed to act as the final control center sending messages to the muscles that control ventilation. Two centers in the pons, the *apneustic* center and the *pneumotaxic* center, appear to control the frequency and depth of breathing. In addition, the chemoreceptors (located on or near the ventral surface of the medulla) respond to the carbon dioxide concentration (PCO_2) in the arterial blood. These chemoreceptors are sensitive to the hydrogen ion concentration, that is, the pH of the arterial blood, and for this reason are also called hydrogen ion receptors. They are, in effect, minute pH meters. When a person *hyperventilates,* too much CO_2 is being exhaled, thereby reducing the CO_2 in the blood. A dangerous condition called *alkalosis* may result. The receptors respond to the decrease in H^+ ion concentration (an increase in pH) and should reduce ventilation.

In *hypoventilation* less CO_2 is exhaled and the CO_2 concentration in the arterial blood is increased. The H^+ ion concentration increases (decrease in pH). A signal is flashed to the medullary center to step up ventilation.

The following is a list of some respiratory rates on the basis of age. These figures express the total number of complete respiratory cycles per minute.

Premature infant	30 to 114
Normal term infant	28 to 48
First month	24 to 100
Adult male	14 to 28

(The female rate is a little faster.)

With position change, the following values occur:

Recumbent	14
Sitting	20
Standing	22

Although it may be believed that a complete exchange of gases occurs during the respiratory cycle, this is not the case. As noted above, normally only about 12 percent of our capacity (500 cc out of 4000 cc) is exchanged, and even with the deepest breathing about 1,000 to 1,500 cc of residual air is present in the alveoli. Consequently, the oxygen concentration varies only from about 20.9 percent by volume in the inspired air to 16.3 percent in the expired air, a difference of 4.6 percent. The mouth-to-mouth resuscitation method is based on the fact that expired air contains a fairly high concentration of oxygen.

PULMONARY FUNCTION TESTING

The great interest in pulmonary function testing expressed today is probably due to the increase in cases of lung cancer and emphysema and other respiratory diseases, as well as to heightened public concern about cigarette smoking, air pollution, and such industrial problems as silicosis and asbestosis.

There are two basic categories of measurements of pulmonary function. In one, the static class, the *volume* (TV, IRV, or ERV) is measured. The other measures the *rate* of flow. For example, the forced expiratory volume (FEV) in one second or two seconds is indicated as FEV_1 or FEV_2. In addition to the mechanical spirometer (see Chapter 9), electronic spirometers are now available. The sensing device is a flow transducer, a pressure transducer, or a hot-wire anemometer. It reacts to pressure or flow of gases and converts the input into an electrical current that can be read out on a meter or recorded on a graph. In the Collins Expirometer,

Figure 6-6.

Marion Spirostat.

Source: Photograph courtesy of Marion Laboratories, 10236 Bunker
Ridge Road, Kansas City, Mo. 64137.

the transducer is a fine-finned turbine blade that revolves on a jeweled bearing; the
bearing is capable of turning at speeds up to 100,000 revolutions per minute as the
patient exhales vigorously. The Collins Expirometer can be calibrated for liters and
tenths of liters to measure vital capacity, and in liters per second to measure peak
flow. The Marion Spirostat (Fig. 6-6) is a relatively simple machine that makes a
tracing on Polaroid film useful for diagnostic screening or as a basis for comparison
of a patient's response to therapy.

The more complex Collins Computerized Modular Lung Analyzer is shown in
Figure 6-7.

Spirometry has recently been recommended to confirm the diagnosis of asthma,
a condition characterized by attacks of dyspnea (labored breathing) and wheezing
caused by spasms of contraction by the bronchi.[2] Asthma can be an allergic response;
it can be attributable to cardiac disease or to other causes.

---

[2]D. A. Mathison, Stevenson, D. D., and Vaughan, J. H., Clinical profiles of bronchial asthma. *JAMA*
May 21, 1973, p. 1134.

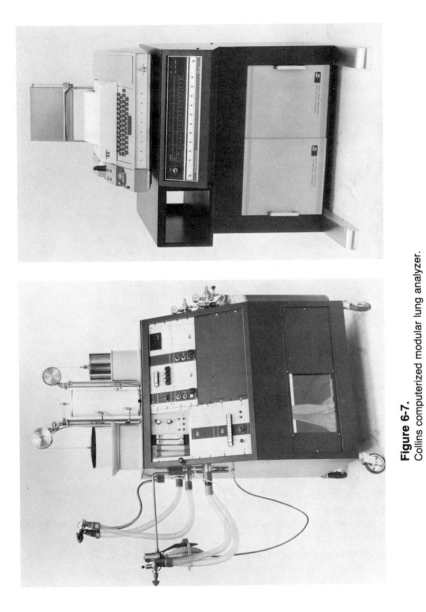

Figure 6-7.

Collins computerized modular lung analyzer.

Source: Manufactured by and courtesy of Warren E. Collins, Inc., 220 Wood Road, Braintree, Mass. 02184.

Either improvement in 1-second forced expiration volume (FEV_1) greater than 20 percent immediately after inhalation of isoproterenol, or greater than 20 percent FEV_1 reduction after inhalation of methacholine confirms the presence of asthma.

THE BRONCHOSPIROMETER

The *bronchospirometer* measures lung function of each lung separately. In cases in which surgery of one lung is contemplated, function of the other lung must first be determined. The bronchospirometer employs a catheter with two tubes and two balloons. The balloons are gently pumped up to close off the natural airways, and the functions of the left and right lungs can be determined separately. In normal persons the right lung absorbs about 55 percent of the oxygen and the left lung 45 percent.

PULMONARY EMPHYSEMA

In *pulmonary emphysema* the alveoli are abnormally dilated and lose their normal elasticity. This causes a marked reduction in gas transfer, with a gradual increase of CO_2 and decrease of O_2 in the blood. In addition, secretions accumulate in the bronchial tree and in the alveoli. The patient becomes short of breath, and has rapid and shallow ventilation to compensate for the lessened capacity. If a normal person inhales CO_2, he will hyperventilate in order to blow off the excess CO_2. In a patient with emphysema, the gradual buildup of CO_2 causes a marked loss in sensitivity to CO_2 concentration as a stimulus to breathing. He is dependent upon the less efficient mechanism of O_2 level to stimulate breathing. If a moderate amount of O_2 is administered, the patient will lose the stimulus to breathe, and hypoventilation may occur. This will cause CO_2 narcosis (stupor) with hypercarbia (excess CO_2 in the blood) followed by acidosis and loss of consciousness. Thus, as little O_2 as necessary to maintain the patient should be administered. A concentration of 24 percent is suggested, but the patient should be watched to see if this is too little or too much. Here we see a good example of how physics, chemistry, and biology are involved in treating a patient.

The measurement of blood gases gives valuable indication as to the physiological state of the patient, since these values usually change before the vital signs. Three parameters are measured: (1) pH, indicating the acid–base balance; (2) Po_2, the partial pressure or tension of oxygen in the arterial and venous blood; and (3) the Pco_2, the partial pressure of carbon dioxide in the blood.

You will recall that it is very important to maintain the pH of the blood within very narrow limits in order to avoid the coagulation of amino acids and proteins. The pH is defined as minus the log of the hydrogen ion molar concentration—pH

$= -\log[H^+]$—or, to be more precise, the $[H_3O^+]$. Consider a solution having a hydrogen ion concentration of 0.0000001 moles, or 1×10^{-7} moles. Taking $-\log(1 \times 10^{-7}) = -(-7) = +7$. If the concentration is increased 10-fold, to 0.000001 or 1×10^{-6}, $-\log(10^{-6}) = +6$. Thus the *lower* the pH value, the *more* acidic the solution.

The pH of arterial blood is about 7.41 and that of venous blood about 7.36. Why is the venous blood slightly lower in pH, that is, slightly more acidic? This is because venous blood has a higher concentration of CO_2, and CO_2 in water can form H_2CO_3, carbonic acid. The P_{CO_2} in the arterial blood is about 40 mm Hg and in the venous blood about 46 mm Hg.

The P_{O_2} refers to the partial pressure of oxygen in the blood, as mentioned above. It is about 95 to 105 mm Hg in arterial blood and 35 to 40 mm Hg in the venous blood. This seems reasonable, as the arterial blood has picked up oxygen from the lungs, whereas venous blood has given up some of the oxygen to the cells. One might ask why the P_{O_2} in the arterial blood is not 0.20×760 mm Hg, or about 150 mm Hg, but only 95 to 105 mm Hg. The reason is that the concentration in the lungs do not attain equilibrium with the outside air.

We mentioned in the section on emphysema that hyperventilation will mean that the patient will blow off too much CO_2. This will reduce the P_{CO_2}, and is called *hypocarbia*. Because there is less CO_2 (an acid anhydride) present, alkalosis may result, with an increase in the pH.

On the other hand, a high P_{CO_2}, *(hypercarbia)* may result if a patient has pain while breathing, or for some other reason limits the breathing and builds up the CO_2 concentration. This would lower the pH.

The above changes in pH illustrate some possible respiratory causes. The body has an effective series of mechanisms to help maintain a fairly constant pH. The main buffer is sodium bicarbonate, with proteins, phosphates, and hemoglobin also contributing some buffering action. The ratio of H_2CO_3 to $NaHCO_3$ is normally 1 : 20. The bicarbonate is called the alkali reserve. The HCO_3^{-1} ion can combine with H^+ to form H_2CO_3 (carbonic acid), which can then break down to CO_2 and water, permitting the CO_2 to be exhaled.

PNEUMOTHORAX

As was mentioned earlier, the pressure in the pleural cavity varies from about 751 to 754 mm. This is less than atmospheric, or, as it is called, negative pressure. Consequently, if the pressure in the pleural cavity is increased to 760 mm or greater, the lung will collapse. The introduction of air into the pleural cavity is called *pneumothorax*. The collapse of a lung is termed *atelectasis*.

Pneumothorax may be:

• *Induced, artificial, therapeutic.* A measured amount of air is introduced into the pleural cavity in order to cause a relaxation of the lung. The amount of air injected is measured by the amount of water flowing from a calibrated bottle into a second bottle and displacing an equal volume of air.

- *Spontaneous*. This is usually caused by the rupture of a tubercular lesion through the visceral pleura or of an emphysematous bleb on the surface of the lung.
- *Traumatic*. This is caused by a perforation of the chest wall or an injury causing rupture of the lung.

Lung collapse can be caused by accumulation of liquid (hydropneumothorax) as well as by pneumothorax. In either case collapse (atelectasis) will occur.

QUESTIONS

1. Why should pneumothorax occur when the pressure outside the lung is less than atmospheric?
2. Why would administering a high concentration of oxygen tend to cause a person with emphysema to stop breathing?
3. What is the difference between volume flow and rate of flow in pulmonary function testing?
4. Would hyperventilation tend to cause acidosis or alkalosis? Explain.
5. What is the difference between tidal volume and inspiratory reserve volume?
6. Explain whether the pH of the blood would increase or decrease in the case of acidosis.
7. Explain why sodium bicarbonate is given to a person suffering from acidosis.
8. Why could the use of a positive-pressure breathing apparatus be helpful in treating a patient who is suffering from respiratory acidosis?

CHAPTER 7

OXYGEN THERAPY

As people are accustomed to breathing air that contains a fixed concentration of oxygen, we may expect that any change in this concentration can have potentially profound effects, both beneficial and deleterious. In addition, both advantages and disadvantages are associated with the various methods of oxygen administration.

The use of oxygen therapy has increased greatly during the last few decades. The rationale for the use of oxygen mixtures above 21 percent O_2—the concentration found in the normal human environment—is that higher concentrations increase the rate of diffusion into the blood.

Consider a person in whom 50 percent of the pulmonary alveolar surface is not functioning: in this case, doubling of the oxygen concentration diffuses more oxygen into the blood, and the possibility of anoxia will be decreased.

THE OXYGEN TANK AND REGULATOR

In most hospitals oxygen is now available from wall outlets. Often, however, oxygen is supplied to the patient's unit from a steel reservoir called the *cylinder* or *tank*, which is transported to the unit on a wheeled carrier.

The large standard oxygen tank is filled with oxygen to a pressure of about 2,200 pounds per square inch. When the tank is installed in the patient's unit, it should be securely anchored to prevent its falling over. Should a tank tip over and its top valve be knocked off, the gas escaping at such high pressure will cause the tank to take off like a rocket with consequent devastation to patients, personnel, and building.

Before an oxygen tank is connected to the regulator, the tank valve should be opened momentarily ("cracked") to blow out dust and dirt which otherwise could damage the regulator.

The oxygen *regulator* is a device that reduces to a safe level the pressure of the oxygen that is delivered to the patient. In addition, it controls oxygen flow rate. Atop the regulator are two gauges; the one nearer the tank indicates the tank pressure, and the other gauge shows the rate of flow in liters per minute. The prescribed rate of flow is adjusted by means of the regulator valve.

There are two types of flow gauges in use: the Bordoun flowmeter and the Thorpe flowmeter. The latter is a vertical glass tube with a float that indicates the flow rate.

A method for determining how long oxygen in a standard large tank will last is based on the approximation that each pound per square inch is equal to three liters of oxygen.

$$\frac{\text{Pressure pounds per square inch} \times 3}{\text{Flow in liters per minute}} = \begin{array}{l}\text{number of minutes} \\ \text{oxygen in tank will last}\end{array}$$

For example: A tank registers 600 pounds and the flow rate is 10 liters per minute; the contents will last about 180 minutes.

$$\frac{600 \times 3}{10} = 180 \text{ minutes or 3 hours}$$

To avoid storage and shipment of oxygen tanks, many hospitals have changed to use of a liquid oxygen source. In the liquid state, oxygen takes up much less space. The oxygen converter makes oxygen available as needed.

Safety Measures. Wherever oxygen apparatus is being used, rigid precautions must be taken to avoid fire. Oxygen itself does not burn, but any combustible material will burn fiercely in the presence of concentrated oxygen. For example, if a piece of steel wool is heated to red heat over a Bunsen burner and then plunged into an atmosphere of concentrated oxygen, it will ignite almost explosively with intense white heat and a shower of sparks. A piece of cloth that merely smoulders in air will burn vigorously in the presence of concentrated oxygen.

No oil should ever be used on any part of the oxygen apparatus, and no one should ever work in the vicinity of oxygen equipment with oil on his hands, clothing, or any other part of the body. The reason for this is that the friction of the oxygen molecules may cause the oil to flare up of its own accord.

No smoking is ever permitted in a room containing oxygen equipment. Ideally, all electrical devices (heating pads, heaters, bell cords, radios, and TV sets) should be excluded from the unit, although some hospitals employ safety devices which permit the operation of electrical equipment in areas where patients are receiving oxygen.

Sparks from electrical equipment are dangerous, and must be avoided. It used to be thought that sparks from static electricity could ignite and cause an explosion, but newer studies have shown that while it is possible to produce fire through a series of sparks, it is not possible to produce a fire with just a single spark from static electricity, even at high oxygen concentrations. Thus it appears that sparks are not a serious danger; however, it still seems prudent to avoid the use of synthetic fibers because static electricity does build up in them.

METHODS OF OXYGEN ADMINISTRATION

FACE MASKS

Face masks come in a variety of types and shapes. They were formerly made of rubber, but plastic masks are widely used. They have the advantage of being disposable, thus eliminating the need for sterilization and reducing the risk of cross contamination which is a perpetual danger in hospitals. Masks are especially useful for short periods of oxygen administration, up to two to three hours, but they tend to be uncomfortable and hot.

Face masks deliver oxygen throughout the therapeutic range from 21 percent up to 100 percent. There are two styles: (1) the nasal style, covering the nose only, which permits the patient to take medication and food while still wearing the mask; and (2) the oronasal style, covering both the nose and the mouth, which makes it easier to maintain a preset concentration of oxygen.

- Simple masks. These are usually made of plastic and are disposable. They are vented by passage of the exhaled air through open holes. The oxygen concentration depends upon the depth of breathing and inspiratory flow rate of the patient. It is impossible to measure the oxygen concentration accurately, but with flow rates of 6 to 10 liters per minute it ranges from 30 to 35 percent.
- Rebreathing masks. These are oronasal masks with a reservoir bag. The masks and bag are part of a closed system (that is, where there is no opening to the outside air) in which a chemical absorber is used for the CO_2 and additional breathing gas is added as needed. These are mainly used for administration of anesthetic gases.
- Partial rebreathing masks. In this type, the first third of the patient's exhaled gas is returned to the reservoir bag. This volume is mostly oxygen, and as the exhaled air is breathed into the bag, and additional oxygen streams in from the oxygen supply, the bag fills and pressure builds up. This opens exhalation ports in the mask through which the balance of the exhaled air, the remaining two-thirds of

the tidal volume, is vented. If the bag is prevented from collapsing, and the rate is about 3 to 4 liters per minute, there will be little CO_2 in the reservoir.

- Nonrebreathing masks. This type is similar to the partial rebreathing mask except for a one-way valve. The oxygen flows into the bag, and during inspiration, also into the mask. But during expiration, the one-way valve prevents the exhaled gas from returning to the bag; instead, it is vented to the atmosphere through another one-way flap valve in the facepiece. This mask allows the most accurate measurement of oxygen concentration, and is used for precise mixtures of oxygen–nitrogen (O_2–N_2) and oxygen–helium (O_2–He), as well as for diluter type or low-concentration type masks discussed below.

LOW CONCENTRATION MASKS

For lower concentrations of oxygen (up to about 40 percent) where the patient can breathe on his own, even though at a reduced rate, a different type of disposable plastic mask is available. These masks are used in cases of emphysema, bronchitis, and other obstructive lung diseases. The Mix-O-Mask, made by the OEM Medical Company, is precalibrated to deliver 24, 28, 35, or 40 percent oxygen. The Accurox made by Med-Econ Plastic Company also has a disposable mask, but has four orifices that are interchangeable on the same mask. Accurox is also calibrated to deliver 24, 28, 35, or 40 percent oxygen. (See Fig. 7-1.)

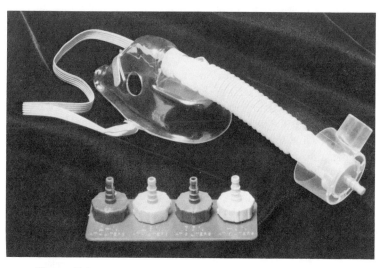

Figure 7-1.
Accurox Low-Concentration Mask with color-coded diluter jets to vary concentrations.

Source: Photograph courtesy of Inspiron Respiratory Division, C.R. Bard, Inc., 8600 Archibald Avenue, Rancho Cucamonga, Calif. 91730.

The masks dilute the 100 percent oxygen by drawing air through a side opening. Because the normal concentration of oxygen is approximately 20 percent, the ratio of air to oxygen is relatively large.

OXYGEN DELIVERY WITH MIX-O-MASK[1]

	Oxygen Flow (liters)	Air/Oxygen Ratio	Total Flow to Patient (LPM)
40% Mix-O-Mask	8	3 : 1	32
35% Mix-O-Mask	8	5 : 1	48
28% Mix-O-Mask	4	10 : 1	44
24% Mix-O-Mask	4	20 : 1	84

The oxygen is not humidified before being delivered; rather, the total volume of air is first humidified by means of either an ultrasonic or an air-operated pneumatic nebulizer.

The masks operate on the principles of the Bernoulli effect and the Venturi effect, discussed in the next chapter. In brief, the pressure exerted by a gas or a liquid flowing through a tube varies inversely with the velocity of the liquid or gas. The masks have a large tube with a calibrated oxygen inlet orifice. As the tube becomes narrower the pressure drops, causing air to be sucked in through side ports. The ports are adjusted to the desired ratio. Evacuation ports get rid of exhaled gases. Since the high flow of mixed gases exceeds the patient's needs, the mask need not fit tightly. The patient's face is kept fairly cool by the escaping gases.

Wall Service Consoles.   Installation of wall service consoles with outlets for oxygen, air, and vacuum as well as ECG and EEG has brought about some improvements in the hospital environment. By eliminating the need for oxygen tanks and vacuum pumps at the bedside, the appearance of the patient's room has been improved; at the same time, the needed services are immediately available. Similarly, the change to plastic, lightweight disposable masks not only has improved patient comfort but also has eliminated a significant source of infection in the hospital. Figure 7-2 shows a typical installation.

NASAL CATHETERS

Nasal catheters may be either *shallow* or *deep*. The shallow type, also called a nasal cannula, is a long plastic tube with two short tips about ¾ inch long, that are

[1]OEM Medical Co., Inc., 29 Meridian Road, Edison, NJ 08817.

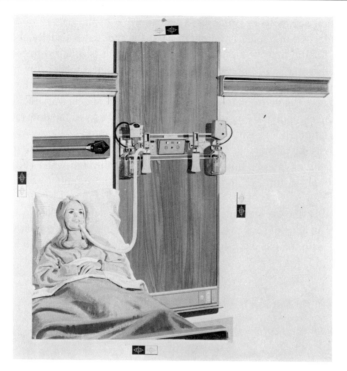

Figure 7-2.
Wall service console.

Source: Photograph courtesy of Chemetron Medical Division, Allied
Healthcare Products, Inc., St. Louis, Mo.

inserted into the nose. The cannula is held in place with a head strap. This kind
of catheter is disposable. The deep type is a long plastic tube with an opening at
the tip. The length is determined by measuring from the tip of the nose to the
earlobe. The tube should be lubricated with a water-soluble lubricant. This procedure
is irritating to the mucous membranes, so the tube should be changed at least twice
a day.

OXYGEN TENTS

Oxygen tents (Fig. 7-3) are also valuable in administering oxygen and, as is the
case with masks and catheters, there are both advantages and disadvantages associated with their use. Tents allow greater freedom and comfort for the patient, and
it is possible to cool the gases being administered. On the other hand, this is a
more expensive method for the patient due to the need for higher flow rates made
necessary by the leaking of oxygen out of the tent.

There often seems to be some misconception as to the degree of oxygen concentration reached in a tent. The flow rate used in the tent is usually about 15 liters per minute for the first 30 minutes; then the flow rate is reduced to 8 to 12 liters, depending upon the concentration desired. Even though a higher flow rate is used, the concentration attained is generally less than with the mask or catheter, because of leakage of oxygen from the tent. Since this leakage is variable, one cannot determine the amount of oxygen concentration on the basis of the flow rate, although the concentration should vary directly with the flow rate.

To avoid drying of the mucous membranes, all oxygen administration procedures require that the oxygen be humidified by passing it through water (Fig. 7-4). The oxygen is bubbled through the water and absorbs the required amount of water vapor.

To humidify the oxygen more efficiently, the water is added by means of a nebulizer or an aerosol generator. The surface area is increased, so that vaporization and mixing with the gas can occur more rapidly, thereby producing a higher vapor pressure.

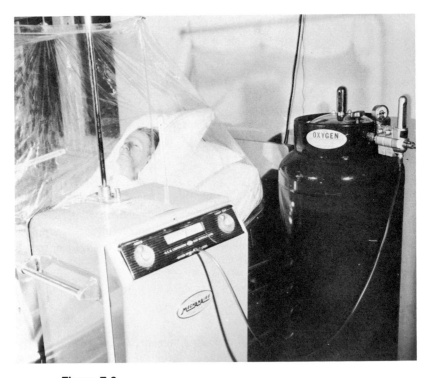

Figure 7-3.
Oxygen tent.

Source: Photograph courtesy of Union Carbide Corporation, Linde Division.

Figure 7-4.
Humidifier attached to oxygen regulator.

Source: Photograph courtesy of Union Carbide Corporation, Linde Division.

Because it is impossible to estimate the oxygen concentration in a tent due to variable leakage, it is important that the oxygen concentration be checked with an oximeter approximately every four hours.

POSITIVE PRESSURE DEVICES

There are many devices by which positive-pressure therapy can be administered. Some masks are so regulated that the oxygen will exert pressure during exhalation; in other masks pressure is exerted during inhalation; still others are designed so that positive pressure is maintained during both inhalation and exhalation. The mask that exerts pressure during exhalation is used in patients with pulmonary edema and increased venous pressure who require oxygen at a pressure slightly higher than atmospheric. The increased pressure of the oxygen pushes the fluid out of the lungs back into the venous end of the pulmonary capillaries and thus increases the oxygen intake of the blood. The positive pressure is controlled by a movable disk with numbered holes that produce pressures approximately equivalent to 1, 2, 3, or 4 cm of water. The resistance to exhalation through any of the numbered orifices in the disk is equivalent to the resistance a patient would experience when exhaling through a tube immersed in the appropriate depth of water. The largest hole in the disk is used when positive pressure is no longer required.

One type of apparatus employing positive pressure is the "cough machine"—exsufflation with negative pressure.

The method consists first of a gradual inflation of the lungs with positive pressures of 30 to 40 mm Hg for a period of time usually set at 2.5 seconds. Inflation is gradual, since a sudden inflation would tend to blow secretions and other foreign bodies farther into the deep recesses of the lungs. The purpose of the initial full expansion of the lungs is to facilitate the passage of air beyond a mucous plug into the alveoli distal to the obstruction, by widening the bronchial tree. After the lungs have been expanded in this manner, the pressure in the upper respiratory passage is suddenly dropped to 30 to 40 mm Hg below the atmosphere in 0.02 second by the swift opening of a solenoid valve (an electrically operated type) and maintained at that level for 2.0 seconds. This is followed by the next inspiratory cycle. A mask or mouthpiece may be used to connect the patient to the apparatus. Normally, five series are given, with six to ten exsufflations in a series, with a 1-minute rest between each series to prevent hyperventilation.

IPPB.   Another type of positive pressure apparatus is the intermittent positive pressure breathing apparatus, often referred to as IPPB. Positive pressure inflates the lungs during inspiration, and expiration is passive by the release or reduction of pressure to zero. IPPB should not be confused with *expiratory* positive pressure breathing in which the patient exhales against resistance to create positive pressure; or with *continuous* positive pressure breathing in which the positive pressure is retained during both inspiration and expiration.

Although the true value of positive pressure devices is still being debated, some studies indicate the value of IPPB for patients with both acute and chronic pulmonary complications, especially where respiratory insufficiency and other difficulties are present. These conditions include emphysema, bronchiectasis, silicosis, asthma, atelectasis, poliomyelitis, pulmonary edema, pulmonary fibrosis, some cardiac conditions, and other conditions involving dyspnea or insufficiency of respiratory ventilation.

Intermittent positive pressure machines are widely used in hospitals and also in some cases for home treatment. They are used as mechanical aids to ventilation, increasing tidal volume and total minute ventilation. These ventilators, which augment or completely take over the patient's ventilation function, can be operated in a *pressure-limited* mode or a *volume-limited* mode. The machines can also be operated on automatic cycling or as demand regulators. The positive pressure delivered to the patient can be set, and is approximately 20 cm of water. A new patient is usually started at about 12 cm; this is increased gradually after some treatment until 20 cm of water pressure is reached. The respiratory rate is usually started at 12 to 13 breaths per minute rather than the normal rate of 15 per minute. The IPPB units can operate with:

1. 100 percent oxygen
2. Oxygen diluted with air
3. Oxygen–helium mixture
4. Compressed air

One hundred percent oxygen produces the highest degree of arterial oxygen saturation; the lack of nitrogen would tend to decrease the total gas volume in the lungs, possibly producing atelectasis. Consequently 100 percent oxygen should be used for relatively short periods of time. The use of oxygen mixed with air can also produce fairly high saturation of arterial blood, but because it does include nitrogen, it would reduce the danger of atelectasis. The oxygen–helium mixture, usually 40 percent O_2–60 percent He, involves an inert gas, and will prevent atelectasis. The low atomic weight of helium, and consequently its high rate of diffusion, improves the transportation of oxygen into the alveoli in the presence of partial airway obstructions. Compressed air is used when the patient requires assistance in breathing without requiring additional oxygen. IPPB machines have now largely replaced tank respirators (iron lungs). For continuous IPPB administration a tracheostomy is usually performed. (A tracheostomy is the surgical creation of an opening into the trachea through the neck.) If a face mask were to be used continuously, the pressure on the facial skin would not only be very uncomfortable, but would damage facial tissue. When a tracheostomy tube is used with an inflatable rubber cuff, gas leakage is reduced to a minimum.

IPPB machines are also used to administer medication in the form of an aerosol (a fine mist spray) (Fig. 7-5). The administration of medication by this route may serve three purposes:

• *Bronchodilation.* Bronchodilators are drugs most commonly used to relieve bronchospasm. Most of them can be given in aerosols.
• *Administration of chemical or antibiotic drugs.* Antibacterial drugs when given by the aerosol route are generally given systemically at the same time. The course of medication is then short and intensive.
• *Softening of secretions.* Surface tension reducers such as tyloxapol (Alevaire) make bronchial secretions less tenacious, thus more easily expectorated.

Water can also be administered effectively by humidification of the inspired air. In breathing, considerable water is eliminated through the drying of the respiratory mucosa. Where breathing is labored or rapid a great deal of water can be lost.

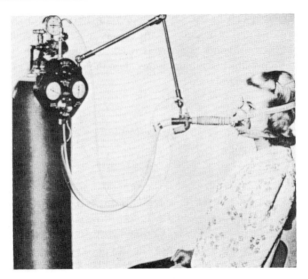

Figure 7-5.

Isoproterenol (Isuprel) administration by IPPB apparatus.

Source: Photograph courtesy of Winthrop Laboratories and Bennett
Respiration Products, Inc.

Replacing it adds to the comfort of the patient and also prevents problems caused
by dehydration.[2]

One report questions the value of IPPB in preventing atelectatic pulmonary complications after surgery.[3] The authors' conclusion is that "there is as yet no objective
evidence of benefit from the use of IPPB in the vast majority of patients with
chronic stable obstructive pulmonary disease."

It has also been found that if the pressure of the administered gas is kept constant,
and relatively low in volume, alveolar collapse often occurs. If the apparatus is set
to a certain pressure, this may be self-defeating, leading to more shallow ventilation.
Any resistance by the patient due to pain, dressings, or incisions makes the volume
even smaller. What is needed is to attempt to return to the same respiratory capacity
that the patient had before the operation. Thus the apparatus should be set on a
carefully measured *volume*, as large as is safe, and the pressure kept for a sufficiently
long time to inflate the lungs.

---

[2]For information on adjusting the machine to the needs of the patient, see J. Tinker and R. Wehner:
The nurse and the ventilator, *Am. J. Nurs.*, July 1974.

[3]R.H. Bartlett, A.B. Gazzaniga, and T.R. Geraghty: Respiratory maneuvers to prevent postoperative
pulmonary complications, *JAMA*, May 14, 1973, p. 1017.

DETERMINATION OF OXYGEN CONCENTRATION

As was mentioned previously, the flow rate is not a sure indication of oxygen concentration. The only way to ascertain the oxygen concentration is with an oxygen analyzer. Some laboratory analyzers measure the oxygen percentage using the principle of chemical absorption. A known volume of gas is injected into a chamber, and the oxygen is absorbed. The decrease in volume is measured on a calibrated scale, giving a reading of the concentration.

There are two types of medical oxygen analyzers on the market. One is based on the difference in thermal conductivity of oxygen and nitrogen. This is called an *electric analyzer* since it uses a battery and a Wheatstone bridge to measure changes in resistance due to small changes in temperature resulting from the difference in heat loss from a platinum wire in oxygen and nitrogen. Oxygen will conduct heat faster than nitrogen; consequently the higher the oxygen concentration the cooler the wire will be, and the lower the electrical resistance will be in the wire. The change in resistance is measured, and the instrument can be calibrated to show the percentage of oxygen. This type of analyzer does not measure other gases.

The other type is called the *physical analyzer*. This type is based on the fact that oxygen is paramagnetic, that is, the molecules are attracted to a magnetic field because they contain unpaired electrons. Most other gases, including nitrogen, are diamagnetic, that is, they do not react with the magnetic field because the electrons are paired. Figure 7-6 illustrates the operation of the Beckman D_2 Oxygen Analyzer. A hollow glass dumbbell is suspended between permanent magnetic pole pieces. The magnetic force exactly balances the torque of the quartz fiber and the dumbbell remains stationary for a given oxygen concentration.

When the oxygen content in the sample gas is changed, the magnetic force is altered. This change in force allows the dumbbell to change its position until it is again in equilibrium. The degree of rotation is indicated on the illuminated scale by the light beam. The degree of rotation is proportional to the change in force: the change in force is proportional to the oxygen concentration in the sample. Since the proportional relationships are exactly linear for the full scale from 0 to 100 percent oxygen, the beam of light will always indicate the exact mm. partial pressure of oxygen, or, as long as the sample is at atmospheric pressure, the exact oxygen concentration in percent. No daily calibration, warm up, or stabilization time is required.

Retrolental Fibroplasia. During the 1940s it was observed that some premature babies became blind as a result of damage to the retina. This condition is called *retrolental fibroplasia*. Oddly enough, it was found more often in the better equipped

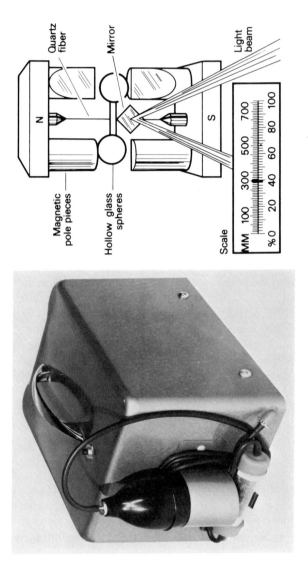

Figure 7-6.
Beckman D₂ Oxygen Analyzer

Source: Photograph courtesy of Beckman Instruments, Inc., Fullerton, Calif. 92634.

hospitals than in the less well equipped ones. Investigation revealed that the cause was the use of high concentrations of oxygen, or else sudden changes in oxygen concentrations in the incubator. Consequently the practice now is to use as low a concentration as possible. This of course means that the oxygen concentration must be monitored closely.

Other Clinical Applications.   There are several apnea/respirator monitors on the market to sense respiratory difficulties in premature and low-birthweight infants. A special transducer mattress pad is used to convert the patient's respiratory movement into flashing white lights. Should respiration stop or become shallow, an alarm will ring. Another infant-monitoring system controls temperature and oxygen and monitors pulse and respiration.

With the increased use of oxygen therapy and mechanical aids to respiration a great deal of clinical information is being gained through respiratory gas analysis and blood gas analysis. This information is useful preoperatively and postoperatively, as well as in medical units where patients have respiratory or cardiac problems. Respiratory gas analyzers are used to determine the composition of exhaled air, and can give information as to O_2, N_2, and CO_2 concentrations as well as the respiratory rate. The blood gas analyzers help determine the degree of alkalosis or acidosis—dangers associated with a number of diseases, and developing as adverse effects of oxygen treatment.

ARTIFICIAL RESPIRATION

Artificial respiration may be administered in several ways. The prone pressure method was suggested in 1893, and still is widely used. Other methods such as rocking, back pressure-arm lift, back pressure-hip lift have also been used. The most commonly accepted method is now the mouth-to-mouth method. This is a form of intermittent positive pressure breathing. Normally, expired air contains approximately 16 percent O_2 and 4 percent CO_2. If tidal volumes of 1000 cc. or more and 12 to 20 respirations per minute for adults are maintained, the patient will receive sufficient oxygen. Mouth-to-mouth airways and masks are sold that will facilitate mouth-to-mouth resuscitation without direct contact and will also help keep the airway open.

HYPERBARIC OXYGENATION THERAPY

Hyperbaric oxygenation, the use of oxygen at pressures above normal, has some important uses in medicine. During the 1920s the British, with their "pneumatic institute," and the French, using a chamber capable of developing pressures above 1 atmosphere, tested it on humans without much success. The use of high pressure air or oxygen was developed by the U.S. Navy and by engineers who built tunnels and subways in the early part of the twentieth century, so when Dr. I Boerema in Amsterdam began the tests for medical purposes in 1959, the basic knowledge was already on hand. Workers at the University of Glasgow also pioneered this technique, and in the United States, research was carried out at Harvard and Duke universities, Minneapolis General Hospital, and Mt. Sinai Hospital in New York to develop hyperbaric oxygenation into a functional medical procedure.

The use of 100 percent oxygen at 3 to 4 atmospheres pressure sends more oxygen to the cells with less strain on the heart. A typical chamber measures 10 feet in diameter and 45 feet in length. It includes an operating room, a small laboratory, a recovery room, and an air lock for entrance.

You will recall that air at normal atmospheric pressure of 760 mm Hg contains approximately 21 percent oxygen. There is a partial pressure of oxygen of 158 mm. Hg in the air. When the air is inspired, it is mixed with the gases in the lung and becomes saturated with water vapor, so the partial pressure of oxygen (Po_2) is reduced to about 104 mm Hg in the alveoli. The red cells in the lungs become 97 percent saturated with oxygen, and they carry most of the oxygen, with a very small amount carried in simple solution in the plasma. Arterial blood contains about 19.4 volumes percent of oxygen (i.e., 19.4 cc of oxygen in 100 cc of blood); there is only 0.24 cc in the plasma. You might ask, "If the hemoglobin is already 97 percent saturated, and carries just about all the oxygen, why go to the effort of using 100 percent oxygen at 3 to 4 atmospheres pressure?" This brings us to Henry's law, which states that the solubility of a gas in a liquid varies directly with the partial pressure of the gas. Therefore, by increasing both the concentration and pressure of the oxygen, it is possible to increase the solubility of the oxygen in the plasma. By using 100 percent oxygen at 3 atmospheres pressure absolute, that is, 44.1 pounds/inch^2 absolute instead of 14.7 pounds/inch^2, the alveolar pressure is then 2000 mm Hg or 20 times normal, and the solubility of the oxygen in the plasma can become the major method of oxygen transport.

The use of hyperbaric chambers raises interesting problems in nursing care. Many procedures have to be modified because of the high pressure. For example, if a vial is open to the atmosphere inside the chamber, and is then closed and taken outside, and a syringe is then inserted into the vial, the high pressure inside would shoot the plunger out of the syringe. Similarly, pH meters take a long time to come to equilibrium, compared with normal operations.

Potential uses of hyperbaric therapy include cardiac surgery for treatment of cases of circulatory insufficiency; decubitus ulcers; poisoning by carbon monoxide, barbiturates, and cyanide; and treatment of gas gangrene or tetanus. High-pressure oxygen also seems to be of value in improving radiation treatment by increasing the sensitivity of the malignant cells to the radiation.

Prolonged exposure to oxygen at high concentrations can produce toxic reactions, and also what scuba divers call "rapture of the deep," a state of euphoria in which the person becomes unable to behave responsibly. For this reason, a person should always be on duty outside the chamber to look for signs of abnormal behavior.

QUESTIONS

1. Name the three principal methods of oxygen administration with the advantages and disadvantages of each.
2. Discuss a very serious complication which can arise when oxygen is administered to patients with chronic emphysema and generalized pulmonary fibrosis.
3. In using a nasal catheter, what effect would opening the mouth have upon the oxygen concentration in the lungs?
4. Why is it necessary to humidify oxygen before it is administered?
5. What complication can occur with prolonged use of 100 percent oxygen?
6. Why is it important to know the oxygen concentration in an incubator?
7. What are two ways in which a ventilator will shut off the supply of oxygen to a patient?
8. Why is it important to set both the tidal volume and the respiratory rate? If volume and rate are too high, would you expect the patient to suffer acidosis or alkalosis?
9. If it hurts a patient to breathe after an operation would you recommend a volume or pressure limiting mode on the IPPB apparatus? Why?
10. Explain the principle behind the electric oxygen analyzer.
11. Assuming an atmospheric pressure of 760 mm Hg, calculate the partial pressure of oxygen.
12. Explain the reason hyperbaric oxygenation chambers are sometimes used with patients suffering from gaseous gangrene.
13. The density of fresh water is 62.4 pounds per cubic foot. Imagine a stack of 33 one-foot cubes of water. What will be the total force of this stack? What will be the pressure in pounds per square inch? (Hint: remember $1 \text{ ft}^2 = 144 \text{ in}^2$.)

CHAPTER **8**

LIQUIDS

Hydraulic brakes may not seem to have much in common with glaucoma, but there is a connection— both illustrate Pascal's principle, one of many statements concerning behavior of liquids. Blood vessels, atomizers, the spinal cord, water beds—all contain liquids whose flow is affected by many of the same factors.

HYDROSTATIC PRESSURE

The pressure exerted by a liquid depends on the height of the liquid and also on its density. The effect of density is shown by the fact that 1 inch of mercury exerts the same pressure as approximately 14 inches of water, because mercury is almost 14 times as heavy as water for an equal volume. That pressure also varies directly with the depth of the liquid, or, to put it another way, with the height of the liquid column, may seem reasonable. However, let us look at Figure 8-1, and ask, "What will the relative pressures be on a person at the bottom of the vessels of water in *a*, *b*, and *c*?" There will probably be some difference of opinion, which is reasonable because the pressure would appear to vary according to the shape of the vessel. Most people would probably think that *a* would have the greatest pressure.

Actually, the pressures are equal in all three, as in each case the *height* of the liquid is the same, and so is the density. This fact is called the *hydrostatic paradox*— paradox because it seems to be contrary to common sense. To summarize then, the pressure of the liquid varies directly with

• Height of the column
• Density of the liquid

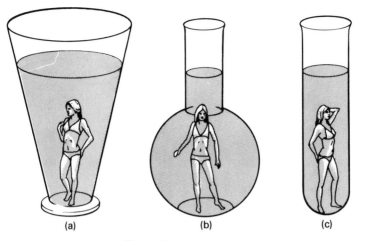

Figure 8-1.
The hydrostatic paradox.

FACTORS AFFECTING FLUID FLOW

PRESSURE

The rate of flow varies directly with the pressure. The difference in pressure between two points in a fluid is called the *pressure gradient*. In a transfusion set, the pressure could be increased in two ways: first, by lifting the container to a greater height, which increases the "head" of the fluid; secondly, by pumping air into the container. In either case the pressure of the liquid will increase its rate of flow (Fig. 8-2).

When an enema is given, the bottom of the container of solution should be no higher than 18 inches above the level of the patient's hips. This will prevent a too rapid administration with a consequent rapid dilation of the rectum, which would stimulate expulsion before the fecal matter was softened. Thus the full benefit of the solution would not be obtained. If the enema is given for an emollient effect, this quick expulsion would defeat the purpose of the treatment. If it were given for a medicinal effect, the rapid return would prevent the proper action of the medication.

LENGTH OF THE TUBING

The rate of flow varies inversely with the length of the tubing. The reason for this is that the flow of the liquid is slowed down by the friction of its molecules with the walls of the tubing. Hence the longer the tubing, the less rapid the flow.

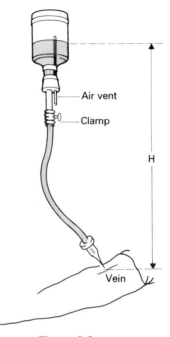

Figure 8-2.
Blood transfusion.

If the tubing is rough on the inside, the liquid flows irregularly in swirls and crosscurrents. The same thing happens in a brook when the water rolls about large rocks with eddies and turbulence. If the flow is irregular, it is called *turbulent* flow. If the tubing is smooth on the inside, the liquid flows evenly in sheets; this is known as *laminar* flow. In this respect plastic tubing is superior to rubber tubing because of its smoothness and consequently the laminar flow through it. The rate of flow through a plastic tube would be greater than through a rubber one of comparable inside diameter and length because in the former there is much less loss of kinetic energy in the liquid.

DIAMETER OF THE TUBING

A French physiologist, Poiseuille (1799–1869), arrived at the following conclusion, now known as Poiseuille's law: *The volume of a fluid flowing through a tube varies as the fourth power of the diameter of the tubing, or $V = D^4$,* other things being equal. This is a surprising fact. If a tube having a diameter of 1 cm has a flow of 1 ml per minute, what will be the flow of a tube having a diameter of 2 cm? $1^4 = 1 \times 1 \times 1 \times 1$, or 1; but $2^4 = 2 \times 2 \times 2 \times 2 = 16$. In other words, doubling the diameter of the tubing increases the rate of flow through an unrestricted tube 16 times. Similarly, a tube with 3-cm. diameter would have 3^4 or 81 times the flow rate of a 1-cm tube.

VISCOSITY OF THE LIQUID

Viscosity is the tendency of a liquid to resist flow. *The rate of flow varies inversely with the viscosity.* For example, physiological saline solution will flow more rapidly than whole blood or plasma.

Blood is approximately 2½ to 5 times more viscous than water. Some diseases cause the blood to increase in viscosity from 2.5 to 14.0 to even 20.0 by dehydration. A decrease in temperature also increases viscosity and thereby slows the flow of blood. Application of heat will make the blood less viscous and enable it to flow more easily to all parts of the body, other factors being equal.

SOME PRACTICAL CONSIDERATIONS

In a transfusion setup, the clamp on the tubing will restrict the rate of flow regardless of the diameter of the tubing. So will the lumen of the needle. Other factors to be taken into account are the blood pressure and the size and condition of the blood vessel, all of which will affect the rate of flow.

In practice, the actual rate of flow will depend on all the factors discussed in this chapter. A shorthand summary of these is the following formula:

$$V = \frac{PD^4}{nL}$$

where

V = volume flowing per unit time

P = pressure

D = diameter of the tubing

n = viscosity of the liquid

L = length of the tubing

In all transfusions, if a rigid, glass container is used, there must be an air vent to let air in to replace the liquid that flowed out. Otherwise, a partial vacuum will result, and no more liquid will flow. On the other hand, if a plastic bag is used as the container, no air vent is necessary because the bag will collapse as the liquid flows out, keeping the pressure constant.

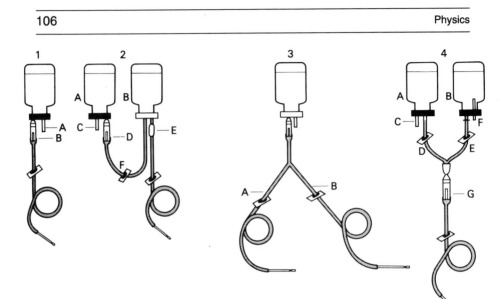

Figure 8-3.

Various means by which intravenous substances can be given by the gravity method.

(1) The solution flows through the drip-meter B (which also may be a filter if blood is given) at a rate that can be controlled by a clamp on the tubing. A provides an air inlet so that air may enter the bottle to displace the fluid that leaves. It has a one-way valve that prevents the fluid from running out. (2) This is referred to as a tandem setup. Fluid will leave bottle B if the clamp F is open. As fluid leaves bottle B, an area of lesser pressure is created in bottle B. This lesser pressure exerts its influence on bottle A and draws fluid from it. The system will operate only if C, the air vent, is open and permits air to displace the fluid that leaves bottle A. D is a drip meter and filter. E is a drip meter. In this setup A will always empty before B. If F is clamped off, no fluid will be able to leave bottle B. (3) This Y arrangement is often used for hypodermoclysis. One bottle of solution provides fluid to two injection sites. A and B parts of the tubing each have a clamp to make separate regulation of flow rate possible. (4) Solution may leave either bottle, depending on the regulation of the clamps D and E. In this setup, A is blood and B is normal saline. If D stops the flow of blood from bottle A, the saline will flow from bottle B if clamp E and the clamp below the filter are opened. The reverse also can be made to happen. Therefore, both bottles must have an air inlet, C and F, so that air can enter to displace the fluid as it leaves. In this setup, G is a filter as well as a drip meter.

Source: Elinor V. Fuerst, LuVerne Wolff, and Marlene H. Weitzel, *Fundamentals of Nursing*, 5th ed., J.B. Lippincott Company, Philadelphia, 1974, p. 337.

IV ADMINISTRATION SETUPS

Figure 8-3 illustrates and describes four setups: the single-bottle arrangement, the tandem arrangement, the Y arrangement, and the two-bottle arrangement.

In the administration of parenteral fluids to pediatric patients or when small volumes of drugs are to be administered intravenously, it is very important to know the precise volume. Calibrations on the large intravenous (IV) bottles are too inexact to be useful. In such a case a small burette chamber is used to ensure accuracy in reading. An example of such a system is shown in Figure 8-4, the Soluset IV Set.

The Soluset comes with either a 100-ml. burette of a 250-ml burette, and with either a microdrip orifice preset at 60 drops per milliliter or a regular orifice preset at 15 drops per milliliter. The sets are labeled to indicate both the size of the burette and the drop size—100 × 60 means the 100-ml burette with 60 drops per milliliter. This is used in pediatric patients. The 250 × 15 is used in adults for repeated measured doses of 1V fluid.

This set also has an interesting additional feature. To prevent the air from entering the system when the fluid is used up, and thereby reduce the possibility of air

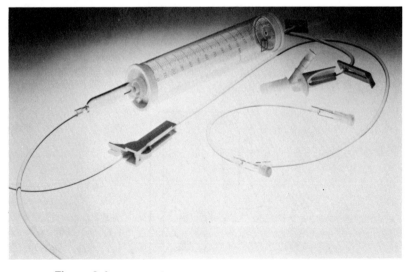

Figure 8-4.
The Abbott Soluset IV set.

Source: Photograph courtesy of Abbott Hospital Products, North Chicago, Ill. 60064.

embolism, the burette has a small hinged rubber flutter valve that closes when the fluid is drained out. This can be seen at the bottom of the burette in Figure 8-4.

As a result of the revolutionary developments in semiconductors and microcomputers, machines are now available from companies such as Abbott Laboratories, Ivac Corporation, and McGaw Laboratories for accurate and constant IV infusion. These IV pumps can be preset for total volumes from 1 to 999 ml, and at variable flow rates. These machines are highly accurate despite fluid viscosity, drop size or height variations. Should air be introduced or occlusion occur so that the pump is unable to function, an audible or visual alarm, or both, will be activated. These machines are especially valuable in administering critical drugs, in pediatrics, or in specialized procedures such as in cancer chemotherapy.

THE SIPHON

The siphon is a device for moving a liquid from one level to a lower one. In Figure 8-5, bottle B must be lower than bottle A. The tube is completely filled with water. The liquid at C falls, leaving a partial vacuum at D. Air pressure (P) forces liquid into the tube to compensate, so a continuous flow results.

BLADDER IRRIGATION

Figure 8-6 shows a setup for *alternate irrigation and drainage* of the bladder. At first, the tube from A is open to the bladder, and the clamp to bottle B is closed.

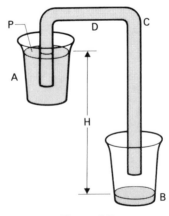

Figure 8-5.
The siphon.

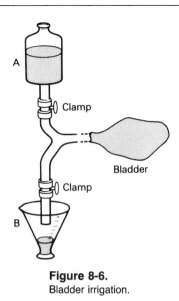

Figure 8-6.
Bladder irrigation.

When the desired quantity of irrigation fluid has reached the bladder, the clamp from A is closed, and the clamp to B is opened. The flow is by straight gravity drainage; the siphon principle does not apply in this case.

TIDAL DRAINAGE

This serves somewhat the same purposes as alternate irrigation and drainage, but it is not necessary to open and close clamps, because, unlike the bladder irrigation apparatus in the preceding diagram, the tidal drainage setup employs the siphonage principle. Filling and emptying of the bladder are accomplished automatically.

Tidal drainage fills the bladder with irrigating fluid at the rate of approximately 40 to 60 drops per minute, and then empties it when the desired intravesical pressure is attained. The system fills by gravity flow and empties by siphonage. The time for a cycle depends upon the rate of flow and also the condition of the bladder. Here are some reasons why tidal drainage is used.

1. It prevents the overstretching of an atonic bladder that results from excessive filling.
2. In a hypertonic bladder, it helps prevent shrinkage of the bladder reservoir caused by mechanical blocking of the emptying impulses.
3. It prevents bed-wetting which, without tidal drainage, would be almost constant.
4. It helps a "cord bladder" to function as normally as possible by exercising the bladder muscles in the normal cycle of slow filling and rapid emptying.

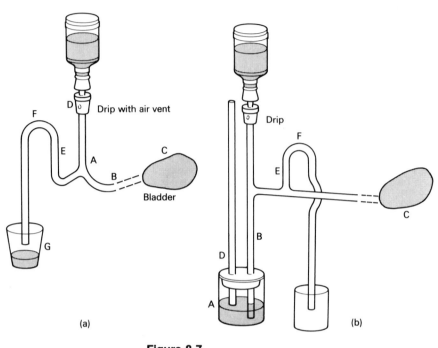

Figure 8-7.
Two tidal drainage arrangements.

In Figure 8-7*a,* the apparatus functions as follows:

1. Fluid flows through A and B, from which air has been eliminated, through a Foley catheter into the bladder C.
2. During filling of bladder C, air in the system escapes through air vent D.
3. Now, as the intravesical pressure increases, fluid rises in the system and ascends tube E until it reaches apex F of the loop. Fluid overflows and starts the siphon which continues to flow into the drainage bottle G until the bladder is empty.
4. At the end of the siphon cycle, air entering D fills the siphon and thereby stops the flow. The tidal drainage is now automatically set up to begin another cycle.

In Figure 8-7*b,* the same cycle is followed except that air is first expelled through tube D. After bottle A is filled, the liquid will rise in tubes B and D, and flow into the bladder C. When the bladder is full, the level will rise in the whole system, including the loop E. When the apex F is reached, siphonage empties the system. Air then enters through tube D when the water level in A drops below the end of the tube.

Figure 8-8 depicts a very neat apparatus called the McKenna tube, which has the siphon loop built right into the glass drip tube.

The intravesical pressure is equal to the difference in height between the liquid in the system and the symphysis pubis. In the early cord injury the bladder is usually atonic, and the pressure should be set low (1 to 2 cm) to prevent overdistention. As the bladder tone increases and reflex contractions begin to appear, the pressure may be increased slightly to maintain a full bladder capacity of 400 to 500 cc. For tidal irrigation of the normal bladder a pressure head of 10 cm is satisfactory; in the spastic bladder it may need to be set as high as 15 to 18 cm.

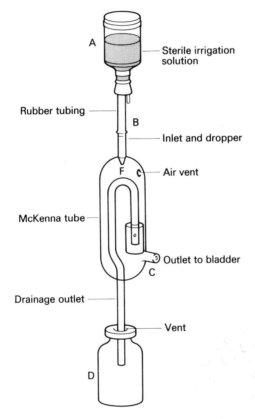

Figure 8-8.
McKenna tube. The solution drips from bottle A through outlet C to the bladder. When the bladder is full, the solution rises in the tube until the level reaches point F, the apex of the siphon loop. At that point the siphon cycle begins, draining into bottle D. The height of the loop is determined by the height at which the McKenna tube is clamped. Bottle A and D are not to scale. The McKenna tube is about 7 inches long.

GRAVITY DRAINAGE

Gravity is used in a number of other drainage methods. Postural drainage is employed in such conditions as bronchiectasis or lung abscess, in which the lungs must be periodically drained of their purulent exudate in order that the healing process may be facilitated.

In this procedure the patient simply flexes his body over the side of the bed, head and shoulders hanging as close to the floor as possible while the legs remain supported on the bed. Evacuation of the lungs is also aided by coughing during the procedure. The frequency of this procedure depends on the patient's condition, but usually it is performed 5 to 10 minutes every 2 to 4 hours during the day, except when there is a sign of fatigue.

UNDERWATER GRAVITY DRAINAGE

This is used chiefly following thoracic surgery. The purpose is to maintain breathing in a manner which is as normal as possible by draining excess fluid from the operative area, providing normal intrapleural pressure, and keeping outside fluids and air from entering this area.

Before the patient leaves the operating room, a thoracotomy tube is inserted in the pleural space and is attached to a drainage tube that is submerged a predetermined distance in a vessel of water.

Whenever the intrapleural pressure becomes greater than the pressure exerted by 1 cm of water (which is the resistance of the water for the distance the tube is submerged), the liquid or air drains or bubbles into the bottle. This can happen spontaneously due to accumulated fluid and air or with coughing, straining, or turning, all of which raise the intrapleural pressure to a positive level. With inspiration, the column of water is merely elevated in the tube to as many centimeters as the intrapleural pressure is less than atmospheric.

THREE-BOTTLE SUCTION

In order to remove air and fluid from the pleural cavity, closed gravity drainage may be used. This was discussed previously. Another method is gentle suction. In this case an electric pump that has a variable suction control may be used. However, quite often the suction apparatus available, whether electric pump, wall suction, or aspirator, has too great a suction.

In this situation a two-bottle or a three-bottle setup is used. In Figure 8-9, the air vent tube in the middle bottle controls the degree of suction (negative pressure) by acting as a "break valve." This tube is usually called the control tube. The control tube is an alternate pathway for air to enter. The deeper the tube is submerged, the greater the water resistance that the incoming air must overcome. No matter how strong the suction exerted by the pump (e.g., 90 or 120 mm), the suction will be only that which is determined by the *submerged* length of the control tube.

If the tube is 3 to 4 inches under water this will result in a negative pressure of 7.5 to 10 cm of water. Remember that it is not the *amount* of water in the bottle that is important, but the distance to which the tube is submerged.

If you want to change 10 cm of water to mm Hg, this is done as follows:

10 cm of water = 100 mm of water. Mercury is approximately 14 times as dense as water, so 100/14 = about 7 mm Hg. That is, 7 mm Hg will have the same weight and will resist the entrance of air through the bubbler tube, as much as will 100 mm water.

The purpose of the safety bottle is to protect the suction apparatus from having water sucked back into it in the event of some difficulty.

The first bottle is to collect the drainage without having it mix with water. This makes it possible to measure accurately the amount of fluid recovered and to observe its characteristics. This bottle is attached to the patient in the operating room under sterile conditions.

As the level in the drainage bottle rises, the length of the tube submerged in it must be subtracted from the length of the tube submerged in the control bottle in order to find the net suction, since the suction must overcome the increasing fluid

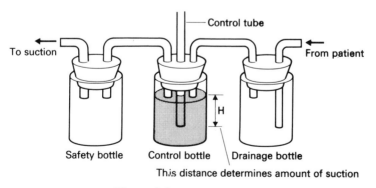

Figure 8-9.
Three-bottle suction apparatus.

pressure in the drainage bottle. When the immersion depths of the tubes in the control and drainage bottles are equal, no negative pressure will be communicated to the patient; the control will break the suction.

WALL SUCTION

The recent use of piped vacuum systems has made for changes in hospital routines. Installation of a large vacuum pump with outlets to many patients rooms has made it possible to eliminate individual vacuum pumps. However, even though the wall suction systems (or ceiling suction outlets) differ from earlier portable systems, the basic principles remain unchanged, and basic techniques and procedures for patient drainage have changed very little. The newer equipment is simply a modification of that described previously, such as the three-bottle suction apparatus.

The piped vacuum systems are usually "dry" systems; fluids and solids should not enter the system, or pipes may become clogged and the pump damaged. Actually the piped systems do have a large safety tank, but every effort should be made to avoid having liquids enter the wall suction. This is especially true where pathological samples are involved.

In order to avoid having fluids enter the wall suction, collecting bottles now come with a float which cuts off further flow when a certain height has been reached (Fig. 8-10). The float operates on the same principle as the float in the water tank in the lavatory. As the liquid rises in the bottle, the float also rises, shutting off a valve and stopping further suction.

SUCTION GAUGES

There are two types of gauges that can be snapped onto the wall suction connector. One is a suction gauge that provides *continuous suction* from 0 to 200 mm Hg. Wall suction usually is approximately 15 or 20 inches of mercury. Since 1 inch is equal to 2.5 cm, 20 inches of Hg line vacuum is equivalent to 50 cm (500 mm) of Hg. The density of mercury is 13.6 g/ml; 50×13.6 equals 680 cm of water. Such high negative pressure is avoided in aspirating the trachea to remove phlegm obstructions because of the hazard of reflex closure of the trachea in the presence of excessively high negative pressure, as well as in removal of air from the trachea of the patient who is already in respiratory difficulty as a consequence of the mucous obstruction. The Suction Regu-Gage permits reduction of pressure to approximately 80 to 120 mm Hg, the pressure usually recommended (Fig. 8-11).

A similar gauge is called an *intermittent vacuum regulator*. This gauge consists of two pressure reduction units in series. The first applies a constant pressure to the second, and in addition, provides the cycling mechanism with its mechanical

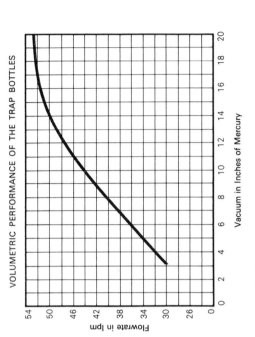

VOLUMETRIC PERFORMANCE OF THE TRAP BOTTLES

Flowrate in lpm

Vacuum in Inches of Mercury

Figure 8-10.

Collection bottle with cutoff float.
Increasing flow rate with increasing suction.

Source: Photograph courtesy of Chemetron Medical Division, Allied Healthcare Products, Inc., St. Louis, Mo.

115

Figure 8-11.
Suction gauge.

Source: Photograph courtesy of Chemetron Medical Division, Allied
Healthcare Products, Inc., St. Louis, Mo.

actuating force. The second regulator adjusts from 90 to 120 mm Hg. The time
cycle for on and off is usually 12 seconds on and 12 seconds off, but the cycle may
be increased if desired.

The purpose of *intermittent* suction is to reduce tissue occlusion by returning
collecting bottle and abdominal cavity to atmospheric pressure during the off cycle.
Backflow of fluid from the elevated bottle during the off cycle tends to break up
clots and solids, reducing "flush" problems.

POSITION OF THE DRAINAGE BOTTLE

If the drainage bottle in a gastrointestinal suction setup is *higher* than the patient,
as in Figure 8-12, some negative pressure will be required to raise the liquid up
to the bottle. If we assume a suction of 120 mm Hg and a distance of 18 inches
between the distal end of the implanted catheter and the liquid level in the collecting
bottle, what would be the reduction in negative pressure? Let us do the calculations:

18 inches of water (assuming a density of 1 g/ml) times 25 mm. per inch equals
450 mm of water. To convert this to mm of Hg, we use the density of mercury,

which is 13.6 g/ml; 450 divided by 13.6 equals 33 mm Hg. Therefore, subtracting 33 mm from 120 mm, we get 87 mm Hg net suction. Placing the collection bottle 18 inches above the catheter means that the suction attainable in the cavity will be 87 mm Hg instead of 120 mm Hg.

If, on the other hand, the drainage bottle is placed on the floor as in Figure 8-13, so that it is *lower* than the catheter, and we again assume 18 inches difference in height, then the siphon effect will facilitate the flow of liquid, and we must *add* to the 120 mm. negative pressure the 33 mm Hg, making a total 153 mm Hg negative pressure in the abdominal cavity. You should notice in these calculations that the important thing is to keep track of the units involved.

If there is a column of water present, the water seal will prevent the stomach from returning to atmospheric pressure during the off cycle.

In pleural drainage, entrapped air and fluids are removed from the pleural cavity to establish the negative pressure outside the lung, which allows for reexpansion of the lung. In this case, very low suction is required, and the collecting bottle is placed below the level of the patient. By having the inlet tube into the bottle extend into the water at the bottom of the bottle, the water seal allows gases to be pulled through into the bottle, but prevents gases from returning to the pleural space. The reason for having the bottle below the level of the patient is to prevent the return of fluid to the pleural space. The three-bottle control setup described in Figure 8-9 has now been improved, and is available in the Micro-Low Pleural Drainage

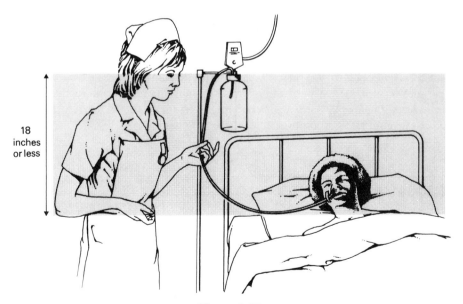

18 inches or less

Figure 8-12.

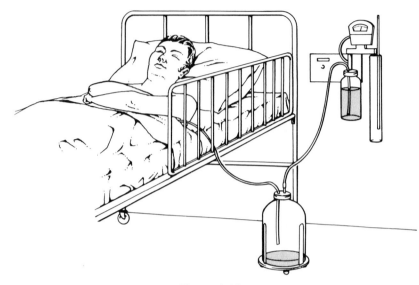

Figure 8-13.

Unit. As will be seen in Figure 8-14, the degree of suction is controlled by the bubbler tube, exactly like the bubbler tube in the middle bottle in Figure 8-9. The amount of negative pressure is measured by the length of the tube submerged in the water. The main advantage of this newer apparatus is the more compact arrangement, which cuts down on clutter at the bedside. (See also Fig. 7-2, photograph of a wall service console with connections for vacuum, air, and oxygen.)

By means of a bubbler tube, a wall suction of 20 inches or 500 mm Hg can be reduced to 80 or 120 mm Hg with the Regu-Gage, which in turn is reduced to 10 cm of water, equal to 100 mm of water or about 7 mm Hg. Because the negative pressure in wall suction will vary appreciably with the number of nearby outlets being used (similar to hot water in a dormitory), for low-level suction the more precise control afforded by the bubbler tube is important.

PASCAL'S PRINCIPLE

Three hundred years ago the great French philosopher-mathematician, Blaise Pascal, stated that *an increase in pressure on an enclosed liquid will be distributed uniformly and undiminished to all parts of the liquid.*

Pascal's principle has many clinical applications. Pressing on the eye will cause pressure on the liquid which will be communicated to the retina, and the person will see flashes of light. The eye disease *glaucoma* is a condition of increased

Figure 8-14.
Micro-Low Pleural Drainage Unit.

Source: Photograph courtesy of Chemetron Medical Division, Allied
Healthcare Products, Inc., St. Louis, Mo.

intraocular pressure due to an obstruction of canals which are normally open for
drainage. This increased pressure affects the entire eye, causing cupping of the
disk, atrophy of the retina, hardness of the eye, and eventual blindness.

Similarly, pressure on the bladder will increase the pressure of the urine, giving
rise to the desire to empty the bladder.

In the same way pressure on the abdomen of a pregnant woman will increase the
pressure of the amniotic fluid, and may cause damage to the fetus.

APPLICATIONS OF PASCAL'S PRINCIPLE

The Queckenstedt test is also a good example of the application of Pascal's law.
The pressure of the cerebrospinal fluid is measured by inserting a needle between

the third and fourth lumbar vertebrae, and by reading the level of the liquid in a glass manometer attached to the needle. If pressure is applied to the jugular vein, the pressure to the liquid should be distributed throughout the liquid and the level in the manometer should rise. If there is movement of the liquid in the manometer, there is no obstruction between the brain and the site of the puncture. On the other hand, if there is no movement, an obstruction exists.

An everyday application of Pascal's principle is the hydraulic jack, used for such purposes as lifting patients and operating tables. In Figure 8-15, the area of the small piston is 1 square inch. When a force of 40 pounds is applied to it, pressure of 40 pounds per square inch will be distributed throughout the liquid. If the large piston has an area of 100 square inches, the piston will exert a force of 40 pounds per square inch × 100 = 4,000 pounds. Thus we can multiply the force manyfold. In this situation it may appear that we are getting something for nothing. But again we have to pay for it in distance. The force has to travel 100 feet to lift the large piston 1 foot.

Hydraulic brakes also work on this principle. The brake pedal pushes on a piston on the master cylinder. The pressure on the brake fluid is transmitted uniformly and undiminished to the four wheel cylinders. There small pistons move to compress the brake shoes. It should be obvious that a leaking brake system is dangerous. Anyone who has had the experience of applying the brakes only to have the brake pedal sink to the floor with no effect knows that the system must be kept filled with the liquid.

Some special mattresses use Pascal's principle. Conventional mattresses are conducive to the formation of decubitus ulcers because the most vulnerable areas of the body, the bony prominences, press hard in carrying a substantial proportion of the body's weight. A water-filled mattress, however, will distribute even pressure throughout the entire area of the body in contact with it, and thereby decrease the tendency for pressure sores to develop. An air mattress is valuable for the same reason (the principle of equal pressure also applies to gases).

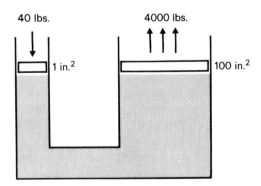

Figure 8-15.
Hydraulic lift.

Figure 8-16.
The Bernoulli effect.

BERNOULLI'S PRINCIPLE

Swiss mathematician and scientist Daniel Bernoulli discovered that when water flows through a tube having a constriction, the speed of the water is greatest at the constriction, and the pressure on the sides of the tube is least at that point (see Fig. 8-16).

Bernoulli's principle, then, is: *the greater the linear velocity, the less the lateral pressure*. At first thought this principle seems to be wrong; one would tend to reason that the pressure would be greatest at the constriction. First of all, we will agree that the constriction reduces the *total* volume of flow through the tube for a definite pressure. In Figure 8-16, let us assume that the flow rate through the tube is reduced from 100 ml per minute to 50 ml per minute by the constriction.

Now, with 50 ml/min coming out of the tube, this same rate must hold true for every part of the tube, including the constriction. How is this done? Imagine a four-lane, one-way highway, and that at one point of the highway there is a construction bottleneck which blocks off every lane but one. When the rush hour comes, traffic must be kept moving at a steady rate, and so a policeman is stationed at the bottleneck. His job is to wave cars through the construction area as rapidly as possible. Now, traffic may be averaging only 10 miles per hour along the highway; but as the cars funnel into the bottleneck, the policeman hurries them through at 25 miles per hour, single file. The same thing happens to the water molecules in the constriction of the tube. Because the molecules move so rapidly in a straight line, they have a diminished tendency to press outward upon the sides of the constricted area.

APPLICATIONS OF BERNOULLI'S PRINCIPLE

Bernoulli's principle was put to work by G.B. Venturi, Italian physicist, to produce suction with the water aspirator (Fig. 8-17).

If we tip the aspirator over, we will have an atomizer (Fig. 8-18).

Squeezing the bulb A forces air through the constriction B. This will result in

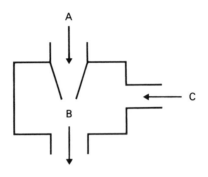

Figure 8-17.
The Venturi water aspirator.

As water flows from A through the constriction B, the lateral pressure
is less than atmospheric pressure and it will exert suction through
the sidearm C.

a lateral pressure that is less than atmospheric, and liquid will be drawn up through
the sidearm C. Opening D is to admit atmospheric pressure to the container.

The same principle is also operative in the lift of an airplane. The cross section
of a wing is not symmetrical at the top and bottom.

Air currents flowing past the wing will separate and rejoin behind the wing. In
order for the divided air stream to converge at the back of the wing, the top stream
must have a higher linear velocity than that of the lower stream since it has to travel
a longer distance. If the top has a higher velocity, the lateral pressure on it will be
less than on the underside. Thus there will be more force up than there is down,
with a resultant upward pressure. This is called the lift (Fig. 8-19).

These examples illustrate again how the same principle may be operative in so
many situations.

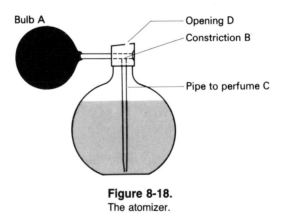

Figure 8-18.
The atomizer.

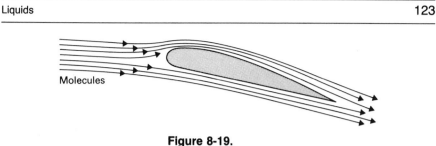

Figure 8-19.
Airplane wing contour.

MEASURING BLOOD PRESSURE

Because blood is constantly being pumped through the circulatory system, it exerts a certain pressure upon the walls of its vessels. This pressure can be measured by different methods. The most direct way of doing it would be to insert a needle into an artery and then measure the height to which the blood rises in a glass tube called the manometer (as is done in determining the pressure of the spinal fluid). This direct method of measuring the blood pressure was first used in 1733 by an English clergyman and scientist, Stephen Hales, who tried it on a horse.

Reverend Hale inserted a brass tube into the carotid artery of the thigh of a horse, and attached this to a glass tube about 13 feet long. It is interesting to note the resourcefulness of the experimenter. Because there were no rubber tubing available at that time, he is reported to have used the flexible trachea of a goose to connect the brass and glass tubing. The blood rose 9.5 feet in the glass tube and oscillated with each beat of the heart. Because the tube was open to the atmosphere, this would indicate pressure above 1 atmosphere. Although a great scientific achievement, it certainly is not a procedure that would be very popular for getting human blood pressure.

With the development of miniature transducers sensitive to pressure, along with recorders to read this pressure from the transducer, it is now possible to get direct readings of both the arterial and the venous pressure, as well as other physiologic pressures, such as intrauterine and cerebrospinal. These meters will record either arterial or venous fast response, or, if desired, mean pressures.

Sphygmomanometer. This instrument provides a more convenient, indirect method of determining the pressure. A rubber cuff is wrapped around the upper part of the arm and is inflated with air by squeezing the bulb. The cuff compresses the brachial artery. When the pressure exerted by the cuff is greater than the pressure of the blood, the circulation is stopped. The pressure of the air in the cuff is transmitted to the mercury manometer or the aneroid gauge. The air is slowly released from the cuff, and the pressure is read at the point at which the blood resumes flowing. This point is indicated by a characteristic sound in a stethoscope held over the brachial artery. The pressure reading here is called the systolic pressure, and is a measurement of the maximum pressure occurring while the heart is pumping.

As the pressure is further released, the sounds change until finally they disappear altogether. The reading made here is called the diastolic pressure, and is the pressure at which the blood is just coasting while the heart is at rest. The difference between the usual 120 mm Hg systolic pressure and the 80 mm Hg distolic pressure (40 mm Hg) is called pulse pressure.

Korotkov Sounds.   Dr. N. S. Korotkov first described his auscultatory method for indirect blood pressure measurement in 1905, and the sounds heard through the stethoscope as the flow of the blood changes are called "Korotkov sounds." However, there are problems associated with hearing these sounds, and as will be seen in Figure 8-20 there are actually five phases to the sounds. The initial sharp "thud" is the systolic phase; phase 4 and phase 5 are both reported as diastolic, and there is some confusion in the literature about whether phase 4 or phase 5 is the one

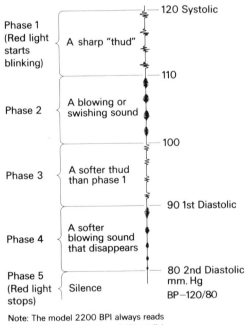

Figure 8-20.
Phases of the "Korotkov sounds."

Source: Courtesy of Picker International, Spring Street, Ossining, N.Y. 10562.

meant. Phase 4 diastolic pressure, sometimes called the muffling diastolic, will differ from the disappearance diastolic phase by approximately 5 mm Hg.

FACTORS INFLUENCING BLOOD PRESSURE READINGS

Variations in blood pressure are caused by two factors—differences that are generally seen from one person to another, and errors in reading the pressure. In the former instance, variations can be due to a number of factors. For example, there are rhythmic variations in the pumping of the heart and the respiratory cycle. Then too, there are changes that result from the subject's emotional state. Sometimes the mere sight of the doctor at the door of a patient's room can cause the diastolic pressure to rise by 20 mm Hg. Blood pressure rises as the bladder fills. Both the systolic and the diastolic pressure are reported to rise by 10 to 33 mm Hg within 15 to 45 minutes after the subject has eaten a meal. Thus, blood pressure is not always stable.

Errors in reading include inaccurate calibration of aneroid gauges, and constrictions that reduce mercury movement in mercury manometers. (In the latter, the time lag will change the reading by as much as 60 mm Hg above the actual pressure.)

Human errors are made, too. The person taking the reading can be influenced by his own expectation about it. There may be difficulty in hearing the soft "Korotkov sounds." These sounds are of low frequency, in the 20- to 80-Hz range; external noises can interfere. Some interesting studies have demonstrated the influence of external factors.[1,2] In one, a movie film was made of the falling mercury column in a conventional sphygmomanometer, while blood pressure was recorded in seven subjects, some of whom had hypertension and some of whom had normal blood pressures. The "Korotkov sounds" were recorded on the sound track. The film was then shown to test observers, who were asked to note the height of the mercury column as they would during an actual blood pressure reading taken to demonstrate response to therapy.[3]

The point is made that many reports analyzing blood pressure response to therapy do not include a statement about the number of readings taken before treatment and following treatment, on which the averages are based. Furthermore, the therapeutic response is usually judged on the basis of subjective, arbitrary criteria, which vary widely from one author to another. It is clear that multiple determinations of blood pressure in each patient, made both before and after therapy, are critical when therapeutic results are being evaluated.

[1]J. Wilcox: *Nurs. Res.* 10:4, 1961.

[2]Rose, Holland, and Crowley: *Lancet* 1:296, 1964.

[3]D. McCaughan. Comparison of an electronic blood pressure apparatus with a mercury manometer, reprinted by Sphygmostat Corporation.

ELECTRONIC BLOOD PRESSURE APPARATUS

Recent developments in electronic blood pressure apparatus have made it possible to reduce the variability from person to person; however, the apparatus must be correctly calibrated. Both the Parke-Davis Model 2200 Blood Pressure Indicator and the Sphygmostat model are based upon a single principle. A sensitive microphone that picks up the "Korotkov sounds" replaces the stethoscope and cuff arrangement. At the beginning of phase 1 through phase 3 or 4, the sound will cause a light to flash; the monitor can also be adjusted to emit a sound through a speaker. At the point of the diastolic reading, the light goes off. The systolic pressure is thus read as soon as the light flashes on, and the diastolic as soon as the light stops flashing.

The Parke-Davis model reads phase 5, the so-called second diastolic, while the

The following analysis traces successive steps in routine indirect measurement of blood pressure.

1. Cuff is inflated above systolic pressure; artery is occluded; no blood is flowing through.

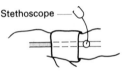
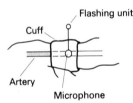

2. Systolic pressure point is reached; blood flows through artery causing first SphygmoStat flash. SphygmoStat light then flashes synchronously with pulse.

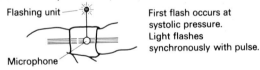

First flash occurs at systolic pressure. Light flashes synchronously with pulse.

3. Cuff pressure continues dropping and reaches first diastolic. SphygmoStat rejects "muffling" after change in sound; flashing ceases at first diastolic. Human ear may not accurately distinguish change in sound due to ambient noise or interference.

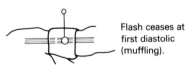

Flash ceases at first diastolic (muffling).

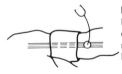

Hearing continues beyond first diastolic; change of sound may not readily be identifiable.

Figure 8-21.
Similarity between electronic blood pressure and conventional sphygmostat.

Source: Courtesy of Technical Resources, Inc., 14 Green Street, Waltham, Mass. 02154.

Sphygmostat model reads phase 4, the first diastolic. It is important, therefore, to report which diastolic pressure is being taken.

Figure 8-21 shows the similarity between the conventional Sphygmostat and the electronic blood pressure monitor.

PLETHYSMOGRAPH

Plethysmographic peripheral vascular studies are an extension of the usual blood pressure measurements. It is frequently necessary to have available accurate objective information about the status of the patient's peripheral circulation before and after therapy. There are two types of measurements. One is the segmental limb plethysmograph, which measures volume changes with each pulse beat of the large arteries of the arms and legs; these are recorded in milliliters per pulse beat. The other type is the digital plethysmograph, which measures the systolic blood pressure at various body areas; it records the pulse waves in cubic millimeters and bloodflow of the toes and fingers in cubic millimeters per second, thereby making possible a study of the circulation through the small terminal vessels. In addition, thermistor thermometers (see Chapter 13) measure the temperature of the second toe and finger on each side.

The graphs in Figure 8-22 show the readings taken before and after vasodilation. It will be seen that the digital pulse rose from 2 cu mm to 5.5 cu mm, and the bloodflow from 3 cu mm per second to 14 cu mm per second, indicating a definite increase in bloodflow.

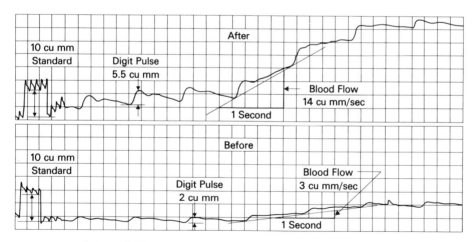

Figure 8-22.
Digital pulse volume and bloodflow before and after vasodilation.

Source: Courtesy of Electro-Diagnostic Instruments, 819 S. Main Street, Burbank, Calif. 91506.

The plethysmograph affords a noninvasive method for acquiring quantitative diagnostic information of the peripheral vascular system.

SCREENING FOR HYPERTENSION

It has been estimated that high blood pressure (hypertension) afflicts 25 million Americans. In about 15 percent of the cases there are definite physiological causes, such as tumors of the adrenal gland producing excess secretion of the hormone aldosterone and leading to fluid accumulation in the kidneys and an increase in blood pressure. Another apparent physiological cause is kidney disease, which produces an excess secretion of renin, also resulting in an increase in blood pressure. The remaining 85 percent of the cases are not clearly understood. Some theories, not definitely established, hold that internalized anger and frustration eventually produce permanent hypertension. Other experiments seem to indicate the possibility of a higher concentration of nitrates, and it is suggested that this be explored. The belief that high salt intake raises the blood pressure seems to be borne out by a number of studies showing a correlation, but no cause-and-effect relationship. Other researchers believe there may be genetic factors involved. For instance, the incidence in blacks is much higher than in white persons, about 28% vs. 17%. However, another study found that among middle-class blacks, the incidence was no higher than among white persons, so this notion is still being questioned.

The paragraph above shows how difficult it is to prove the etiology of a disease, especially when there may be multiple factors involved. In short, there is no definite answer as to the cause or causes of high blood pressure.

As mentioned above, in about 85 percent of the cases of hypertension, no specific causes can be found. This type is called *essential hypertension*. There is now some support for the hypothesis that the kidney is involved in hypertension by releasing excess renin. This in turn stimulates a chemical called angiotenin I, which circulates to the tissues where an enzyme, angiotensin II, is released. This chemical is a strong vasoconstrictor, and due to the smaller volume, pressure increases rapidly. In addition, over a period of several hours, salt retention also occurs, with the attendant retention of water and an increase in blood volume, and an increase in blood pressure.

It is estimated that one-half of those having high blood pressure do not know it. Because hypertension dramatically increases the risk of heart disease, stroke, and kidney disease, the need for better screening of the population is evident.

PREVENTING STASIS OF BLOOD

One example of the use of pressure to prevent stasis of blood in the sinuses of the calf veins is the inflatable surgical boot. Prolonged bed rest and general anesthesia reduce the ability of the calf muscles to pump blood back toward the heart. It has been estimated that 15 to 50 percent of all surgical patients experience clotting in the deep venous system of the calf during and after surgery. This clotting can lead to thrombosis—the formation of a clot that occludes or shuts off a blood vessel.

To avoid this, patients have had their legs wrapped in surgical bandages, anticoagulants have been used, and electrical stimulation of the calf muscles has been tried. The inflatable surgical boot attempts to relieve or eliminate deep venous stasis by exerting pressure on the foot and the lower leg. Intermittent peak pressure of 40 to 60 mm Hg is used in a double-wall plastic boot, consisting of a soft plastic inner boot and a stiff outer boot which is attached only at the top of the boot and the heel. The foot and leg will experience the intermittent pressure which will force the blood out of the leg toward the heart. There are two types. One is a pneumatic device designed for use in the operating room or in areas where there may be explosive gases. A second device is electronically controlled, and uses a compressor. This type is suitable for bed use. The inflatable boot significantly decreases the frequency of deep venous thrombosis. Venograms demonstrate that the veins and

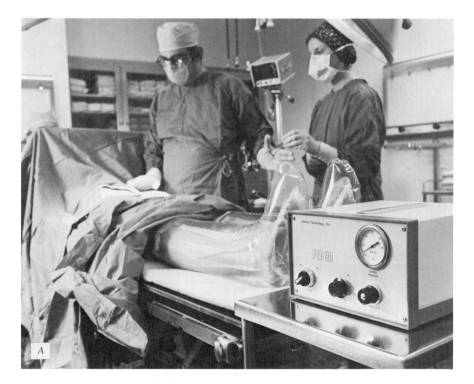

Figure 8-23(a).
The PED-90 inflatable surgical boot.

Source: Photograph courtesy of Clinical Technology International Inc., P.O. Box 92, Randolph, Mass. 02368.

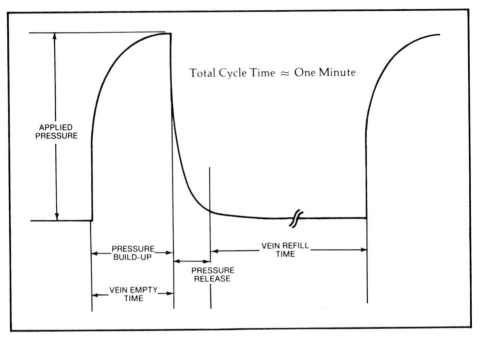

(b)

Figure 8-23(b).
The pressure cycle of the PED-90, showing emptying and filling of vein.

Source: Courtesy of Clinical Technology International Inc., P.O. Box 92, Randolph, Mass. 02368.

venous sinuses empty and refill on each cycle. The inflation pressure and cycle time can be adjusted according to individual need (Fig. 8-23).

ATHEROSCLEROSIS AND ARTERIOSCLEROSIS

The heart pumps about 7,200 liters of blood in 24 hours (assuming 70 ml per contraction and an average of 72 contractions per minute). The elasticity of a normal blood vessel allows it to expand as the blood is forced into it, thereby preventing the pressure from becoming too high. When the heart is at rest, the blood keeps flowing due to its inertia and also because of the contraction of the expanded blood vessels, which in effect squeeze the blood along.

In the case of arteriosclerosis (hardening of the arteries) there is a lack of flexibility of the artery caused by a loss of elasticity in the middle layer of the walls. The pressure in the system rises, sometimes to a dangerous level, because the decreased "give" in the blood vessels keeps their volume relatively small (Fig. 8-24).

Another disorder of the blood vessels is atherosclerosis. The vessels become coated with a deposit of a fatty substance called cholesterol. As the lumen of the blood vessel becomes smaller, the pressure in the whole system increases. This is quite logical, because the blood is being pushed through a pipe of smaller inside diameter. This problem has been studied intensely during the last two decades. It appears that there is some connection between the amount of saturated fats in the diet and the amount of cholesterol in the blood. However, no definitive *causal* relationship has been found as yet.

Fats that are fluid at room temperature are classified as oils, and those that are solid at room temperature are called fats. Both are mixtures of triglycerides, with a number of fatty acids in various proportions. These fatty acids are classified as saturated or unsaturated, according to the amount of hydrogen they contain. Those that have two fewer hydrogen atoms than a saturated fatty acid are called monounsaturates; those that have four, six, or eight fewer hydrogens are called polyunsaturates. There is good evidence that the unsaturated fatty acids lower the blood cholesterol level, whereas saturated fatty acids tend to raise it. Even though final proof is not in as to the causal relationship of saturated fats and atherosclerosis, it does seem sensible to reduce the amount of saturated fats in the diet. It is also important to realize that the body manufactures cholesterol in amounts up to four to six times those found in the diet, and that emotional stress might cause the cholesterol level to rise.

It is interesting to note that for every 100 cases of coronary thrombosis reported, about 97 of them occur in men.

Various corrective methods of a mechanical nature have been devised for use in treating cases of arteriosclerosis and atherosclerosis, the main principle being to increase the volume of the vessels and therefore to reduce the pressure. One treatment for atherosclerosis involves scraping the lumen of the blood vessel to remove cholesterol deposits; attempts to dissolve the deposits chemically have also been made.

Artificial arteries are being used extensively as replacement parts for occluded vessels. A number of synthetic materials are available. These are often treated with silicone plastics to reduce damage to the blood as it is passing through the artery.

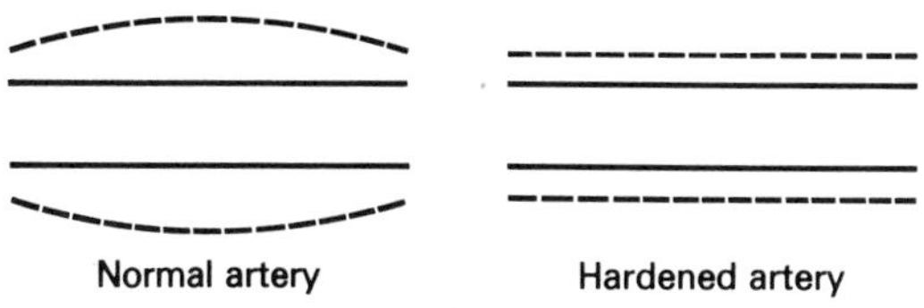

Normal artery   Hardened artery

Figure 8-24.
Loss of elasticity in arteriosclerosis.

As scientific research continues, substitutes for diseased parts are invented, and we can probably expect that replacement organs will be used with increasing frequency.

THE HEART–LUNG MACHINE

Heart–lung machines, also known as pump oxygenators, were developed so that open-heart surgery of several hours' duration could be performed. Without a machine that pumps the blood and exchanges oxygen for carbon dioxide, the heart could be stopped for only a few minutes without causing destruction to the oxygen-sensitive tissues of the brain, other parts of the central nervous system, and the kidneys.

The possibility of operating unhurriedly on a dry heart was very attractive. It would allow surgeons to repair valves and chambers that formerly were inoperable.

It is estimated that one in every 100 children is born with one or more congenital heart defects that cause premature death.

The heart, although small (the size of a fist), pumps more than a million gallons of blood per year. Before the use of a machine, a "bypass" method of circulation was developed. The circulation of the donor was tied into that of the patient. Thus the donor did the breathing and pumping for the patient. This was fairly successful.

As far back as 1930, work was begun to develop a machine to take over the work of the heart and the lungs for a period of time. In order to oxygenate the blood, it was run in a thin film over stainless-steel screens, or picked up in a film on rotating disks, both in an atmosphere of oxygen.

In 1953 John Gibbons developed the helix-reservoir machine. This oxygenates the blood by bubbling the oxygen through the blood in a vertical mixing chamber. The oxygen is dissolved in the blood, and the carbon dioxide is removed. (See Fig. 8-25).

At the top of the mixing chamber is a debubbler chamber. It is filled with stainless-steel sponges coated with an antifoam silicone to reduce surface tension and thereby to break up the bubbles. It is important to remove gas bubbles to avoid embolism. The carbon dioxide and excess oxygen escape through a vent.

The blood then flows by gravity to the top of a helix-reservoir, which is a coil of plastic tubing surrounded by a warming bath. Here the blood flows through the tubing, being heated to body temperature by the bath.

A "finger-type" pump forces the blood into the large femoral artery or into the subclavian artery.

The machine, which can deliver up to five liters of blood per minute, has made possible extensive reconstruction of heart valves and arteries and veins. At many hospitals these operations are performed on a routine basis, numbering in the hundreds in each hospital.

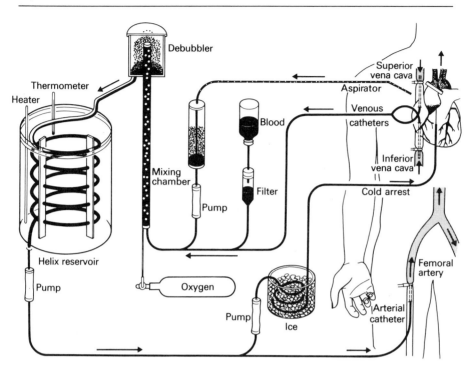

Figure 8-25.
Helix-reservoir machine.

SPECIFIC GRAVITY

Specific gravity is a figure which compares the density of a liquid with the density of water at the same temperature. The density is the mass per unit volume ($D = M/V$).

The density of water is taken to be 1 g/cc (at 4°C). Thus a liquid having a density of 2 g/cc would have a specific gravity of 2, that is, twice as heavy as water. In the metric system the numerical value of the specific gravity is the same as the density.

A device used to determine the specific gravity of a liquid is called a *hydrometer*. A hydrometer is a weighted glass tube that floats in the liquid. The greater the specific gravity of the liquid, the higher the hydrometer will float in the liquid (Fig. 8-26).

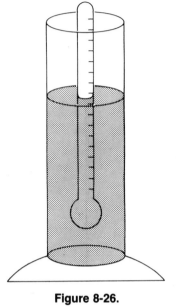

Figure 8-26.
Hydrometer.

One test for the composition of urine is the measurement of its specific gravity. Water has a specific gravity of 1.000. Urine contains a number of dissolved salts, and normally has a specific gravity of between 1.015 and 1.030. In cases of illness, there may be an increase of salts in the urine, and the reading on the hydrometer (for this purpose called a urinometer) will be correspondingly higher.

The antifreeze in an automobile radiator is checked the same way (a different hydrometer must be used for the alcohol and the ethylene glycol type antifreeze).

QUESTIONS

1. Calculate the pressure in millimeters of mercury that is equal to 14 cm of water.
2. Suppose that a transfusion were being given. Explain what effect warming the blood slightly before injecting it would have on the rate of flow. What effect could injection of cold blood have? What could be one reason for not allowing the blood to be warmed before injecting?
3. A sphygmomanometer reads 120 mm Hg. What is this in pounds per square inch? Are these readings on the absolute or gauge scale?

4. A hydraulic jack chair has a piston of 360 square inches to lift it, and a piston of 12 square inches under the foot pedal. How far does the pedal have to move to raise the chair one foot? What is the mechanical advantage?

5. What is the purpose of an air vent in a blood transfusion set? Why is there no vent in the plastic ones?

6. How does an inflatable boot using intermittent air pressure reduce the danger of thrombosis?

7. How does an electronic blood pressure apparatus determine the systolic and diastolic pressure? How are the results shown to the operator?

8. Discuss the reproducibility of blood pressure readings. What factors contribute to the results?

9. How does the location of the drainage bottle affect the amount of suction?

10. The tube in the middle bottle in a three-bottle suction apparatus is immersed in 21 cm water. The wall suction is set for 200 mm Hg. What will be the negative pressure in millimeters of mercury experienced by the patient? What would your answer be if an electric pump were set at 100 mm Hg suction?

11. The specific gravity of a urine sample is determined to be 1.050. Is this value indicative of excessive salt retention or elimination? Suggest possible clinical causes.

12. A patient is scheduled to receive a blood transfusion at the rate of 120 ml/hr. How would you determine how many drops per minute is required if you do not know the size of the drop for that particular setup?

CHAPTER 9

HEAT

If you were asked to define "heat," how would you do it? Many common phenomena are difficult to define. In days of yore heat was considered to be some invisible fluid that could flow from one substance to another. Today, we can use the kinetic theory to explain many natural phenomena, including heat. From this basis, we are able to see many situations—involving both patients and equipment—in which heat is a factor.

Heat is related to the average kinetic energy of the molecules of the material. The faster the molecules are moving, the greater their kinetic energy and the more heat the material possesses.

MEASUREMENT OF THE INTENSITY OF HEAT

It is important to distinguish between the *quantity* of heat and the degree or *intensity* of heat. A cup of water at 100°C possesses a relatively high heat intensity but a small quantity of heat. A bathtub full of water at 60°C contains a lower degree of heat, but a much greater quantity of heat. The bathtub contains many more molecules in motion.

The intensity of heat is measured by a thermometer. There are many types of thermometers; in the one in common use the liquid expands as the temperature rises. The liquid is usually mercury or colored alcohol. However, the property of expansion of solids or of gases also may be used in the design of thermometers.

In addition, electronic methods are used, such as thermocouples and thermistors. (These are described in Chapter 13.) The advantage of electronic methods is that the temperature is automatically recorded on a chart.

MEASUREMENT OF THE QUANTITY OF HEAT

A *calorie* is the amount of heat required to raise the temperature of 1 g of water 1°C. In the British system a BTU (British Thermal Unit) is the amount of heat necessary to raise the temperature of 1 pound of water 1°F. Since there are 454 g in a pound, and 1 °F is equal to 5/9°C, the BTU is equal to approximately 252 calories.

The kilocalorie is equal to 1,000 calories, and is the amount of heat required to raise 1 kilogram of water 1°C. In medicine the calorie is used in the measurement of the basal metabolic rate and in dietary calculations.

EXPANSION OF MATERIALS WITH TEMPERATURE

We have already learned in the study of Charles's Law that the volume of a gas increases with an increase in temperature. Liquids and solids also generally expand on heating. The following table lists the expansion in length of various solids. We call this the thermal coefficient of linear expansion. It is the fractional increase in length per degree change in temperature.

THERMAL COEFFICIENTS OF LINEAR EXPANSION
FRACTIONAL EXPANSION/°C

Aluminum	0.000023 or 2.3×10^{-5}
Brass	0.000019 or 1.9×10^{-5}
Glass	0.000009 or 9.0×10^{-6}
Pyrex glass	0.0000036 or 3.6×10^{-6}
Steel	0.000011 or 1.1×10^{-5}

This table states that a 1-inch aluminum rod would increase 2.3×10^{-5} inches if the temperature is increased by 1°C. That is, it is the fractional change in length per degree Celsius. If the rod is measured in meters or centimeters, the same *fraction* would be involved. We can write this in the following formula:

$$\frac{\Delta L}{L} = \frac{k}{°C} \Delta T$$

where ΔL is the change in length, $k/°C$ is the linear coefficient of expansion of the substance per degree Celsius, and T is the change in temperature in degrees Celsius.

For aluminum, $k/°C$ is given above as 2.3×10^{-5}.

How much would a 10-meter aluminum rod expand if it is heated to 45°C from 25°C?

$$\frac{\Delta L}{L} = 2.3 \times 10^{-5}/°C \times 20°C \qquad 10 \text{ meters} = 1,000 \text{ centimeters}$$

$$\frac{\Delta L}{1,000 \text{ cm}} = 2.3 \times 10^{-5} \times 20$$

$$\Delta L = 46 \times 10^{-5} \times 1 \times 10^{3} = 46 \times 10^{-2} = 0.46\text{-cm expansion}$$

Because $1°F = \frac{5}{9}°C$, the expansion per degree Fahrenheit will be $\frac{5}{9}$ of $k/°C$. For example, to calculate the $k/°F$ for glass, we multiply $9 \times 10^{-6} \times \frac{5}{9} = 5 \times 10^{-6}/°F$.

How much would a 5-foot glass rod at 32°F expand if it were heated to 62°F? Five feet equals 60 inches.

$$\frac{\Delta L}{L} = k/°F \times \Delta T \qquad \frac{\Delta L}{60 \text{ inches}} = 5 \times 10^{-6} \times 30$$

$$\Delta L = 9 \times 10^{-3} = 0.009 \text{ inches expansion}$$

The effect of thermal expansion is apparent in many situations. A long pipeline has expansion loops at intervals throughout its length. Railroad tracks would buckle in the summer if they were continuous, so small gaps are provided between each length of track to take up the expansion. The characteristic "click-click" of the wheels is attributable to these gaps. Lately, steel alloys with a lower coefficient of expansion are being used so that these gaps can be reduced in size or eliminated.

Expansion joints in concrete are also placed to allow temperature change without cracking the concrete. In recent years you may have noted roads being poured in continuous strips. Air bubbles incorporated into the concrete allow for the necessary expansion.

If a syringe is heated (for instance, in sterilization) with the needle attached, the syringe may crack. As can be seen from the table on this page, glass has a smaller coefficient of expansion than steel. Thus the metal will expand more than the glass when heated, and contract faster than the glass when cooled.

What makes a glass crack if hot water is poured into it? In Figure 9-1, the inside of the thick-walled glass will expand appreciably because of the higher temperature.

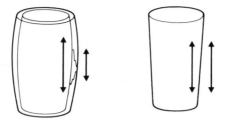

Figure 9-1.
Cracking of glass caused by uneven thermal expansion.

Glass does not conduct heat very well, so the outside of the glass will remain relatively cool for a little while, and thus will not expand as much. It is the uneven expansion of the walls of the glass that causes the cracking. From this we see that a thin-walled glass if less likely to crack, as both sides of the glass will expand at approximately the same rate.

Another application of the expansion of solids is in the bimetallic thermostat. If we join two strips of dissimilar metals, such as aluminum and steel, as the temperature is increased, the metals will expand unequally. From the table of thermal linear expansion, we can see that aluminum will expand twice as much as steel. This will cause the strip to bend (Fig. 9-2). Such bending is used to control furnaces, ovens, electric frying pans, and other kinds of equipment in which a switch is turned off or on.

METHODS OF HEAT TRANSFER

There are three methods by which heat is transferred from one place to another: *conduction, radiation,* and *convection.* Let us examine each in turn.

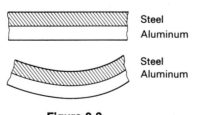

Steel
Aluminum

Steel
Aluminum

Figure 9-2.
Bimetallic thermostat.

CONDUCTION

The Latin *ducere* means "to lead." Heat is transmitted through a conductor. This implies that a substance is required through which the heat can be conducted. The mechanism of conduction can be pictured by imagining an iron rod being heated at one end. The molecules will pick up energy from the flames, and will move faster. The adjacent molecules will also move faster and thus become warmer. Thus, by a step-by-step transmission of the heat throughout the rod, the other end will also become warm.

Following is a list of some good conductors and some poor conductors of heat. The units are in calories per second passing through a plate 1 cm. in area and 1 cm. thick with a temperature differential of 1°C. between the faces of the plate. A poor conductor is called an *insulator,* and can be used to inhibit or prevent the conduction of heat.

$$cal/sec \cdot °C \cdot cm^2 \cdot cm$$

Conductors		*Insulators*	
Silver	0.96	Air	0.000057
Copper	0.92	Flannel	0.00023
Aluminum	0.50	Water	0.0013
Zinc	0.27	Brick	0.0015
Iron	0.12	Glass	0.0025

It will be noticed that copper conducts better than steel. This explains why steel saucepans are sometimes copper clad, in order to distribute the heat more evenly and more quickly. On the other hand, Pyrex cooking vessels are not as efficient as metal ones in transferring heat (although the ease of cleaning Pyrex vessels may offset this disadvantage).

The very low conductivity of air makes it a good insulator. It should be noted that this applies to dry, trapped air. The usual type of insulation against conductivity losses of heat, such as rock wool or glass wool, consists of fluffy mats of fibers producing thousands of pockets of trapped air. If fur coats were worn for functional reasons, they would be worn inside out. This would result in an outer windbreaker surface with the trapped air spaces inside. Many coats, of course, do use this arrangement.

The reason why we feel so much colder in the wintertime when the humidity is high is that the water vapor in the air conducts away the heat of the body at a faster rate than would occur if the air were dry.

Hot-water bags are made of rubber instead of metal. If they were of metal, they not only would fail to conform to the body contours, but also would conduct heat so rapidly that the patient would be burned. Even rubber conducts heat too rapidly, so hot-water bags are wrapped in a flannel cloth (this should be prewarmed so that heat conduction to the patient will begin immediately). In the table of conductors and insulators, it can be seen that flannel is a good insulator. This is because of its many air traps.

What would be the effect upon the hot-water bag if the flannel cloth were to become wet? You will notice in the conductivity table that water is a much better heat conductor than air. This means that a bag wrapped in a wet cloth will lose heat much faster.

The application of hot moist packs and compresses is beneficial in some situations. However, in their use certain precautions have to be observed to prevent burning the patient. Before a hot moist pack is applied, the skin area should be lubricated with petrolatum, which acts as an insulator and slows down the transmission of heat.

The ideal method of preventing heat conduction, of course, is to eliminate the conductive medium altogether. The vacuum bottle (often called the Thermos) works on the principle that unless there are gas molecules to conduct the heat, heat will not be conducted. Therefore, the bottle is made as a double-walled vessel with a vacuum between the walls. If the vacuum is broken the bottle's effectiveness is destroyed.

Most children living in a cold climate have at one time or another put their tongues on the cold doorknob on the outside of a door, and some have even had the tongue freeze to the knob. If you touch an outside knob, it feels colder than the wooden door. Actually both must be at the same temperature. Why does the knob feel colder? Because the metal is a better conductor, and conducts the heat away from the skin faster than the wood. The same situation holds true when we compare two tumblers, one of aluminum and the other of glass. Most people will say that the aluminum will keep a drink colder than the glass, and as proof will say that the aluminum feels much colder. Again, the fact that it feels colder actually indicates that the aluminum conducts the heat away from the hand faster, and it therefore warms up more rapidly. We find quite often that the psychological impression is exactly the opposite of the physical fact.

Metal bedpans should be warmed before being offered to the patient; otherwise, the metal will remove heat so rapidly from the area of contact that the patient will feel uncomfortable.

RADIATION

Radiation is the transfer of energy in the form of waves. It has been found that these can be transmitted even through "nothingness." The sun, for instance, radiates heat, light, and other forms of energy called electromagnetic radiation (this is discussed in more detail in Chapter 11). Part of this energy from the sun is transferred to our planet. How did it get here? The answer is that nobody really knows. Between the sun and the outer limit of the earth's atmosphere, there is essentially an absolute vacuum. So far as we know, there is no medium through which these forms of energy can be conducted.

Hold your hand near a light bulb and turn the light on; you will feel heat almost immediately. Here again, light and heat are radiated.

If we were to ask, "Is there any heat in the vacuum of outer space?" there probably would be a variety of answers. It seems logical to assume that the closer we are to the sun, the more intense the radiation. This is true. However, the vacuum itself has no heat because there are no molecules to become agitated. Yet if you were to go out there with a thermometer, it would register a high temperature because now there are molecules (the thermometer's) which can absorb the radiation and move faster, thus producing heat. In other words, radiation *becomes* heat when it is absorbed by a substance.

Radiant heat, like light, can be reflected or absorbed by a surface, depending upon the lightness or darkness of the surface. A white suit will reflect a large percentage of the heat while a dark suit will absorb more. The astronauts also found problems with radiant heat, and the space capsule had to be air-conditioned to make the temperature comfortable.

A silvery surface is an even better reflector of heat than a white surface. The silvered interior of a vacuum bottle reflects radiant heat back upon its origin (the hot coffee) and thus minimizes heat loss by radiation. The aluminum paper used in attics reflects radiated heat.

CONVECTION CURRENTS

As air above a radiator becomes warm, it will expand and rise. Cooler, more dense air will come in to take its place. Thus a current of air (i.e., a flow) will result, and the heat absorbed by the air is transferred elsewhere. The unequal heating of the earth's surface produces similar convection currents and is the cause of winds and storms. Convection currents also may occur in liquids if one side of the container is heated more than the other. This is one cause of ocean currents.

SPECIFIC HEAT

It was mentioned earlier that the molecules or atoms are continually moving in random motion in solids, liquids, and gases and that the higher the temperature, the faster they are moving and the higher the average kinetic energy of the molecules. In solids and liquids the bonds of attraction holding the particles together are appreciable and differ from substance to substance. Thus different substances will usually have different internal energies, even though they are at the same temperature.

If we add a certain amount of heat, usually denoted by Q, to two different substances, they will not have the same increase in temperature. The number of calories required to raise 1 gram of a substance 1°C is called the *specific heat*. This is 1 cal/g °C for water, since we originally defined a calorie as the amount of heat necessary to raise 1 gram of water 1°C. Water has a very high specific heat. It is only 0.5 cal/g °C for ice at 0°C, 0.22 cal/g °C for aluminum, and 0.09 cal/g °C for iron.

In other words, it takes only about $^1/_{11}$ as much heat to raise 1 g iron 1°C as it takes to raise 1 g water. Or, to turn it around, a certain amount of heat will raise the temperature of 1 g iron 11 times as much as that for 1 g water. This is the reason the temperature of land close to large bodies of water tends to have less fluctuations than in places more remote from water.

We can now put the above information into an equation.

$$Q = \text{s.h.} \times m \times \Delta T$$

Q is the number of calories of heat added, s.h. is the specific heat, m is the mass in grams, and ΔT is the change in temperature in degrees Celsius.

If we want to know how many calories it will take to raise 50 g water from 20° to 50°C, we can use the above equation, remembering that water has a specific heat of 1.0 cal/g·°C.

$$Q = 1.0 \text{ cal/g °C} \times 50 \text{ g} \times 30°C$$

$$Q = 1,500 \text{ calories, or 1.5 Cal, or 1.5 kcal}$$

Please note that 1,500 calories raised the temperature of 50 g water 30°C. How much would the same number of calories raise the same number of grams of iron?

$$1,500 \text{ cal} = 0.09 \text{ cal/g °C} \times 50 \text{ g} \times \Delta T$$

$$333 °C = T$$

The next section brings up an additional factor where there is a change of state.

CHANGE OF STATE

Matter is considered to exist in three states: solid, liquid, or gas (vapor). Most substances can exist is any one of these states. Even in the solid phase the molecules or atoms are in continuous motion, vibrating about a mean (average) position in the substance. As heat is applied to the solid, the molecules vibrate more vigorously. Finally a point is reached where the energy is sufficient to break the bonds holding the molecules in the crystal lattice, and the solid melts. This requires a definite amount of heat per gram of a particular substance. The amount of heat needed to melt one gram of a substance at its melting point is the *heat of fusion*. Heat of fusion could also be called the heat of melting; for example 80 calories would have to be added to melt a gram of ice, and 80 calories would have to be removed to freeze a gram of water at 0°C.

The change of state such as from a solid to a liquid or the reverse occurs without any change in temperature. If some water with ice cubes in it is warmed, the temperature will remain at 0°C until all the ice is melted. After all the ice is melted, the liquid water at 0°C will rise in temperature, 1°C per gram of water per calorie. It will thus take 100 calories to raise 1 gram of water from 0° to 100°C.

To change one gram of liquid water to vapor requires 540 calories. This energy is called the *heat of vaporization*.

Which would be better for an ice bag, 100 g liquid water at 0°C or 100 g ice at 0°C? If we assume a temperature of the patient of 40°C, the liquid water would rise 40 degrees, requiring $100 \times 40 = 4,000$ calories that could be taken from the patient. The ice, on the other hand, would take on $100 \times 80 = 8,000$ calories to melt plus 4,000 calories to go from 0° to 40°C. Thus the ice would be three times more efficient than the liquid water.

When a liquid evaporates, whether at the boiling point or at room temperature, it takes on the heat of vaporization. The body keeps cool, to a large degree, by the evaporation of perspiration. If 100 g water evaporates, it will take on $100 \times 540 = 54,000$ calories. The rate of evaporation varies directly with the vapor pressure of the liquid (vapor pressure is the pressure of the vapor above the liquid that would exist if the liquid were in a closed container). At room temperature (20°C) water has a vapor pressure of 17.4 mm, ethyl alcohol 44 mm, ethyl ether 442 mm, and ethyl chloride 988 mm. Alcohol thus will vaporize fairly quickly and cool the body when used as an alcohol rub. Ethyl chloride, with its very high vapor pressure, will vaporize so rapidly that it will freeze the skin; it is sometimes used as a local anesthetic for minor surgery. The following simple experiments will serve to demonstrate the cooling effect of evaporation:

COOLING BY EVAPORATION (EXPERIMENTS)

1. Blow across the back of your wrist.
2. Moisten the back of your wrist with *water,* and blow as above.
3. Use *alcohol* instead of water and repeat step 2.
4. Use one or two drops of *ethyl chloride* instead of water and repeat step 2.

Letter 1, 2, 3, and 4, as follows: (A) the procedure that made your wrist feel coldest, and (B), (C), and (D) in order of how cold your wrist felt.

Explain results.

1. Wet a washcloth in warm water and wring it out.
2. Hold it against your face and feel its warmth.
3. Open it and wave it back and forth about 25 times.
4. Hold it to your face and compare its warmth with step 2.

Waving the washcloth hastened evaporation (by removing it from the vicinity of the more humid air). The same result would have been achieved if you had held it still, but it would have taken longer.

Explain the result.

Another experiment will clarify the meaning of heat of fusion.

HEAT OF FUSION (EXPERIMENT)

This experiment is designed to compare the effectiveness of cooling a patient by using liquid water at 0°C and a liquid–ice mixture at 0°C. Which would be more effective for reducing fever?

Prepare a pitcher of ice water as follows: put ice in the pitcher up to about the half-way mark; add cold water up to the three-quarters mark. Stir gently with the thermometer and note the temperature about every 10 seconds. Add as many ice cubes as necessary to keep several ice cubes in the pitcher. When the temperature stops falling and there are still ice cubes in the pitcher, you have ice water; the water and the ice are at the same temperature.

Pour ice water, including at least two lumps of ice, into a small beaker (about 250-ml capacity). Pour water (without ice cubes) from the pitcher into an exactly similar beaker, up to the same level. Note the temperature of the water in the second beaker *immediately* and then the temperature in the first beaker.

Warm each beaker in your hands for about a minute and then stir the contents of each and note the temperatures.

Warm, stir, and note the temperature two more times.

Readings	1st	2nd (after warming)	3rd (after 2 warmings)	4th (after 3 warmings)
1st beaker (with ice)				
2nd beaker (without ice)				

What can you conclude about the effectiveness of an ice cap containing ice water with cracked ice in it, compared with one containing only ice-cold water?

Do both ice caps start at the same temperature? Does the body give heat to both ice caps? (Did your hands give heat to both beakers—that is were your hands cooled in both cases?)

The heat that your hands gave to the second beaker was used to raise the temperature of the water in it. What happened to the heat that was given to the first beaker?

DISTILLATION

Distilled water has many uses in the hospital, particularly in the preparation of medications. Distilled water is that which has been rendered chemically pure by boiling and then condensing the steam back into liquid. The distillation of a liquid is a cyclic process involving evaporation (*taking on* the heat of vaporization) and condensation (*giving off* the heat of vaporization).

REFRIGERATION

The cycle of evaporation and condensation is also employed in a refrigeration system.

In the system outlined in Figure 9-3 the circulating element is a liquid that vaporizes readily. Formerly ammonia and sulfur dioxide were used; but since these are toxic, they have been supplanted by fluorocompounds called Freons.

A compressor (really a motor-driven pump) compresses the gas. This causes heat to be given *off* through a set of coils located near the motor compartment of the refrigerator. In this manner the gas loses the heat of vaporization and becomes a liquid.

The liquid rises in the system to the level of the refrigerator box (i.e., the cooling unit). At this point the liquid passes through a small opening, where it changes from the liquid to the vapor state, taking on the heat of vaporization. Immediately following the constriction are copper coils which line the walls of the cooling unit. Heat is taken from these coils, and in turn is removed from the cooling unit. Then the vapor is pumped back to the compressor to repeat the cycle. The refrigerator system is really a heat pump, taking *on* heat from the cooling unit and pumping it *out* at the back of the refrigerator.

Refrigeration devices, stationary and portable, have a variety of clinical uses. Oxygen tents, for instance, are cooled by refrigeration. A variant of the refrigerator has been used to produce hypothermia, which is the lowering of the body temperature to a point well below the normal. Hypothermia is sometimes employed in operations such as open-heart surgery, in which the blood circulation might be interrupted for

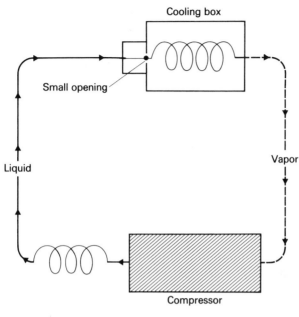

Figure 9-3.
Refrigeration cycle.

some time. The oxygen need decreases with the temperature. Patients have been cooled to a body temperature of as low as 5°C, and a number of cardiac operations have been performed in the 10° to 15°C range. Cooling of the patient has been accomplished by immersion in cold water, application of ice bags, irrigation of the stomach with cold water, and more lately with electric refrigeration machines that pump an antifreeze solution through a blanket.

Since the development of the heart–lung machine, hypothermia is used less often in open-heart surgery.

THE FREON PROBLEM

As mentioned above, the Freons—a series of chlorofluorocarbons, with formulas such as Freon-12 (CCl_2F_2) and Freon-112 ($C_2Cl_3F_3$)—are widely used as coolants in refrigerators and air-conditioning systems.

Another use of the Freons, as propellants in pressure cans, has increased tremendously in the last decade. Aerosol cans are used for spraying literally thousands of substances ranging from hairsprays and deodorants to paints, insect sprays, and lubricating oils. About 800,000 tons are produced annually worldwide. Of this amount, about 60 percent is used for spray cans, 25 percent as a coolant, and the remainder as a foaming agent in the production of cushions and insulation.

In 1974, several scientists working in the field of photochemistry discovered that the relatively inert Freons could react with ozone (O_3). Dr. Frank Sherwood Rowland at the University of California at Irvine suggested that the large amount of Freon released to the atmosphere by spray cans could reach higher altitudes, where it could possibly react with the ozone layer, and slowly reduce the O_3 concentration. Why should we care about this possibility? The ozone layer absorbs a great deal of the ultraviolet radiation from the sun, and thus protects us from excessive radiation. It is believed that an increase of 2 to 5 percent in UV radiation may increase the incidence of skin cancer in 50 to 100 years.

The use of Freon in aerosol cans has been drastically reduced, but no equally satisfactory substitute has been found for its use in refrigeration. Scientific panels appointed by the U.S. government are still studying this problem, and when more facts are known, rational decision can be made as to what needs to be done.

RELATIVE HUMIDITY

A liquid in a closed container will evaporate. As the amount of vapor above the liquid increases, the vapor will begin to condense again. After a while a point of equilibrium will be reached where the rates of evaporation and condensation will be equal. We say that the system has reached a point of dynamic equilibrium. Please note that the action does not stop; but that since the rates are equal, the net result is no change. The pressure of the vapor at that point can be measured with a manometer. It is found that the vapor pressure of a liquid varies directly with the temperature of the liquid. The following temperature–vapor pressure table is for water. Liquids that vaporize more readily than water have a correspondingly higher vapor pressure for a given temperature.

Temperature		Vapor Pressure (mm)
°F	°C	
32	0	4.6
50	10	9.2
59	15	12.7
68	20	17.4
77	25	23.6
86	30	31.5
212	100	760.0

What has this got to do with relative humidity? Relative humidity is the percent of water vapor in the air compared with the maximum that the air can hold. This maximum is the vapor pressure at a particular temperature.

$$\text{Percent of relative humidity} = \frac{\text{amount in air} \times 100}{\text{maximum possible}}$$

For example, if the air at 25°C has a vapor pressure of 17.4 mm, what is the relative humidity? Using the above equation, and looking up the vapor pressure at 25°C (23.6 mm), we get

$$\frac{17.4 \times 100}{23.6} = 73.8 \text{ percent}$$

Now, if we take the same sample of air, and lower the temperature to 20°C, what will the relative humidity be? At 20°C the vapor pressure is 17.4 mm, and our sample has 17.4 mm. Thus at 20°C the air is 100 percent saturated; that is, it has all the water that it can hold.

EFFECT OF HUMIDITY ON COOLING

We saw previously that the body dissipates heat by evaporating perspiration. If the air is very dry, with a low humidity, the water will evaporate rapidly, and the body will cool off rapidly. On the other hand, if the air is very humid, evaporation is reduced, and the person feels very warm and oppressed. In air conditioning it is just as important to reduce the humidity as it is to reduce the temperature of the room. A person can feel quite comfortable at 80°F with a low humidity, and uncomfortable at 70°F with a high humidity.

STERILIZATION

Sterilization is the process by which *all* living organisms (including bacterial spores) upon an object are killed. The only reliable way to achieve complete sterilization is through the employment of a physical agent, such as exposure of the object to be sterilized to steam or dry heat, or by the use of a chemical agent.

The so-called boiling water sterilizer cannot be depended on to kill spores. Because it is open to the atmosphere, the water will attain a temperature of only 100°C (212°F), which is insufficient. Therefore, this device is not a sterilizer in the absolute sense of the term.

True sterilization can be accomplished by means of a dry-heat sterilizer, operating on the same principle as the electric oven. Electric heating elements within the cabinet raise the temperature to the desired level. While the literature generally discusses exposure of the material to be sterilized by dry heat for 1 hour at 160°C (320°F), several manufacturers recommend cycles of 30 minutes at this temperature. The World Health Organization Committee on Hepatitis recommends that if there

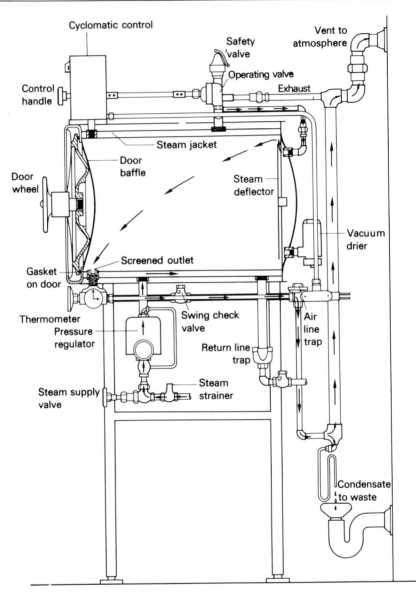

Figure 9-4.

Cross section of autoclave.

Steam from the main source enters the jacket via the supply valve and the pressure regulator, which maintains a constant pressure. The operating valve admits steam from the jacket to the interior of the autoclave.

Source: Adapted from The Surgical Supervisor, American Sterilizer Co., Erie, Pa.

is evidence that all parts of the (dry heat) chamber are equally heated, a temperature of 170°C (338°F) for ½ hour is sufficient. One advantage of dry-heat sterilization is that it does not damage the fine cutting edge of surgical instruments. Dry heat is recommended for the sterilization of syringes because it will not erode the ground-glass surfaces.

The most common method of sterilization is autoclaving (Fig. 9-4). The autoclave is essentially a closed cylinder in which the materials to be sterilized are exposed to high-pressure steam. Remember that it is the *heat* that kills the organisms, not the steam. As one company puts it, "The bacilli are not squeezed to death." The autoclave is basically a pressure cooker. You will recall that the boiling point of water varies directly with the pressure, that is, the higher the pressure, the higher the temperature.

We can see in Figure 9-5 that at 15 pounds per square inch gauge pressure, the temperature of the steam is 121°C; at 25 pounds the temperature is 130°C; and that at 40 pounds the temperature is 142°C. Please note that at 0 pounds pressure (atmospheric pressure) the temperature is 100°C.

These temperatures apply to pure steam. If there is air mixed with the steam, the temperature will be lower. On the newer models there is an automatic valve controlled by a thermostat to remove the air. As steam enters the autoclave, the air is forced out. When the steam begins to escape, the temperature rise causes the thermostat to close the valve automatically.

Most autoclaves have a double wall, called a steam jacket. What is its purpose? In a single-wall autoclave the steam will be cooled by the walls and will condense,

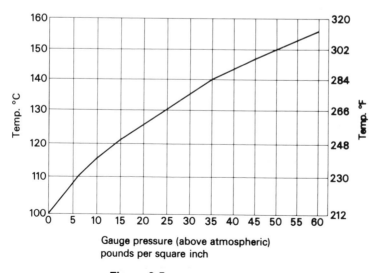

Figure 9-5.
Temperature–pressure relationship.

producing a great deal of water. But if a second wall is added, and the space between them is filled with steam, the inside wall is kept warm, and little condensation will occur. The condensation within the jacket does not affect the material being sterilized; it is either run off through the drain or else is reused.

In order to check whether the desired temperature has been attained, indicator tapes are usually used. A plain tape looking like masking tape is put on the outside of the pack. When the temperature for which it is calibrated has been attained, such as 250°F, a series of black lines appears. However, to make sure that the heat has penetrated into the center of the pack, another type of tape is used, in which a color change is produced, such as from purple to red. Tapes can be purchased for either steam or gas sterilizers.

In order to insure that everything is working satisfactorily, hospitals run tests about every 30 days. A tape containing a spore culture is used. One is inserted into the center of a large pack, another is placed in the sterilizer outside the pack, and a third is not heated at all. All three strips are sent to the laboratory for culture testing. This is an experimental way to insure that the spores are killed in the sterilizer, with the control strip showing that the culture strip is functioning.

To summarize: it is the *heat* that kills the bacteria. The time required depends upon the type of microorganism involved and also upon the penetration rate into the pack. One advantage of the steam sterilizer over the dry-heat sterilizer is that the steam will penetrate more rapidly. If a steam sterilizer rather than a dry-heat sterilizer is used with cutting instruments, the cutting edges may become dulled due to oxidation. This may be avoided by submerging them in a solution of diethylene glycol during autoclaving. This chemical shields the metal from the water and the oxygen that cause corrosion.

Instead of heat, it is possible to use a sterilizing gas. There are on the market gas sterilizers that use gas cartridges of ethylene oxide. The gas is packed under pressure; the can is punctured by pushing down on a needle, and the gas fills the sterilizing chamber. Since the container is under pressure, it must be kept away from fire and flame, and the user must avoid both breathing the vapor and allowing it to contact the skin or eyes.

BODY TEMPERATURE

HEAT PRODUCTION

Food materials, after being reduced to simple forms by digestion, are absorbed and then are conveyed to the tissues by the bloodstream. The food then is oxidized in the cells. As is well known, carbohydrates and proteins produce approximately 4 calories of heat per gram of food, and fats produce 9 calories of heat per gram. Actually, not all of the food that is ingested reaches the cells. It has been estimated that on the average about 98 percent of the carbohydrates, about 95 percent of the fats, and about 92 percent of the proteins ingested are absorbed and reach the body cells. For precise calculations some deductions for "losses in digestion" are made, and the fuel values of foods are estimated as follows.

Carbohydrate (1 g):	98 percent of 4.10 cal = 4.0 cal
Fat (1 g):	95 percent of 9.15 cal = 8.7 cal
Protein (1 g):	92 percent of 4.35 cal = 4.0 cal

REGULATION OF BODY TEMPERATURE

Heat is being produced constantly in the body, and at the same time is being lost by it. Both heat production and heat loss occur at rates that vary greatly; but in spite of these variations, certain body mechanisms ensure the maintenance of a temperature that normally is held within fairly narrow limits.

All parts of the body are maintained at a nearly uniform temperature by the circulation of the blood. In this case the principle of heat transfer is conduction, and can be compared with the operation of a hot-water heating system in a house.

The structure of the brain called the hypothalamus is the body's "thermostat." It receives impulses from the peripheral temperature receptors and transmits impulses to both the visceral efferent system and the somatic efferent system. For instance, if the body becomes overheated, the peripheral blood vessels dilate to bring more blood to the surface of the body so that more heat can be removed by radiation and conduction. This is why the skin appears flushed when the body is overheated. Conversely, if the body is exposed to cold, the blood vessels contract in order to decrease the amount of blood at the surface and thereby to minimize the heat loss.

A high outside temperature will cause the sweat glands to excrete perspiration; as this evaporates, it removes body heat. Since most of the body heat is generated in muscle tissue, the muscles remain as lax as possible in hot weather.

If the outside temperature is low, and heat is being removed from the body too quickly, a number of reflexes are initiated. Besides vasoconstriction, there is inhibition of the sweat glands. The pilomotor muscles contract, producing "gooseflesh" (which reflex is somewhat obsolete in humans, incidentally, since the original purpose of it was to allow the body hairs to become erect, thus providing a large number of "dead-air" traps for insulation). Finally, the movement known as shivering begins, which causes the rapid generation of heat in the skeletal muscles.

The rate at which tissue metabolism is carried on varies according to the degree of outside temperature. Increased temperature, with consequent vasodilation, causes the blood to flow more rapidly and thereby increases the oxygen supply and hence the metabolic rate. Decreased temperature will slow down metabolism because of the reduced circulation.

LOCAL APPLICATIONS OF HEAT AND COLD

Heat and cold are sometimes applied to various parts of the body for their therapeutic effects. The application of heat will usually reduce pain and muscle spasm (by relaxing the muscle); promote healing of a part (by increasing the metabolic rate); and reduce congestion in one body area by increasing the blood flow in another. The temperature of any application of heat should never exceed 110°F because of the danger of burning the patient.

Cold applications are used to check hemorrhage, prevent edema, and produce anesthesia. Hypothermia blankets are used to reduce high fever. Cooled water circulates through plastic blankets placed between the patient and the bed.

Remember that these various effects of heat and cold applications do not hold true if the applications are prolonged. The immediate effect of heat is vasodilation; but if the hot application is maintained for an hour or so, the blood flow is actually reduced. On the other hand, a prolonged application of cold eventually will cause vasodilation. The reasons for this reversal of effect are not yet clear.

Would an application of a cold pack be helpful immediately after a sprain or strain has occurred? Yes, because the vasoconstriction will decrease the accumulation of fluid in the area. However, if edema is already present, the application of cold will decrease the circulation, and it will take longer for the fluid to be reabsorbed; thus at this stage cold is not indicated.

Both the hot and the cold receptors become adapted to new temperatures. This is sometimes a danger because a person may be exposed to temperatures that will cause tissue damage without his being aware of it.

BASAL METABOLIC RATE

Basal metabolism is the energy production required to maintain the life of a person. The basal metabolic rate (BMR) of the subject is measured with the subject awake, as nearly as possible in a state of complete muscular and mental rest, and in a postabsorptive state (i.e., 12 to 14 hours after a light meal).

Nurses and other health practitioners are well aware of the clinical significance of the BMR, but relatively few are aware of what a figure such as +20 represents. Is it calories, liters of oxygen, or some other unit? The BMR is measured as *the percent of variation of the basal metabolism from the expected normal value.*

There are several methods of measuring the BMR. The most basic one is called the direct method, in which the subject is placed in an insulated chamber of known heat capacity, and the increase in temperature is used as a measurement of the heat evolved. This requires very elaborate equipment and is seldom used.

More commonly employed are various indirect methods, which have a number of modifications. The closed circuit forms use compressed oxygen, whereas the open circuit forms use oxygen from the air.

For precise work, the amount of carbon dioxide evolved is measured in addition to the oxygen consumed. However, since usual diets do not vary a great deal, and the body oxidizes carbohydrates, fats, and proteins in amounts that are physiologically predetermined and which do not vary a great deal, it is generally sufficient to determine only the oxygen consumption.

Through careful calculations it has been found that one liter of oxygen consumed is equivalent to 4.825 calories.

The BMR increases in proportion to body surface rather than weight. Charts such as the DuBois chart have been developed that make it possible to get an approximate value for the surface area from the height and weight of a person.

The apparatus used is called a *spirometer,* from which pure oxygen is administered to the patient through a mask (Fig. 9-6). The expired air returns to the spirometer through an efficient carbon dioxide absorber. The entire closed gaseous system is kept saturated with water vapor and the oxygen consumption is measured by the progressive decrease in the volume of oxygen in the spirometer. This takes about 6 to 8 minutes, depending on instructions received with the apparatus.

The result is expressed in milliliters of oxygen used per minute. After the observed value has been reduced to standard temperature and pressure (760 mm Hg pressure and 273° absolute temperature, i.e., 0°C), the number of liters per hour is calculated. Because 1 liter of oxygen is equivalent to 4.825 calories, the number of liters of oxygen consumed per hour is multiplied by 4.825 to get the number of *calories* produced *per hour*. This is then divided by the number of square meters of body surface, to get the number of *calories per hour per square meter* of body surface.

Because, on average, a person produces about 40 calories per hour per square meter, the above calculated value is compared with the average. Let us assume that a patient produces 50 calories per hour per square meter. This is 50–40, or 10 more than the average; that is, $10/40 \times 100 = 25$ percent above average. This is reported as $+25$.

A person producing 35 calories per hour per square meter would be 5 less than 40. This is reported as $5/40 \times 100$ or -12.5.

The age and sex of the patient must also be considered. The heat production per square meter of body surface diminishes progressively from infancy to old age, being about 50 calories per square meter per hour at the age of 10 or 12 and about 32 calories at age 90. Female patients have a metabolic rate a little lower than that of male patients of the same age group. Summarizing the procedure: In computing the BMR of a patient we need:

1. To correct, that is, to reduce, the oxygen consumed to standard temperature and pressure.

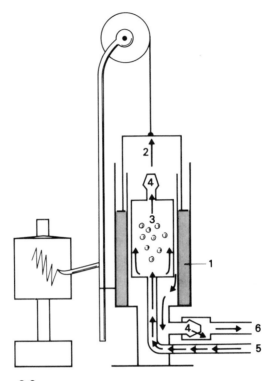

Figure 9-6.
Spirometer.

1, water seal in tank; 2, movable chamber; 3, soda-lime canister which absorbs carbon dioxide from expired air; 4, one-way valves through which air passes in direction indicated by arrows; 5, expiratory tube; 6, inspiratory tube. The upper chamber may be filled with oxygen or room air. Its movements are recorded on the revolving drum. As the oxygen in the chamber is used up, the record assumes a downward slope.

2. To calculate the number of liters of oxygen consumer per *hour*.
3. To multiply the number of liters of oxygen consumed by its caloric value (4.825 calories per liter).
4. To divide the result (calories per hour) by the number of square meters of body surface.
5. To subtract the quantity (calories per hour per square meter) from the basal rate considered normal to the subject in question. (The basal rate considered normal takes into consideration the sex and the age of the patient.)

This figure is customarily expressed as normal or as a percentage above (+) or below (−) normal.

Let us work a BMR problem in detail. A patient having a body surface of 2 square meters used 2 liters of oxygen calculated at 0°C and 760 mm Hg pressure, in 6 minutes. The volume is already at standard conditions.

$$\frac{2 \text{ liters}}{6 \text{ minutes}} = \frac{x \text{ liters}}{60 \text{ minutes}}$$

$$6x = 120 \qquad x = 20 \text{ liters of } O_2/\text{hr}$$

$$4.8 \text{ cal/l} \times 20 \text{ l/hr} = 96 \text{ calories/hr}$$

This is for the whole patient. Dividing by the body surface, we get

$$\frac{96 \text{ cal/hr}}{2 \text{ m}^2} = 48 \text{ cal/hr·m}^2$$

As the average value is 40 cal/hr·m^2, the difference will be $48 - 40 = 8$ cal/hr·m^2. $8/40 \times 100 = 20$. Because the patient is above the average, this would be reported as $+20$.

A main purpose of the BMR test is to estimate the activity of the thyroid gland, the function of the thyroid being to control the rate at which the tissues metabolize.

The spirometer test for thyroid function is now being replaced by various tests using radioisotopes. The radioactive iodine uptake tests are discussed in Chapter 14, and the in vitro radioimmunoassay (RIA) tests for T-3 and T-4 are discussed in Chapter 15.

QUESTIONS

1. Is it correct to say that there is no heat in a vacuum? How does radiant energy become heat?
2. Why is it unreliable to test the temperature of water by keeping your hand in it for some time?
3. What would be the difference, if any, in the effects on the patient of two ice bags, one filled with 100 g liquid water at 0°C and the other filled with 100 g chopped ice at 0°C? Calculate the amount of heat that each bag could absorb from a patient, assuming the final temperature of each to be 38°C.
4. How many calories of heat would be removed from the body by the evaporation of 20 g perspiration?
5. Why are some cooking utensils coated with copper on the bottom?
6. Explain the difference between quantity and intensity of heat.
7. Using the coefficients of thermal conductivity, show whether a dry or a wet flannel wrapper on a hot-water bag is more likely to burn the skin.

8. Some recipes have the instruction to add 5 minutes to the boiling time at a high altitude. Explain.

9. How many calories will be required to change 50 g ice at 0°C to steam at 100°C?

10. How many calories will be required to raise the temperature of 100 g water from 10° to 30°C? How many calories for 100 g aluminum?

11. If 1 kcal heat is applied to 80 g water at 20°C, what will be the resulting temperature?

CHAPTER 10

SOUND

*Sound, like heat, is another common phenomenon
that is difficult to define. However, an understanding
of characteristics of sound, such as speed, pitch,
and frequency, is basic to the study of hearing and
the methods used to test and improve it.*

If a tree falls in a forest, and there is no one to hear it, is there any sound? This problem has been debated with much vigor and heat in many philosophy classes. What would be your answer?

As is always the case in a discussion, there is a need to define terms. One definition of sound is: vibrations in a medium which are *capable* of being received by the ear and interpreted by the brain as sound. Let us call this the *physical* definition. According to this definition, then, there *is* sound if the tree falls, even though there is no one to hear it, because vibrations which are potentially audible were created. Sound can also be defined as the *auditory response* or the *interpretation of the disturbance*. We can call this the *psychological* definition. If this definition is accepted, it means that there must be a hearer before there is sound, and thus there is no sound when the tree falls without a hearer, even though there are vibrations in the air. In fact, if we leave a room where the radio is playing, if there is no one to hear it, there is no sound.

If you ask which definition is right, the answer is that they both are, with one placing the emphasis on the physical manifestations and the other on the subjective aspects of the phenomenon.

THE SPEED OF SOUND

Sound can be transmitted by solids, liquids, and gases. Its speed is greatest in solids, least in gases. The seismographic apparatus used by oil companies in prospecting for oil wells is based upon the differences in the speed of sound in liquids and in solids. Small explosive charges are set off, and picked up by microphones. The time it takes for the sound to reach the apparatus is recorded on a graph. If an underground pool of oil or water is present, the sound will be slower than through the solid ground.

Sound travels through a liquid about four times as fast as through the air. In dense solids (such as iron) the speed is about 15 times as great.

The velocity of sound in air is 1,089 feet per second at 0°C, plus 2 ft/sec faster per degree Celsius. Thus at 10°C, the velocity will be 1,089 plus 10 × 2 = 1,109 ft/sec. In other words, the speed varies directly with the temperature.

Because the speed of light is so much greater (186,000 miles per second) than the speed of sound, which is only about $1/5$ of a mile per second, we see an event before we hear it. For example, if it takes 1 second after we see the steam coming out of the whistle of a ship before we hear it, the ship is approximately 1,100 feet away. Similarly, if there is a time lapse of 10 seconds between seeing the lightning and hearing the thunder, the storm is 2 miles away (5 seconds equals 1 mile). If the time lapse is 0 second, it is time to worry.

An echo is the reflection of sound from a surface. If it takes 10 seconds before the echo of, say, a yodel is heard, how far away is the mountain? Ten seconds equals 2 miles. Since the sound has to go back and forth, the mountain is only 1 mile away.

The speed of sound at sea level is approximately 760 miles per hour, decreasing with lower temperatures. The speed of sound at any temperature is called Mach 1. As a plane goes faster than Mach 1, a great deal of vibration will be set up in the plane, and a loud noise is created. This is usually called the "sonic boom." The noise seems associated with the compression of air molecules in front of the plane and their coming together behind the plane. Passengers in a plane going faster than the speed of sound would not hear the plane's own loud exhaust noise, because the sound could not catch up with the plane.

THE WAVE CHARACTERISTIC OF SOUND

You are familiar with waves in a liquid. When a stone is dropped into a calm pool of water, the waves or ripples spread out from the source of the disturbance. Sound waves behave the same way as they travel through their medium.

As sound waves spread out in all directions, their energy is quickly expended. The stethoscope largely overcomes this rapid loss of intensity by receiving the sound waves from a small area (such as the heart region) and transmitting them through a tube so that they are able to move in but one direction, arriving at the listener's ears at very nearly their original strength.

How do sound waves travel through their medium? Remember that although

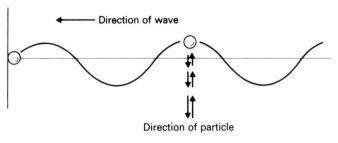

Figure 10-1.
Transverse waves.

other forms of energy (light, heat, etc.) also move in waves, their type of wave motion differs according to the nature of the energy source. For instance, sound waves are said to be *longitudinal* and light waves are *transverse*. Let us compare these two wave characteristics.

Transverse waves can be pictured by considering a rope, fastened at one end, being waved up and down.

Imagine a certain particle on the rope. This does *not* travel along the rope; it merely rises and falls. But the wave moves along the rope. Therefore, the particles of the substance in which the waves move travel *transversely* (perpendicular to) the direction in which the wave is going (Fig. 10-1).

Sound waves, as we noted, are *longitudinal* (or *compressional*). This means that the particles of the medium that carries the wave move in the same direction as the wave. Let us consider a tuning fork. As the prongs move to the left, the air will be compressed on the left side of the fork. When the prongs a moment later move to the right, the air on the left of the fork will undergo rarefaction (decompression), while the air on the right of the fork will become compressed. Thus a series of compressions and rarefactions will be produced (Fig. 10-2).

The disance from one compression to the next is one wavelength. This is usually represented with a sine curve. The frequency of a tone becomes higher as the wavelength becomes shorter (Fig. 10-3).

The shorter the wavelength, the more waves per second would pass a given point, and therefore their frequency—the number of vibrations per second (vps), also called cycles per second (cps) or Hertz (H_z)—increases. The soprano tone above

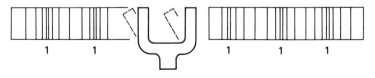

Figure 10-2.
Longitudinal (compression) waves.

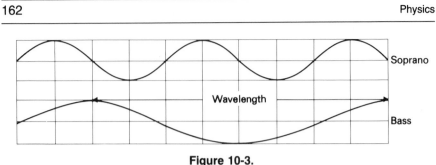

Figure 10-3.
Wavelengths.

would have a shorter wavelength and a higher frequency than the bass tone. The tone thus can be characterized by either the frequency or the wavelength.

Pitch and Frequency.   The intensity or loudness of a sound is dependent upon how much energy is used in the generation of the sound. If a tuning fork is hit harder, the prongs will move a greater distance back and forth, but still with the same frequency. Thus the tone will sound the same, but will be louder. This is shown by the dotted line in Figure 10-4. The loudness of a sound depends on the amplitude of the wave.

Sometimes the words "pitch" and "frequency" are confused. Pitch is *not* synonymous with frequency. It is the subjective interpretation of frequency, and is dependent not only on frequency, but also on the intensity and the composition of the sound. The composition of a tone is the result of its overtones. These will be described later.

Similarly, the loudness of a tone is an interpretation of the intensity of the sound by the brain. It varies directly with the amplitude, but is also dependent upon the frequency and the composition of the tone. This is the reason why in hearing-range curves the "loudness" is not plotted against frequency, but rather in intensity units such as decibels, pressure in dynes per square centimeter, or watts per square centimeter. In other words, terms like "loudness" and "pitch" are psychological terms rather than physical units, although this does not decrease the importance of them.

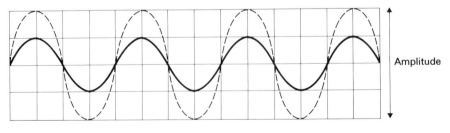

Figure 10-4.
Amplitudes of two waves with the same frequency.

Analysis of Sound. A sound may be analyzed by picking it up on a microphone, changing it to an electric current, amplifying (magnifying) the current, and making it visible on an oscilloscope, an electronic device similar to a television set in that it has a fluorescent screen. The impulses are made into sine waves and the length and shape of the sound waves are rendered visible on the screen. What determines the difference in quality of tone in human voices or in musical instruments? Each

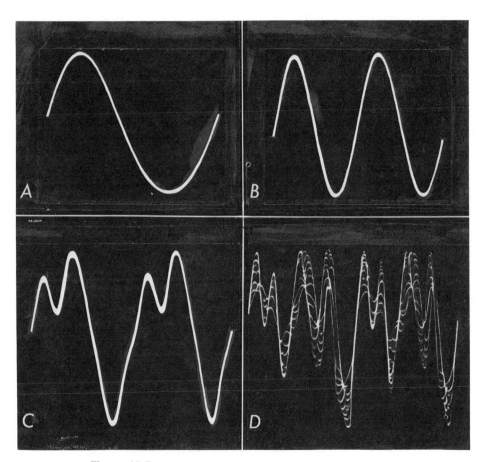

Figure 10-5.

Sound waves from various musical instruments as they look when projected on the fluorescent screen of an oscilloscope.

A is a fundamental sine wave. B is its second harmonic; note that there are twice as many peaks and troughs occupying the same space, which represents an interval of time. C is the pure note of a French horn. D is the note of a trumpet. The small ripples intertwined with the fundamental wave are overtones. They are much higher in frequency than the fundamental and serve to lend a characteristic quality to the note of a particular instrument.

person's voice has a fairly characteristic tonal quality. The same holds true for instruments. The sound of a French horn differs from that of a trumpet or a violin. Each tone contains not only the basic wave, but other higher frequencies than the fundamental one as well. Simple multiples of the fundamental tone are called *harmonics* (Fig. 10-5). *Overtones* are more highly pitched tones that "ride along" on the sine waves; they determine the quality or *timbre* of the tone. Figure 10-5 shows the oscilloscope patterns of various instruments. The integration of the fundamental tone with various overtones can be seen. Note the relatively simple pattern of a French horn and the ragged mixed quality of a trumpet. Here is a good example of a visual representation of the feeling that we have when we hear these sounds. We consider the French horn a clean, simple, pure tone and the trumpet brassy and complex.

The *timbre* of an instrument depends on its shape and the material of which it is made. The attempts to reproduce a Stradivarius violin have not been completely successful because even the composition of the varnish will affect the timbre. Some concert violinists use two violins, one for brilliant passages and one for more melodious music.

The human voice as well as reed instruments can be considered from the point of view of physics to be composed of a power supply, an oscillator, and a resonator that can be modified in order to change the sound. The lungs supply the air stream; the length, thickness, and tension of the vibrating vocal folds produce the sound; and the larynx, pharynx, mouth, nasal sounds, and nose act as resonators. In order to change the sound, the articulators—the lips, jaw, tongue, and larynx—can change position or volume. Each instrument and voice thus has its own characteristic timbre or quality. You can easily recognize the voice of someone you know well, even over a low-fi instrument, such as the telephone.

The human voice has small fluctuations, or microtremor, and these vary with the degree of stress. A lie-detector apparatus, the voice-stress analyzer, is based on this phenomenon. This machine is different from the polygraph machine which records breathing, blood pressure, and skin conductivity as a function of perspiration under stress. The voice-stress analyzer picks up the sound on tape, converts it to an electrical current, isolates the microtremors from the rest of the voice pattern, and plots the graph on a tape. By asking nonstressful questions at first, the baseline pattern is established for the person, and when more important questions are asked, the response may indicate increased stress, unless the person is a pathological liar.

MUSIC OR NOISE?

Music is distinguished from noise in that a musical note has a regular frequency and noise has no definite frequency or regularity. During the last few years many people think it has become increasingly difficult to distinguish between music and noise. Music used to have easily recognized tempo, melody, and harmony. With some modern music, such as atonal music, it may seem that neither melody nor harmony is very evident. There is a great deal of conditioning involved in musical taste. The sound of bagpipes or of Oriental music may seem beautiful or unpleasant depending upon the previous experience of the listener.

Why do some combinations of notes produce harmony while others are discordant? If the frequencies of these various notes are measured, we will find definite numerical relationships between them. Mathematics, then, is one of the technical bases of musical composition.

HEARING

Hearing involves both mechanical and electrochemical processes. The source of the sound sets up compressional waves in the medium. A person speaking sets up waves in the air. These will be gathered by the outer ear, and will impinge upon the eardrum (tympanic membrane) producing a vibration similar to the original sound wave. The movement of the tympanic membrane is transmitted to the three bones, the malleus, the incus, and the stapes, and then to the fenestra ovalis. The size of this oval window is about 1/30th as large as the eardrum, and the force is thus applied to a smaller area with an increase in pressure. The ossicular chain appears to be a first-class lever with the fulcrum between effort and resistance. Movement of the stapes may be hindered by *otosclerosis,* a disease involving the temporal bone, in which there are various bone changes including hardening of the stapes. In such cases an operation called *stapes mobilization* is possible. Sometimes gentle pressure breaks the hardened membrane about the stapes and restore movement; with the bones free to transmit the vibrations, the patient hears again. If this simple loosening of the bone is not sufficient, the stapes is removed, leaving the stapes footplate attached to the inner eardrum. The footplate is then punctured, and two small polyethylene prostheses are used to reconnect the bones.

Another operation, known as *fenestration,* is used in cases where, as a result of otosclerosis, there are progressive osseous changes around the oval window. A new window is made by pulling a flap of skin over a new opening that has been made into the inner ear. If the bones are intact and movable, the vibration is transmitted to the inner ear. Figure 10.6, depicting the human ear, shows how the soundwave is transmitted.

The inner ear contains the semicircular canals, which are involved in the sense of equilibrium, and the cochlea, which is filled with a liquid. As the stapes pushes on the oval window, the liquid is under pressure. Liquid is not compressible to any significant extent, but provision is made in the inner ear with the round window to have the required "give." In other words, as the oval window pushes in, the round window moves out an equivalent amount. The net result is a "ripple" or compression wave in the liquid which stimulates the cilia of the hair cells of the organ of Corti. The high frequencies are picked up near the oval window, and the lower frequencies are picked up toward the thicker, apical end. So far the process of hearing has been entirely mechanical. Now we shall investigate other factors involved in this process.

There are an estimated 23,000 hair cells that pick up the vibration. We usually say that this impulse is transmitted to the brain where it is "interpreted." That covers a lot of ignorance! Through the work of anatomists, physiologists, neurologists, and psychoacousticians much has been learned, but much remains to be learned.

How do these nerve cells pick up the stimulus? It is possible that they exhibit a so-called piezoelectric effect (the effect whereby mechanical pressure on a certain type of crystal produces electricity—the principle used in crystal microphones). As yet there seems to be no definite answer to the question whether the electrical potential developed in the hair cells is directly responsible for resulting impulses

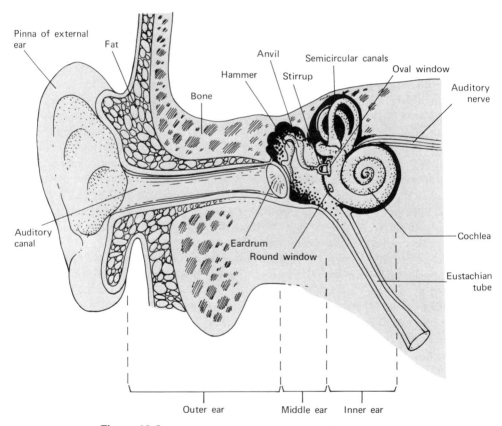

Figure 10-6.
Anatomy of the human ear.

Source: Shirley R. Burke, *Human Anatomy and Physiology for the Health Sciences*. John Wiley & Sons, New York, 1980, p. 270.

in the auditory nerve fibers, or whether these cells cause some chemical reaction in the intervening ganglia, which in turn stimulate the nerve fibers.

Another problem is whether there is selectivity of the hair cells. The resonance theory (also called the Helmholtz theory) suggests that each hair cell acts as a resonator; that is, it has the ability to respond to only one frequency, and this is connected to a specific nerve fiber. Another possibility is the frequency theory, which suggests that action in the cochlea is not restricted to one fiber, but is widespread. The pitch is determined by the frequency of the nerve impulse transmitted to the brain, and analysis is thus the function of the higher centers.

Whatever the action involved, the sense of hearing is a wonderful gift. When we think of the many tones we hear even when one note is played on the trumpet (as we saw in Fig. 10-5) then it becomes even more difficult to understand how the sound of a symphony orchestra can be transmitted as depolarization waves over nerve fibers.

THE HEARING RANGE

The normal human ear is responsive to a wide range of sounds, extending from the deep bass tones with 16 to 20 cycles per second (cps) to the high-pitched tones of 16,000 to 20,000 cps. There is much individual difference in sensitivity. Tones below 16 cps cannot be heard, but may be felt; those above 20,000 cps are called ultrasonic. These will be discussed later.

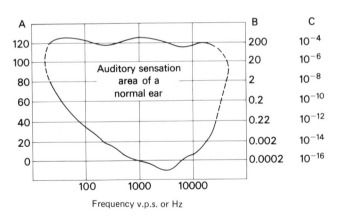

Figure 10-7.
Hearing range of the normal ear.

A: Intensity level in decibels; B: Pressure in dynes per square centimeters; C: Intensity in watts per square centimeter.

Source: H. Stearns, *Fundamentals of Physics and Applications*, 2nd ed. MacMillan, New York, 1956, p. 164.

The hearing range curve looks a little like a map of the United States without California and Florida. On the vertical axis is plotted the intensity and on the horizontal axis the frequency in vibrations per second, now called Hertz (Hz). (Fig. 10-7.)

What does this curve show? If you place a ruler horizontally at the bottom of the graph and move it slowly toward the top, you will notice that the center of the frequency range will be heard first. We are most sensitive to sounds in the region of 1,000 to 4,000 cps. In order to hear the tones of 50 cps or 12,000 cps the volume must be increased about a millionfold. This is one reason "hi-fi" fans usually keep the volume so high. They want to be able to hear the highs and lows that they have paid so much money to be able to reproduce.

The hearing range decreases with age, and a person of sixty usually can't hear above 8,000 cps. However, this usually is not too much of a handicap, as the human voice range is about 85 to 1,050 cps, differing of course with the sex of the speaker.

THE INTENSITY OF SOUND

In Figure 10-7 we have used on the vertical axis three different units of sound intensity: A, decibels; B, pressure in dynes per square centimeter; and C, intensity in watts per square centimeter. These are all measurements of the *intensity* or power of the sound, which is to be distinguished from the *loudness* of the sound, which is subjective measurement, and is reported as *phons*. Since the subjective loudness varies at different frequencies, it is better to use the sound intensity; this is done by comparing the intensity of one sound with another at 1,000 Hz.

The ear has an incredible range of sounds that can be heard, from the most quiet sound to one a million-million times louder without immediate pain. If we call the intensity of the first detectable sound 1, we can set up a logarithmic scale that will indicate how many times more intense another sound is. The $\log_{10}$ of 1 is 0, because $10^0 = 1$. If we have a sound that is 10 times more intense, the $\log_{10} 10 = 1$, as $10^1 = 10$. The log of the intensity compared with the threshold intensity is called the *bel,* after Alexander Graham Bell, the inventor of the telephone. The following diagram shows the relationship between bels and relative sound intensities.

Intensity	Bels	Decibels (dB)	
1 : 1	0	0	$10^0 = 1$
1 : 10	1	10	$10^1 = 10$
1 : 100	2	20	$10^2 = 100$
1 : 1,000	3	30	$10^3 = 1,000$
1 : 10,000	4	40	$10^4 = 10,000$

As the bel unit is rather large, the bel is divided into one-tenths, the decibels.

A sound having an intensity 1,000 times threshold would be 3 bels, or 30 decibels. A sound 100,000,000, or 10^8 times more intense would be 8 bels or 80 decibels.

Let us try one additional value. How much more intense would a sound be that has a value of 60 decibels? This would be 6 bels, or 10^6, which is 1,000,000 times greater.

Normal breathing sounds would be about 10 decibels, the rustle of leaves 20, normal conversation 55, a power lawnmover 6 feet away 90 dB, and a jet plane at takeoff 140 dB.

We are suffering from all types of pollution, including *sound pollution*. Many times in the streets we are exposed to sounds of 80 to 100 decibels, and in many places of employment this level may be exceeded. It appears that continuous exposure to high noise levels will damage hearing acuity; many companies now test the hearing of employees at various intervals. This is done both for humanitarian reasons and for protection against damage suits for hearing losses. The extremely high level of electronically amplified music that many young people are exposed to for hours daily may also be causing hearing damage. Much sound approaches the pain level of 130 decibels. Some hi-fi sets can reach 140 to 145 decibels, definitely within the pain threshold area.

The Occupational Safety and Health Administration (OSHA) is for the present time setting the noise standard for employees at 90 dB, approximately the noise made by a heavy truck. This means that on the basis of an 8-hour day all U.S. businesses will have to limit noise levels to 90 dB. At higher noise levels, the time of exposure will have to be reduced proportionately by one-half for every 5 dB over 90. Thus, at a noise level of 95 dB, the worker could be exposed for only four hours per day, and of 100 dB, for 2 hours per day.

A number of critics feel this level is too high, especially because most other countries have set 85 dB as a goal and are working at attaining it. However, the cost of meeting the 90-dB level is expected to be $13.5 billion, and if the 85-dB level were set, $32 billion. OSHA is setting a more realistic goal for the present time, and will press for attaining it before lowering the limits. If the noise level cannot be reduced in some industries, workers can wear ear protectors. It is a step forward that we are using phrases like "noise pollution," and it is important to be aware of the fact that loud, prolonged noise, whether in a factory or at a rock music concert, can cause deafness and irritability. The National Bureau of Standards reports that prolonged exposure to sounds over 75 dB can result in stress symptoms, such as a faster pulse, shallow breathing, a rise in blood pressure, and tinnitus, a ringing in the ears.

AUDIOMETERS

An audiometer is an apparatus which produces sounds of definite frequencies and intensities for the purpose of testing hearing. It employs an electronic circuit to produce tones of known frequencies and loudness. The time-honored sound producers used in detecting hearing losses were the conversational voice, the whisper, the clicking of coins, or the ticking of a watch. These all suffered from the lack of standardization; one clock might tick louder than another. There are two types of audiometers, the step-type and the sweep-type. The step-type has a limited number of specific frequencies, while the sweep-type has a continuous range of frequencies. For ordinary purposes the step-type of audiometer is as suitable as the sweep-type.

The hearing is often tested at 125, 250, 500, 750, 1,000, 2,000, 3,000, 4,000, 6,000, and 8,000 cycles per second. The percentage of loss is plotted on a graph, such as that shown in Figure 10-8.

An audiometer amplifies the sound up to a maximum of 110 decibels. This means that the sound is 110 dB or 11 bels, which is equal to 10^{11} or 100,000,000,000 times louder than the normal threshold of hearing; a person who could not hear that sound would have very severe hearing loss.

Instead of plotting the percent hearing loss, one can plot the sound level in decibels required before the person can hear a sound of a particular frequency. Figure 10-9a,b shows such hearing threshold level graphs for two patients.

It will be seen in Figure 10-9a that there is a moderate to severe general hearing loss in both ears throughout the frequency range from 250 to 8,000 Hz. In addition, there is a general bone conduction loss shown for the right ear. Figure 10-9b shows a patient with hearing losses in both ears in only the higher frequencies. The air conduction symbols in Figure 10-9a,b indicate that no masking was used.

MASKING IN PURE-TONE THRESHOLD TESTS

It has been found that if a severe hearing loss is present in one ear, and the sound has to be raised above 50 dB in order to be heard, part of the sound will cross the head and will be heard by the other ear. However, because the sound from the "poor" ear is heard through the head in the "good" ear, this gives an incorrect

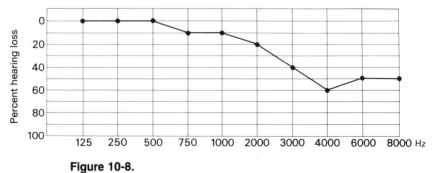

Figure 10-8.
An audiometer curve, showing hearing loss at higher frequencies.

AUDIOGRAM — CYCLES PER SECOND (HERTZ)

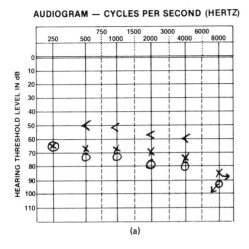

(a)

AUDIOGRAM — CYCLES PER SECOND (HERTZ)

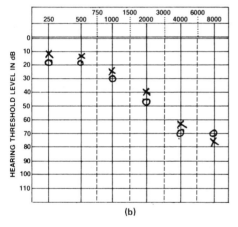

(b)

KEY TO SYMBOLS

		No Response
Right ear—red		
Air (unmasked)	○	○
Air (masked)	△	△
Bone (unmasked)	<	<
Bone (masked)	⊏	⊏
Left ear—blue		
Air (unmasked)	✕	✕
Air (masked)	▢	▢
Bone (unmasked)	>	>
Bone (masked)	⊐	⊐

Figure 10-9.

(a) General hearing loss in air and bone conduction. (b) High-frequency losses.

Source: Courtesy of Isidor Haitkin, Davis Hearing Aid Center, 851 Forest Avenue, Staten Island, N.Y. 10310.

reading for the "poor" ear. Putting a masking tone into the "good" ear so that it is occupied by that stimulus makes it possible to get a valid measurement of the hearing loss in the "poor" ear. Therefore, it is important that where significant differences exist between the two ears, a masking sound be used to avoid having the "good" ear pick up some of the sound.

This type of crossover is also found in bone-conduction testing, in which the mastoid bone conducts the sound to both ears with almost equal intensity.

HEARING AIDS

A hearing aid is an electronic device that assists a hearing-impaired person to hear. It consists of a small microphone to pick up the sound and convert it into a faint electrical current. This current is then amplified so that it will actuate a small receiver (actually a miniature loudspeaker) that fits in the ear. Hearing aids fall into two classifications, the *air conduction* type and the *bone conduction* type. The air conduction type is used in all cases except where the ear is not functioning because of some condition such as a perforated eardrum or immobilization of the ossicles. In these cases, conduction of sound by the bone may be the answer. If the damage is in the nerve fibers, a hearing aid will not restore hearing.

Before audiometry, the companies manufacturing hearing aids often used word lists in an attempt to determine the type of hearing loss. Ordinarily the vowels are pronounced at low frequencies and the consonants at high frequencies. If words were missed because of the consonants, the indication was that hearing was deficient at the higher frequencies. Of course, this was only an approximation.

Unfortunately, even though a person may have had an audiometer test, the fitting of the hearing aid is often done by untrained people. The aid should be matched to the hearing loss by trained technicians. If there is a linear (uniform) deficiency in hearing, all frequencies should be amplified equally. Otherwise, only those frequencies to which the patient has an inadequate response should be amplified.

A hearing aid contains a microphone to pick up the sound, an amplifier to magnify the sound, and a small speaker that will transmit the amplified sound to the ear.

Various types of microphones are used. These are the magnetic type, also called the dynamic type, the ceramic type using piezoelectric ceramic material, and the electret condenser type. The microphones can be either omnidirectional or directional. The latter type gives the patient a better feel for the direction from which the sound is coming.

There are four different ways of placement of the hearing aids.

- *Conventional.* Also called the body or pocket type, generally used for cases of severe hearing loss in the range of 65 to 80 dB.
- *Behind-the-ear.* Also used for moderate to severe hearing loss.
- *The Eye-glass.* Components of the hearing aid are fitted into the temples of the glasses.
- *In-the-ear.* Least powerful type, used for hearing losses in the 30- to 65-dB range.

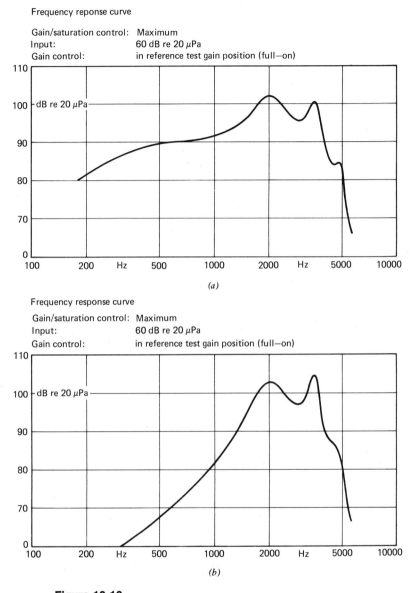

Figure 10-10.
(a) Oticon I-11V. (b) Oticon I-11H.

Source: Courtesy of the Oticon Corporation, 29 Schoolhouse Road, Somerset, N.J. 08873.

There are approximately 21 companies, both domestic and foreign, that sell hearing aids in the United States. About 33% are the behind-the-ear type, and 40% are of the in-the-ear type.

The fitting of a hearing aid is of critical importance in helping a person achieve the optimum value from the apparatus. As was mentioned earlier, the hearing aid should amplify those frequencies that the patient has difficulty hearing.

Figure 10-10a shows the frequency response curve for the Oticon Company hearing aid I-11V. It will be seen that this unit has a broad range of amplification over the range from 200 to 5,000 Hz. Figure 10-10b, on the other hand, amplifies the higher frequencies (I-11H).

Which model would be better to use with the patient shown in Figure 10-9b in whom there was a high frequency loss but lower frequencies were heard satisfactorily? It would be I-11H, because this would amplify the frequencies that were not detected well.

It is also important that the earmold be fitted correctly. If air escapes from the earmold, the lower frequencies will be attenuated, and these lower frequencies will not be amplified as much as they normally would.

Most hearing aids can be adjusted so as to attenuate either the lower or the higher frequencies. A typical setting would be N for normal, or full range of amplification. A setting of the switch to -12 position would reduce the amplification 12 dB at 500 Hz and 8 dB at 1,000 Hz, thereby reducing the lower frequencies somewhat.

SELECTIVE LISTENING

Experiments have shown that we are able to listen independently with either ear and to connect either to the mind, so to speak.

This ability to ignore sounds that we are not interested in is quite complex. Consider the following discussion:

> Most of us are familiar with the "party situation" a room full of people all talking, shouting, and even singing at the same time. Yet you can easily listen to a joke being told by someone on your right and the next moment "switch" to a couple of giggly girls behind you, without even moving your head. Your ears were getting the same sounds all the time, but you directed your attention now here, now there. If you had made a tape recording of the party with a single microphone at the spot where your head was, you would not be able to make any sense of the playback; you could not "direct" your attention anywhere but to the loudspeaker. The same sounds that were such fun to you at the party are now a garbled racket. Obviously listening, that is, concentrating on a particular sound, is not as "simple" as it seems.[1]

[1]W. A. Van Bergeijk, et al.: *Waves and the Ear,* New York, Doubleday, 1960.

Inability to select sounds seems to be involved in hearing losses in older people. The sounds are heard, but the masking noise is too distracting. Inability to understand words, even though the sound is heard, shows that to some extent hearing is done by the brain. Often older people are sold hearing aids that *will* amplify sounds so an individual tone can be heard in a quiet room, but the patient may *not* be able to distinguish among sounds to get the meaning. Nurses and other health personnel can do much to help their patients by explaining the need to match hearing aids to the specific hearing loss, and explaining that amplification alone will not necessarily enable them to understand speech if the deterioration is in the nerve or the brain. In fact, the person should be advised that he may only be wasting money by buying hearing aids if such conditions prevail. Countless numbers of people have done this, and after a few weeks, have put the expensive equipment in a drawer. On the other hand, where the difficulty is in transmission of the sound through the ear, hearing aids can be a great blessing.

"HI-FI"

"Hi-fi" means high fidelity, that is, a high degree of faithfulness in reproducing sound. As might be expected, the term hi-fi has been misused. Many of the phonographs and sound systems so labeled are not. One way to check up on a sound system is to find out its frequency response range. A phonograph that reproduces sound in the range from 60 to 12,000 cps is not truly hi-fi. It is the two extremes of the sound spectrum that cost so much to reproduce. To get tones of 20 or 30 cps requires a large speaker appropriately called a "woofer." For the high frequencies (12,000 to 20,000 cps) a small speaker is used, called a "tweeter." It is no wonder that hi-fi addicts keep the volume high, as was mentioned earlier. They want to hear the low bass rumbles and the high tweets on the piccolo.

High fidelity is often confused with stereophonic (usually referred to as "stereo") sound. Stereo sound is produced by using two microphones separated by some distance to pick up the sound. The sound is then recorded on two tracks on the record or tape. The two tracks are then reproduced and transmitted to two speakers. In this way the hearer receives the sound coming from the same directions as the original sounds. For instance, the violins in an orchestra may be picked up more distinctly on the right microphone, and the horns on the left microphone. If the sounds are then played back over two speakers placed some distance apart, the separation will be heard, and the sound will appear more natural. It is thus possible to have stereo sound reproduction with varying levels of high fidelity, depending on the type of speakers and amplifiers involved.

Recently four-channel reproduction has been developed. The listener is, in effect, surrounded by sound coming from four speakers (quadraphonic). There are some records and tapes being produced for quadraphonic systems, and usually these can be added to current high fidelity stereo systems.

ULTRASONICS

Ultrasonics is the term used to designate sound above 20,000 Hz. People who have dogs can use a high-frequency whistle that the dog can hear, but which they cannot.

Bats send out waves of very high pitch. An insect flying nearby will reflect the sound to the bat, and this will guide the bat to the insect. This is the principle of radar.

Ultrasonic waves may be produced by either *magnetostrictive* devices that depend upon a change in length of some metal such as nickel or invar when magnetized, or by crystals or ceramics exhibiting a piezoelectric reaction (that is, they move when a current is put into them). The vibrating head of the device is called a *transducer,* and will produce a beam of ultrasonic waves in air or a liquid.

Ultrasonic waves are used in cleaning metal parts. The vibrations shake the solution and produce cavitation or bubbles that help the cleaning action. This process is used in many industrial applications. There are also ultrasonic cleaners available for use in cleaning hospital equipment such as syringes and needles. At low intensity ultrasonic waves are used as direct therapy in the treatment of arthritis and bursitis, and with high intensity beams carefully controlled brain lesions have been produced to treat Parkinson's disease. Ultrasonic energy has also been used to bore into kidney stones to break them up as well as to dissolve obstinate scars. Ultrasonic dental probes are used to break up tartar accumulations on teeth.

A very important medical use of ultrasound waves is the ultrasound scanner or the ultrasonograph. This is discussed in Chapter 15.

ACOUSTICS

It was pointed out earlier that noise is a hazard because it is capable of causing deafness, and that it is very important in industry to be able to determine whether a worker suffered a hearing loss on the job or whether he had it previously. Consequently, audiometer testing is becoming an important part of employment procedures. Deafness may develop slowly and insidiously in an environment with a noise level above 90 or 100 decibels. At first the worker may experince ringing of the ears which will disappear some hours after leaving work. It is believed that this is a type of nerve fatigue, and that eventually acoustic trauma may result.

Allowance must be made for simple atrophy and degeneration of nerve fibers of the eighth cranial (acoustic) nerve that occurs naturally with old age.

Excessive noise also can bring about adverse psychological reactions. The tenseness, stresses, and strains of daily life may to a significant extent be due to all the noise around us. These particular effects are undergoing extensive study at the present time. How can we reduce noise? One efficient way is to use sound-absorbent materials in building construction. Acoustical ceiling tiles absorb much of the sound (up to 85 percent) instead of bouncing it back; this can make a striking difference in the comfort and the quiet of the environment. More emphasis is being placed on sound conditioning in hospital construction and, if the personnel are aware of the value of sound reduction, much improvement can be made in old buildings. Many hospitals are far from quiet and conducive to rest. Terrazzo floors, ceramic tile walls, and hard plaster ceilings may be easy to keep clean, but are acoustically poor.

QUESTIONS

1. Why is a man's voice lower in pitch than a woman's?
2. If you are standing by a railroad track and a fast train goes by with its whistle blowing, why does the pitch seem to change according to the distance of the train from you?
3. Many blind people continually tap their canes while negotiating a stretch of sidewalk. One reason for this, of course, is to probe for unseen obstacles. Can you think of another reason?
4. It is said that a thin wineglass can be shattered by a violin note of a certain pitch. Why?
5. If a patient's audiogram indicated a poor response in the frequency range of 500 to 2,000 cps, which hearing aid button do you think should be recommended, I-11V or I-11H (see Fig. 10-10)?
6. Why is hearing sometimes temporarily impaired by a cold or an attack of sinusitis?
7. Compared to a sound of 1 bel (10 dB), how many times more intense would a 5-bel (50-dB) sound be?
8. The treble A note has been set at 440 Hz. What effect would it have on a vocalist if the symphony orchestra tuned to 448 Hz?
9. An older person experiencing hearing loss amplifies the sound with a hearing aid but still has difficulty understanding speech. Would you expect that this is the result of a conductive hearing loss or sensorineural loss?

10. How do you explain that if two singers of the same sex sing the same note of middle G (256 Hz) it is possible to tell the difference between them?

11. Would a very hard, smooth surface be a good or bad acoustical remedy in a noisy factory? Why?

12. Why are "woofers" in speaker systems so large?

13. What is the difference between the terms hi-fidelity and stereophonic? Can a monophonic system be hi-fi?

14. A person hears the thunder 2 seconds after seeing the lightning. How far away is the storm? What assumption does this make about the speed of light?

CHAPTER 11

LIGHT, COLOR, AND VISION

One answer to the question, "What is light?" is: "Light is a form of electromagnetic radiation having a range of wavelengths from 4,000 to 8,000 Angstroms." But we do not explain things simply by putting a name on them. In fact, the final answer to the question is that we don't know what light really is. We can, however, observe its effects and apply them to a study of many aspects of vision.

THE ELECTROMAGNETIC SPECTRUM

The energy radiation from the sun covers a broad spectrum, and is called *electromagnetic radiation*. It extends from very long electric and radio waves, whose wavelengths can be measured in miles, to the very short cosmic, gamma, and x-rays, with wavelengths in the range of 1/1,000,000 to 1/1,000,000,000 of a millimeter (Fig. 11-1).

What is this radiation? It seems to be composed of electric and magnetic fields at right angle to each other moving through space. You may conclude that the above explanation is not very clear, and you will be right because we do not know what radiation really is. The radiation appears to be in the form of waves, and the wavelengths can be measured very accurately. On the other hand, many experiments seem to show that light is composed of small particles. These have been given the name "quanta." The use of quantum mechanical equations has somewhat reconciled the two aspects of radiation, but there is no complete model or picture of what radiation really is. Some have suggested the word "wavicle" to combine the words "wave" and "particle." It has also been facetiously suggested that we use the wave

Long wavelength   Short wavelength

Electrical waves	Radio Local broadcast Short wave	Infra- red	Red Blue Visible	Ultra- violet	X-rays	Gamma rays	Cosmic rays

Figure 11-1.
The electromagnetic spectrum.

theory on Monday, Wednesday, and Friday, and the particle theory on Tuesday, Thursday, and Saturday. Actually, we use whichever one is more suitable for a particular problem.

In fact, the choice of an experiment will actually determine whether the radiation will appear to be a wave or a particle.

If we think of the radiation as waves, we may wonder how waves can go through a vacuum. About 70 years ago it seemed unthinkable that we could have waves if there is no medium to wave or distort. How can you have wiggles in nothingness? To resolve the dilemma the "luminiferous aether" was invented. The coining of such a name is a good example of the scientific habit of presuming the existence of something undetectable by known methods in order to explain a phenomenon. Eventually, of course, the reality of this impalpable "something" has to be either proved or disproved.

All attempts to detect the presence of the aether have failed, and contradictory properties would have to be assigned to it; so the concept of the aether had to be abandoned. Where does that leave us? We have energy transmitted through a vacuum in the form of waves and/or particles.

The retina is sensitive to only a small fraction of the total electromagnetic spectrum —the range of electromagnetic radiations registered by the senses as light.

PLANCK'S EQUATION

At the turn of this century, German physicist Max Planck suggested a simple equation:

$$E = hV$$

where E is the energy of the electromagnetic radiation, h is a constant (6.6×10^{-27}), and V is the frequency.

This is a very useful equation that applies to the whole electromagnetic spectrum including, of course, visible light. In the case of electric or radio waves, the higher the frequency, the more energy. As will be discussed in a later chapter on x-rays, the higher the frequency, the shorter the wavelengths; hence, the greater the energy and the more penetration of the rays. Also, we shall see that the energy of light waves varies according to the color.

LIGHT

The speed of light has been found to be 186,000 miles per second, or 3×10^{10} cm/sec. This is so fast that we usually think of seeing an event instantaneously. Actually, it does take a definite time for the light to travel to our eye. In the case of the 93 million miles to the sun, it takes about eight minutes for the light to reach the earth. That means that we can see the sun eight minutes after it has actually dropped below the horizon. The 200-inch telescope at Mount Palomar has picked up light from galaxies 2,000,000,000 light-years away. (A light-year is the distance that light can travel in a year. How many seconds are there in a year?) This means that we see today something as it looked two billion years ago but which, for all we know, may no longer exist.

Sunlight has approximately 5 percent ultraviolet rays, 45 percent visible rays, and 50 percent infrared rays. The infrared ray produces heat when absorbed, the ultraviolet ray causes tanning of the skin, and the short ultraviolet rays are bactericidal.

SUNLIGHT AND SUNBURN

It has been estimated that 1.5 billion-billion kilowatt hours of energy (1.5×10^{12} kwh) reach the earth annually. Fortunately, the upper atmosphere has a belt of ozone that filters out most of the ultraviolet rays, so only about ¼ of 1 percent reaches the earth. The ultraviolet light that does reach us will produce a tan. The skin contains a pigment called melanin that is drawn to the skin's surface by sunlight.

Excessive ultraviolet light can also produce skin cancer over a period of years. It is estimated that 100,000 people annually develop skin cancer of one type or another, and about 5,000 deaths occur annually from this cause. There are three types of skin cancer. The most common, found in about 60 percent of the cases, is called basal cell carcinoma, which is isolated and rarely spreads. The second is squamous cell carcinoma, which accounts for approximately 38 percent, and spreads slowly. Finally there is a small number of cases of a more virulent type called melanoma.

Certain drugs such as tranquilizers, sulfonamide drugs, and antiseptics, as well as some perfumes, can increase the photosensitivity of the skin; this effect varies from person to person.

It is a strange phenomenon that in our culture so many people bake themselves in the sun in order to develop a tan that will make them look healthy and athletic. What is really happening is that the aging process of the skin is greatly accelerated. One beneficial effect of "suntanning" oneself is that the chemical ergosterol in the skin changes into vitamin D, which is helpful in preventing rickets.

The amount of melanin produced in the skin by ultraviolet light varies greatly from race to race. Among the white races, olive skin contains a great deal of melanin, while blondes and redheads have very little, and therefore burn very easily and have difficulty developing a tan.

In 1978, a Food and Drug Administration advisory panel recommended that all sunscreen products include sun-protection-factor information in the package. You probably have used some tanning lotion labeled with a number such as SPF 3 or SPF 6 on the bottle. The higher the number, the better the protection against the ultraviolet radiation. The sunscreen preparations work in two ways, either by setting up a physical barrier to the light, such as zinc oxide creams, or by chemically absorbing part of the ultraviolet portion of the light. Chemicals such as paraaminobenzoicacid (PABA), benzophenones, and cinnamates are used.

REFRACTION

Light traveling from one medium to another of a different density will be bent if it enters obliquely. Bending of a light ray is called *refraction*. Light travels faster in air than in glass or in water.

In Figure 11-2, the light beam will bend toward the perpendicular. It can be seen also that as the wave front hits the water, it is slowed down and thus the direction is changed. This is analogous to a group of soldiers marching around a corner. The ones on the inside mark time and the ones on the outside walk faster. The principle of refraction is the basis of the eye, of lenses, and of many optical instruments.

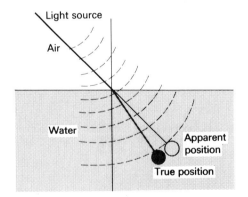

Figure 11-2.
Refraction of light.

If you were standing on the bank of a stream and were to try to spear fish, they would not be where you saw them. The brain is actively engaged in vision. Since we "know" that light travels in a straight line, we tend to continue the direction in air, and "see" the object where it is not. The shimmering of light above a radiator is partly caused by the refraction of light as it travels through air layers of different densities.

LENSES AND THE EYE

Lenses are pieces of glass shaped to utilize the principle of refraction. A *convex* lens has more glass in the middle than at the edges, and thus will pull the beams of light into the middle. In other words, they are converging. On the other hand, *concave* lenses (which are "caved in") cause the light beams to diverge (Fig. 11-3).

The eye is similar to a camera, but it is only relatively recently that cameras have been built to perform some of the functions of the eye. The lens of the eye is a *biconvex* lens (bulging outward on both sides) that focuses the light on the "film," the retina. The iris has an opening called the pupil, which opens or closes depending upon the intensity of the light. This is similar to setting the f-number on a camera. On a focusing camera the distance from the lens to the film is set by using a coupled range finder. The eye automatically adjusts for far and near vision thousands of times a day. How is this done? See Figure 11-4 and note the suspensory ligaments of the lens.

The curvature of the lens is changed by varying the tension on the suspensory ligaments. When the ligaments are tight, the lens is flat and thin, and has less converging power. When the ligaments are loose, the lens becomes more curved and has more converging power. One would think that the ciliary muscles would tighten in order to tighten the ligaments. Actually the reverse is true. As the ciliary muscles tighten, they become thicker and less elongated. This relaxes the ligaments, and the lens becomes rounder and more convergent.

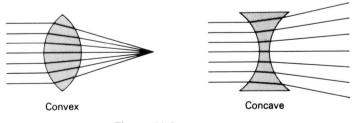

Convex   Concave

Figure 11-3.
Convex and concave lenses.

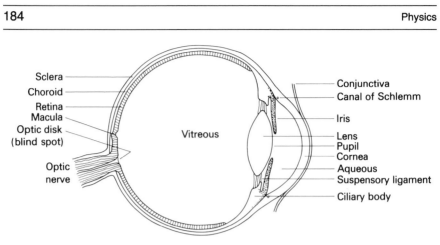

Figure 11-4.
Horizontal section of a right eyeball.

Source: Courtesy of the American Optometric Association, 243 North
Lindbergh Boulevard, St. Louis, Mo. 63141

The ability of the eye to change its focus is called *accommodation*. The eye at
rest normally is focused for objects beyond about 25 feet. Then as the ciliary
muscles contract and the ligaments relax, the lens becomes rounder especially in
the anterior portion. Figure 11-5 presents a graph showing the loss of accommodation with age—really a depressing picture. Note the continued decrease in accommodation with age, and the rapid drop from 40 to 50 years.

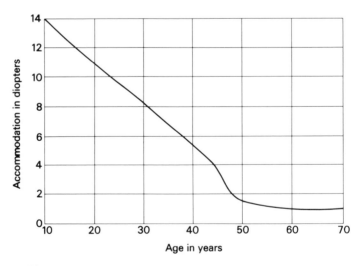

Figure 11-5.
Loss of accommodation with age.

Source: H. Stearns. *Fundamentals of Physics and Applications,* 2nd
ed. Macmillan, New York, 1956, p. 164.

What can be done when there is not sufficient accommodation to bring both near and far objects into focus? Some people use two pairs of glasses with different corrections. Others prefer bifocals, which are two lenses in one pair of glasses with the greater correction in the lower part used for reading. Others may require trifocals (three lenses) to secure sufficient correction for various distances.

Whenever the light is not focused on the retina, or on the film in a camera, every spot of light appears as a circle. This is called by the descriptive name "circle of confusion," familiar to everyone who has looked at a blurred, out-of-focus picture.

REFRACTIVE DEFECTS AND THEIR CORRECTION

Myopia or Nearsightedness.   The myopic person can see well at close range, but not at a distance. This condition is caused by a lens that is too "round," that is, too refractive. An elongated eyeball can also cause myopia. The light from a nearby object is diverging, and the extra round lens is able to focus it on the retina (Fig. 11-6a). Light rays from a distance, however, come in parallel, and are focused in front of the retina (Fig. 11-6b). What type of lens would be used in the correction of this defect? Since there is too much converging power, a diverging lens would be used, that is, a concave one (Fig. 11-6c).

Hyperopia or Farsightedness.   The farsighted person can see well at a distance but not nearby. Hyperopia can be caused by too "flat" a lens (i.e., insufficiently refractive) or else by too short an eyeball. The parallel light coming from a distant object can be focused (Fig. 11-7d), but the diverging light from a close object will be focused behind the retina (Fig. 11-7e). What type of lens would be needed? Because more converging power is required, a convex lens is used (Fig. 11-7f).

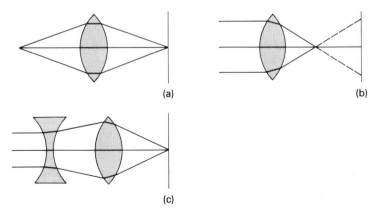

Figure 11-6.
Correction of myopia with diverging lens.

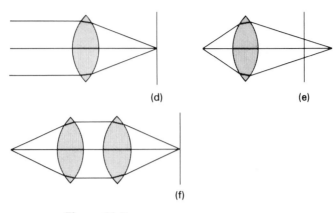

(d)   (e)

(f)

Figure 11-7.
Correction of hyperopia with converging lens.

Astigmatism.   Light vibrates in all planes. Let us consider the two main planes, the vertical and the horizontal. Try to picture half the light from an object coming toward you in vertical waves, and the other half the light in a horizontal plane. Astigmatism is a condition in which the cornea of the eye is not spherical; and because the cornea normally refracts the light rays to a certain extent before they reach the lens, the total effect is that of an asymmetrical lens. This may be pictured by thinking of an orange that has been flattened so that the curvature vertically is different from the horizontal curvature. Now, if the vertical curvature of light is focused on the retina, the horizontal half will be out of focus. The eye will adjust so as to bring the horizontal half into focus, but now the vertical component is out of focus. No matter what the eye does, half of what it sees will be out of focus. The eye keeps trying to focus, the muscles become fatigued, and a headache usually results.

The correction is to grind the lens with a different curvature in the two planes to compensate for the difference. If you have glasses which correct for astigmatism, it is important to keep them on straight; otherwise the wrong correction will be made. If you wear glasses and wish to check to see whether they are corrected for astigmatism, look through each glass separately, rotating the lens, while looking at a vertical object. If astigmatism is present, the object will seem to wave back and forth as the lens is rotated in one direction and then in the other direction.

The chart in Figure 11-8 is used to check for astigmatism. Look at it with one eye at a time. If any of the lines appear darker than the others, astigmatism is present.

This device is not infallible because it might not indicate small irregularities that nevertheless can produce severe symptoms.

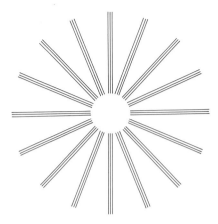

Figure 11-8.
Astigmatism chart.

CONTACT LENSES

Contact lenses are similar to eyeglasses and work on the same principle of refraction. They are simply small lenses that are ground with the same inside curvature as the eyeball, and are placed right on the eye.

At first, contact lenses can be very irritating and the wearer has to develop a tolerance to them. Initially they can be worn for only an hour or two at a time. Later, some people may be able to wear them all day, whereas some who are very sensitive may never become accustomed to them.

Contact lenses are made of soft plastic and have little danger of breakage. However, they are so small that the user has to be careful not to lose them and must also be sure which is the left lens and which is the right lens. The lenses are corrected for the individual eye, and are not interchangeable.

The statement above about becoming accustomed to wearing contact lenses applies mostly to so-called *hard* lenses. The second generation of lenses, the *soft* lenses, are much more comfortable and can be worn for longer periods of time. An estimated 10 million Americans wear contacts, and the soft lenses are growing about 25 percent annually. In 1981, the third generation of lenses was introduced. They are the so-called *extended wear* lenses. They are being developed by a number of optical companies, such as Continuous Curve Company, San Diego, and the Revlon Company, with Hydrocurve II. The principal feature of these lenses is that they can be worn all day and night for up to 2 weeks without being removed. The advantages according to Revlon are, "One does not have to remove, clean and disinfect contact lenses on a daily basis." They also mention the advantage of "being able to read the alarm clock immediately upon awakening and enjoying an overnight camping trip without bothering with lens removal, cleaning and storage."

These lenses are very thin: at the center, they are about half the thickness of a human hair. The main reason these lenses are so comfortable is that they are very hygroscopic. They absorb 50 to 90 percent of their weight in water. This usually makes them compatible with a person's tears, and permits the cornea to absorb oxygen from the air.

The goal is to make disposable lenses. Dr. O. A. Battista, an American chemist and inventor, has licensed a French company to test lenses he has made from edible proteins. It will be interesting to see future developments in this field.

RADIAL KERATOTOMY

We noted in the discussion on myopia that this is the result of too much convergency of the eye. This may be caused by an eyeball that is too elongated or a lens that is too convex. A surgical procedure, called radial keratotomy, is being tested in this country. The operation was originally developed in the Soviet Union, where an estimated 4,000 cases have been performed. In the United States, an estimated 1,000 cases have been tested. The operation consists of cutting 8 to 16 hairline slits, one-thousandth of a millimeter deep, in the curved surface of the cornea. These cuts will allow the cornea to flatten, thereby reducing the converging power of the eye and, it is hoped, making it possible to focus distant objects on the retina instead of before it. This operation is still experimental, and there is still much debate about its safety. A National Eye Institute study is under way to determine the safety as well as the permanence of this procedure.

THE RETINA

The retina, on which the image is focused, contains two forms of receptors known as rods and cones. There are estimated to be about 130 million rods and 7 million cones. The outer part of the retinal screen contains mostly rods; going toward the center, the cones are more numerous; and in the center of the eye (the fovea) there are only cones. The rods are much more sensitive to light than the cones, and thus are used in dim light and night vision. However, they do not give as sharp a focus as the cones. The cones respond in fine detail to color as well as to white light and so are used in daylight vision. The fovea is about 1/16th of an inch in diameter, and is used in bright light.

Associated with the rods is the pigment known as visual purple. In bright light the visual purple decomposes; in dim light it normally is rapidly restored, permitting the eye to adapt itself readily to a wide range of light intensity. One constituent of this pigment is vitamin A. If the diet is deficient in vitamin A, the eye has increasing difficulty in adapting itself to dim light or darkness; night blindness may result.

The pigment involved in vision is the organic molecule rhodopsin. Recent experiments show that humans have four different kinds of rhodopsin in their retina. Three of them are in the cones and respond to either blue, green, or red. The fourth form is in the rods and are sensitive of low-level light intensities. All rhodopsins are composed of a compound called 11-*cis*-retinal and a protein portion called opsin. Changes in these molecules have now been demonstrated to occur in 3 to 10 picoseconds, that is, 10^{-12} seconds, or 1 million-millionth of a second. The integration of chemistry, biology, and physics in the study of the human body is becoming increasingly important, and we can certainly appreciate that "we are fearfully and wonderfully made."

The eye has more than 10 million electrical connections and can handle 1.5 million messages simultaneously. It is much more complicated than any television camera. The eye is cleaned by the steady production of tears, and the eyelids act like "windshield wipers," blinking normally three to six times a minute, or more often when tired. The blinking also gives the eye a brief chance to rest. The signal from the retina is transmitted to the brain by the optic nerve at the speed of about 300 miles per hour. The brain then takes this electrical stimulus and interprets it. The whole electrochemical sequence takes only about 0.002 second or 2 milliseconds.

The spot on the retina where the optic nerve enters has no light receptors and is called the "blind spot." To detect its presence, look at Figure 11-9.

Hold this page about 18 inches away from your eyes. Close your left eye and look at the cross with your right eye and you will see the circle. Now move the page slowly toward you. At a distance of about 6 to 10 inches from your eye, the circle will suddenly disappear and then reappear.

When looking at a scene, why do we not see a blank spot? The brain seems to fill in with whatever is in the environment. This and other subjective aspects of vision will be discussed later in the chapter.

Occasionally small tears occur in the retina, and these may develop into a retinal detachment. For small breaks the sclera is treated with a laser beam which irritates the choroid so that after the subretinal fluid is removed, adhesions form between the retina and the choroid in the region surrounding the break, sealing off the break. The word laser is an acronym of *l*ight *a*mplification by *s*timulated *e*mission of *r*adiation. Or, to put it differently, the laser transforms the light into a series of coordinated pulsations producing a very intense beam that is focused onto a small spot. This creates scar tissue that forms the bond.

Figure 11-9.
The blind spot.

For greater tears an operation called "scleral buckling" is possible. A piece of polyethylene tubing is inserted under sutures on the sclera. This seals off the breaks, and also will decrease the volume at that point so that the vitreous fluid will help hold the retina against the choroid.

OPHTHALMOSCOPE

The ophthalmoscope is an instrument used to visualize the interior of the eye. It was invented more than a century ago by Helmholtz. Light from a lamp is reflected by a mirror into the patient's eye. It is then reflected from the eye to the observer through a small hole. This instrument indicates to the examiner not only the general condition of the eye but also something of its refractive power (Fig. 11-10).

GLAUCOMA

Glaucoma, an eye disease, gives us a good illustration of fluid pressures. In this disease, the pressure within the eye (intraocular) increases. Over a period of years this increased pressure will cause damage to the optic nerve, resulting in reduction of side vision (peripheral vision) and loss of visual acuity.

There are two types of glaucoma: The *narrow-angle,* acute type causes blurred

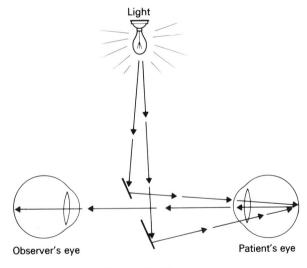

Figure 11-10.
The ophthalmoscope.

vision and halos around light; prompt medical treatment is necessary. The second type, chronic, simple, *open-angle* glaucoma, is usually asymptomatic for many years until the elevated pressure has already damaged the optic nerve. In the United States, about 2 percent of the population over 35 years of age have increased intraocular pressure, and this percentage will increase with age. An estimated one to two million Americans suffer from glaucoma, which is often called "the sneak thief of sight," because of the lack of symptoms during the initial stages.

Normal intraocular pressure is about 15 mm Hg. Fluid is continually produced in the eye and is continually draining out. If the drainage opening is clogged for some reason, the liquid will accumulate, and pressure will build up to 25 to 40 mm Hg or higher. In some cases operations are performed to open up the drainage path, but during the last few decades the use of drugs such as pilocarpine, epinephrine, and a recent drug called Timolol, have been used to reduce intraocular pressure. More recently, argon lasers have been used to "focus on a spot smaller than a pinhead where the fluid is supposed to flow out."[1] Another laser, called the *Q*-switched laser, emits pulses of nanosecond (10^{-9} second) duration that may avoid inflammation and scarring while opening up the duct.

To measure the intraocular pressure, an instrument called a tonometer is used. The patient tilts his head back and looks at the ceiling. A drop of local anesthetic is applied to the cornea. The footplate of the tonometer is placed on the cornea. A slight pressure is applied to the central plunger, causing the central cornea to be displaced inward. The pressure within the eye exerts a force that moves the indicator, which records the amount of pressure.

A newer type of tonometer is an electronic one, in which a graphic presentation of the intraocular pressure is made for a test period of four minutes. There is also available a noncontact tonometer, in which a brief, controlled air pulse "senses" the intraocular pressure.

One simple and inexpensive tonometer utilizes a 5.5-gram weight to measure ocular depression. The *higher* the intraocular pressure, the *less* the deformation that will result, and the lower the units will read. The reading of the instrument is converted to millimeters of mercury pressure by a conversion scale. To measure higher pressures, it is possible to use additional weights up to 15 grams.

Routine tonometry takes less than one minute. Yet it can help save people from blindness.

CATARACTS

A cataract is an opacity of the crystalline lens or its capsule. Cataract removal used to be a lengthy operation; but in recent years the operation has been so modified

---

[1]Carl Kupfer, Director, National Eye Institute, interview in *U.S. News and World Report,* Oct. 20, 1980, p. 63.

that it is now a fairly simple one. An incision only 2 mm long is made in the lens, and a small probe is inserted. This probe has a titanium tip that moves 0.0015 inch forward; it vibrates 40,000 times per second against the lens and thus emulsifies it. At the same time sterile isotonic fluid is introduced through one opening in the tip of the probe, and the liquid and fragments are aspirated through the center of this tip. Thus, with a single instrument that simultaneously emulsifies the cataract, irrigates, and aspirates, the actual procedure requires only about 3 minutes.

STEREOSCOPIC VISION

Why do you have two eyes? You might answer rightly that having two eyes affords a wider angle of vision and also acts as a safety measure in that if one eye is damaged, a person would still have sight in the other.

There is another value in having two eyes (binocular vision). It affords *stereoscopic* vision, the perception of depth. The closer an object is to the eyes, the greater the angle between the images. By experience, we have learned to correlate this angle with the distance (Fig. 11-11). This is the same principle used in some range finders that are attached to cameras. It is another example of the involvement of the brain in seeing.

True three dimensional (3-D) pictures must be taken by a camera with two lenses separated by a distance in the same manner as are a person's eyes. This arrangement will produce two pictures that are slightly different. When a person looks at each of these pictures through a viewer, each eye will see the respective picture as it would in viewing the actual scene. The two pictures then are merged or integrated

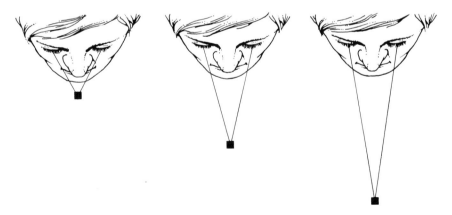

Figure 11-11.
Variation of angle of vision with distance.

by the brain into one picture having the added dimension of depth. Some motion pictures advertised as 3-D are not truly three-dimensional. They are shown on a wide screen, so that the viewer seems immersed in the picture, and thus the realism is heightened; but nevertheless the picture is "flat."

If the two images do not form the same size on the retina, there is difficulty in space perception, a condition called *aniseikonia*. If the two images are not on the same place on the retina because of crossed or divergent eyes, the person sees double, a condition called *diplopia*.

SUNGLASSES

The eye adjusts to varying intensities of light by opening or closing the iris. If the light is very strong, too much light may enter the eye and possibly damage it. Ordinary sunglasses are colored glasses that will absorb some of the light and thus decrease the intensity of the light reaching the eye. A specially treated transparent plastic is available in Polaroid sunglasses. These sunglasses absorb the glare by absorbing the light vibrating in a horizontal plane while transmitting the rest of the light that is vibrating in a vertical plane (Fig. 11-12). Try to picture light vibrating in either a vertical or horizontal plane.

In Fig. 11-12, part A represents a vertical wave, part C a horizontal wave. The Polaroid filter has crystals lined up in an orderly fashion, and thus can serve as a filter for light.

In A we see a vertical wave coming to a filter oriented vertically. Think of a vertical picket fence. The vertical wave would pass through the vertical filter, but if we rotated the filter to a horizontal axis B, the light would be stopped. Similarly, in C a horizontal wave would pass through a horizontal filter, but would be stopped by a vertical filter such as in D.

Since the glare component is usually horizontal, Polaroid sunglasses use vertically oriented lenses and thus stop the horizontal component while transmitting the vertical light.

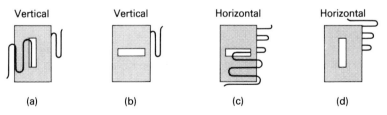

Figure 11-12.
Polarization of light.

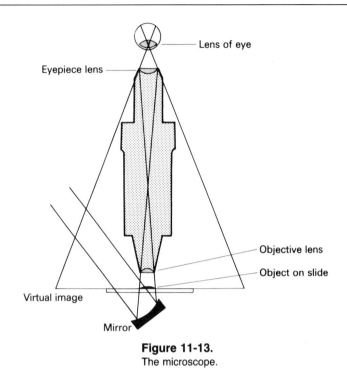

Figure 11-13.
The microscope.

THE MICROSCOPE

The microscope employs the principle of refraction to magnify an object (Fig. 11-13). The light from a lamp is reflected by the mirror through the object on the slide. The light passes through the objective lens, and the image is magnified. This image is further magnified by passing through the eyepiece lens; and as can be seen from the diagram, the eye sees a large "virtual" image.

What is a virtual image? It is something we see where it is not. This may sound peculiar, but can be explained by the following example.

If we look at a light bulb in a mirror, as in Fig. 11-14, we would "see" the bulb at point B, not at point A where it really is.

In other words, the image of the lamp at point B is a virtual image.

By varying the power of the microscope lenses, useful magnification of more than 1,000× can be achieved.

When the oil immersion lens is used in microscopy, the space between the lens and the slide is filled in with a drop of oil having an index of refraction similar to that of the lens, so that there is less bending of the light and less distortion than there would be with an air gap. The oil used is cedar oil or xylene.

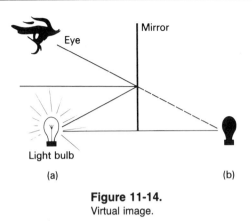

Figure 11-14.
Virtual image.

The Electron Microscope. If the object is so small that it approximates the wavelength of light, it cannot be seen with visible light. Then the electron microscope is used. This device employs an electron beam instead of a light beam. The wavelengths of electrons are much smaller than the wavelengths of visible light. The electron microscope uses magnetic fields instead of lenses to focus the electron beam. The image formed by the electron beam is rendered visible by being projected onto a fluorescent screen.

COLORS

A piece of glass which is triangular in cross section is called a *prism*. When a white light is passed through a prism, the light is bent (refracted). Each wavelength of light is bent differently from the others so that the beam of light is dispersed into the various colors: red, orange, yellow, green, blue, indigo, and violet. Red light

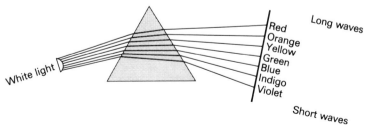

Figure 11-15.
Dispersion of light by a prism.

is refracted least, and violet light most (Fig. 11-15). There is a handy mnemonic (memory) device used to recall the colors of the spectrum: Roy G. Biv.

When another prism is used to refract the light the other way again, the colors blend to become white light once more. This can be done with pure light but not with pigments. The colors of the rainbow, dewdrops, diamonds, and cut glass are all due to the dispersion of the light. Each color is thus a particular wavelength of light. A synthetic gem made of titanium dioxide is also brilliantly colored because of its high refractive index.

A device used to analyze the chemical composition of substances by means of their light is called a *spectroscope*. The light emitted from a heated substance is passed through a prism and either viewed through an eyepiece against a calibrated scale or else photographed. Each element has its own characteristic wavelengths, that is, characteristic colors. Even the light from a distant star can be analyzed in this manner, and the elements determined. In fact, helium was found on the sun by this method before it was discovered on earth.

COLOR VISION

The mechanism by which the eye "sees" colors has been under discussion for many decades. The most commonly accepted theory was that of Helmholtz, who suggested a tricolor system of receptors, one red, one yellow-green, and one blue. Other suggestions such as a four-color system have also been made. The color was thought to depend only on the wavelength of light, and the stimulus was picked up by the appropriate color cone. Some years ago the whole theory was questioned by Dr. Edwin H. Land of the Polaroid Corporation, who demonstrated that if two black-and-white transparencies are made with different filters, and two shades of one color (such as yellow light) are used to project the pictures so that they are super-imposed on the screen, the viewer will see the scene in full color. All that is required is that the two shades of the colored lights used be somewhat different in wavelength. This is a tremendously interesting discovery. It emphasizes the subjective aspect of vision. The brain seems to analyze the light from an object in relationship to the whole picture. It appears that the eye can recreate a whole range of colors from just two wavelengths selected from almost anywhere in the visible spectrum. The implication of this discovery is that we see colors that are not there, so far as a certain frequency is concerned. Whatever the process involved, it is a further demonstration of the brain's involvement in the seeing process.

COLOR BLINDNESS

Color blindness is the total or partial inability to recognize colors or to distinguish a certain color from another. In the usual type of color blindness red and green are indistinguishable; both appear gray. In much rarer instances blue and yellow likewise register as shades of gray. Also, it is possible to be "blind" to but a single color.

Men are more susceptible than women to color blindness. It is estimated that 10 percent of the male population has some difficulty in distinguishing red from green.

Color blindness can be detected by means of charts consisting of colored dots arranged in recognizable patterns which cannot be detected unless the colors composing them are perceptible. Such tests are given in industry, driver's license tests, etc.

RETINAL FATIGUE

If one looks at a bright object for some time, the retina becomes fatigued and will not respond to the light for some time. If you look at a bright light bulb for about 30 seconds, and then look at a light wall, the shape of the bulb can be seen in black. If you stare at a red piece of paper for some time, and then look away to a white surface, the complementary color should be seen. Try the same thing with a yellow object.

The complementary colors are those which will add to produce white light. They are the opposites on a color wheel. Thus it is said that if the retina becomes fatigued to a certain color, when a person looks at a white surface (which reflects all the colors), the retina will respond to all the colors except the one to which it is fatigued, and the person will "see" the complementary color. To demonstrate this for yourself, color in the flag (Fig. 11-16) as indicated. Stare at the flag for a half-minute or so and then look at a white surface.

SUBJECTIVE ASPECTS OF VISION

There is often a very great difference between the physical process of seeing a thing—that is, focusing its image on the retina—and the psychological process of

Figure 11-16.
Retinal fatigue experiment.

interpreting the image perceived. You will recall that, theoretically at least, the blind spot of the eye should leave a hole in the center of every image; yet the brain seems unable to accept this lack of continuity and compensates accordingly. There are other indications that we see what we have a need to see. If you ask a dozen people to report on an incident they have all viewed simultaneously, the impressions should differ materially from one another. This is the reason why it is so difficult to secure a true account of the happenings from eyewitnesses to an accident. Similarly, a man and a woman might notice quite different aspects of a person coming into the room. Women can often, after a quick glance, describe in detail the color and texture of another woman's dress and accessories, while the man may not even remember the color.

Some interesting experiments have been performed which show the effect upon vision of such influences as experience and immediate needs. For example, a group of people were kept without food for some time until they were hungry. Then they were asked to look at some objects behind a gauze screen that blurred the scene. What did these hungry people see? That's right—they saw food. Similarly, when children from different economic groups were asked to estimate the size of coins, the children from poor homes "saw" the coins as much larger than did the rich children.

An interesting test used "aniseikonic" glasses which are designed to distort the image. However, some people while wearing these glasses saw no distortion of the images of those whom they loved or respected. Newlyweds saw no distortion of

(a)   (b)

Figure 11-17.
Experiments showing integration of stimuli.

each other's images, whereas older couples did. Navy enlisted men let the images of junior officers go out of shape much faster than those of admirals.

Two simple experiments which you can perform yourself are illustrated in Figure 11-17. The purpose is to show the integration by the brain of two simultaneous stimuli.

Look through the tube at a distant object with one eye and look at your hand with the other (Fig. 11-17a). Your brain will combine the images and you will appear to see through your hand.

Put two pennies each ½ inch from the line (Fig. 11-17b). Hold the card upright on the line. Put your nose against the top of the card. Look with both eyes and your brain will combine the images.

COLOR AND EMOTION

People tend to think of colors as being "warm" or "cool" or somewhere in between. That is, some colors seem to embody more energy than others. If you were to ask 100 people which color is warmer, red or blue, 99 probably would say that red is the warmer color and that blue is a cool color. This seems to be generally accepted, at least in our culture, and may be due to the association of red with fire and blue with cool water. Whatever the reason, the psychological verdict is that red is warmer, more exciting, and more stimulating.

If we look at Planck's equation, $E = hV$, and remember that red is a relatively long wavelength with low frequency, it should mean that a wave of lower frequency should have less energy. Thus red light is actually (physically) less warm (has less energy) than blue light. Blue light has a shorter wavelength, that is, higher frequency. Since energy and frequency vary directly, blue has higher energy than red. This is another case where the psychological response and the physical facts are completely opposite.

Colors are increasingly being employed to establish moods. Some mental hospitals are experimenting with color therapy, placing manic patients in "cool" rooms, as one example.

Thus, color is not only a matter of wavelengths, but involves subjective interpretation of the stimuli by the brain.

Actually, the importance of color to human beings can scarcely be overstated, even though there may often be a lack of a conscious appreciation of it. Most people have an emotional response to their surroundings, and color is one of the main attributes of these surroundings. Warm-colored surfaces appear to approach the observer, so a room that is finished in warm tones seems smaller than it really is. This conveys an impression of coziness and intimacy.[2]

---

[2]General Electric, *Light and Interior Finishes,* TP-129.

LIGHTING

The unit of light intensity is the footcandle, the amount of illumination on a surface one foot distant from a source of one candle. For *general* lighting a level of five to ten footcandles is recommended. For *local or functional* lighting, the following levels are advisable:[3]

Recommended Footcandles (Minimum)	*Visual Tasks*
10–20	Card playing
20–30	Casual reading—good type on white paper
	Easy sewing, such as basting with contrasting thread
	Facial makeup
	Simple musical scores
30–50	Household activities in kitchen and laundry
40–70	Prolonged reading, or study, or both
	Sewing on medium-colored fabric
	Machine stitching
	More difficult musical scores
	Shaving
	Benchwork
100–200	Fine sewing
	Hobbies with small details

A third use of light is for *accent or decoration,* such as using spotlights on a picture.

The amount of reflection of light from a surface is also important in planning for visual comfort. For example, a flat white paint will have about 82 percent reflection, whereas a dark-colored paint, such as a blue or green, will have from 8 to 12 percent reflection. Wood finishes will range from 10 to 40 percent depending on how dark they are.

[3]General Electric, *The Light Book,* 146-1220, p. 4.

QUESTIONS

1. What is a virtual image?
2. What color is a pure blue object in pure red light? In white light?
3. Why are older persons more likely to use bifocal glasses?
4. What is meant by the statement that we do not see things instantaneously?
5. A shore fisherman must allow for refraction when spearing fish. Must a skin diver do the same? Explain your answer.
6. Explain how colors can be used to set a mood.
7. Discuss three uses of lighting.
8. What is the relationship between Pascal's law and the tonometer?
9. If the plunger in a tonometer moves farther than normal, would this indicate intraocular pressure above or below normal?
10. In the experiment in Figure 11-17, in which the hole appears to be in the hand, would this occur if one were to look alternately through each eye?
11. If you stare at a red object for awhile and then look at a white surface, what color would you see? Why?
12. The index of refraction r is approximately equal to the speed of light in air divided by the speed in water. If $r = 1.3$ for water, and 1.8 in some crystal glass, what would this indicate about the speed of light in glass compared with that in water?
13. Explain the reason for a physiological "blind spot." How is this term descriptive psychologically?
14. Explain the reason radial keratotomy could help a patient who is nearsighted. Would this also be of help for farsighted people? Explain.
15. What is the cause of astigmatism? Would the wearing of eyeglasses crooked have more or less effect on an astigmatic person or one who is simply farsighted?
16. The speed of light equals frequency times wavelength ($C = \gamma \times \lambda$). The speed of light is a constant, 3×10^{10} cm/sec. Explain from the equation how frequency varies with wavelength.

CHAPTER **12**

MOLECULAR PHENOMENA

*Many clinical procedures related to promoting and
maintaining health involve molecular reactions.
Knowledge of molecular phenomena has additional
practical applications, such as in the use of
cleansers and disinfectants.*

SURFACE TENSION

The molecules of a liquid or a solid are held together by electrostatic and other
types of attractions, just as are the atoms or ions of the molecules themselves. In
a vessel of water, for instance, all the molecules except those at the surface are in
a state of equilibrium because their mutual attraction is the same throughout. But
the molecules at the *surface* of the liquid are subjected to unequal forces. There
is a very slight attraction between these surface molecules and those of the air above
them, but a strong attraction to the water molecules below. The net result is that
all the surface molecules are pulled inward, and so any body of liquid tends to form
a sphere. This tendency is demonstrable by scattering drops of water—or better
yet, mercury—over the surface of a table.

As the surface molecules are attracted toward the center, they become densely
packed together as a film. The toughness of this film of molecules is known as
surface tension (Fig. 12-1).

The effect of surface tension can be shown by carefully placing a needle on a
piece of tissue paper and floating the paper on water. As the paper becomes wet,
it will sink, leaving the needle floating on the water even though the iron is much
more dense than the water. If you fill a container to the limit with a liquid, you will
notice that the surface tension causes the top of the liquid to bulge. This effect
sometimes can be a nuisance, as when one is attempting to pour out precisely one
drop from a graduated cylinder. The liquid tends to hang out over the edge of the

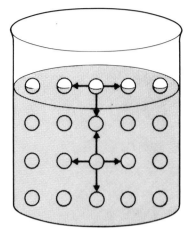

Figure 12-1.
Surface tension.

lip until the surface tension breaks, whereupon many drops are released simultaneously.

Certain fabrics are waterproof because of the action of surface tension. The spaces between the threads of the fabric used in tents and umbrellas are much larger than the water molecules that the fabric is supposed to keep out. To understand why the fabric remains waterproof, first consider the structure of a water molecule (Fig. 12-2).

Hydrogen Bonding. Note that the water molecule is not linear, but forms an angle of 105°. You might say, "What is so important about that?" Because one end of the water molecule is more negatively charged and the other is more positively charged, the molecule acts like a dipole, and "hooks up" into large units.

The attraction of the positive end of one water molecule to the negative end of another molecule is called *association*. More recently, it has been called *hydrogen bonding* (Fig. 12-3). The more positive hydrogen atom bonds to the more negative oxygen atoms; in this way water molecules join together into units or chains.

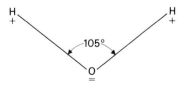

Figure 12-2.
Nonlinear shape of a water molecule.

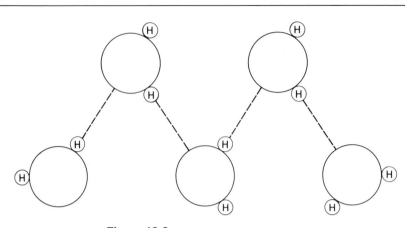

Figure 12-3.
Association or hydrogen bonding of water.

Now we can see why water will not run through the relatively large openings in a tent or an umbrella. The large aggregations of water hanging together prevent it from coming through (Fig. 12-4). However, if you have ever camped in the rain, and touched the inside of the tent wall, you have made the discovery that once the surface tension is broken, the fabric is no longer waterproof.

Measuring Surface Tension. The surface tension of liquids varies according to the nature of the liquid; for example, mercury has a higher surface tension than

Threads in a tent or umbrella

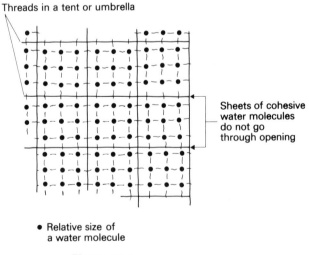

Sheets of cohesive water molecules do not go through opening

• Relative size of a water molecule

Figure 12-4.
Water molecules retained by cloth.

water. The surface tension of any liquid can be measured by an apparatus called a *tensiometer*. This instrument has a fine metal ring suspended in the liquid by a wire. As the wire is pulled up with a calibrated force, the amount of resistance of the surface tension is measured. The following table depicts a comparison of the surface energy (i.e., tension) of various liquids exposed to air at 20°C. The surface energy is measured in ergs per square centimeter.

Ethyl alcohol	22.3
Phenol	40.9
Water	72.8
Mercury	465.0

The efficacy of any cleansing agent depends on its surface tension. The lower the surface tension the better, because the cleansing agent is thereby more able to penetrate the small crevices in and about dirt particles so that they can be loosened and washed away. How do we lower the surface tension of a liquid to make it a better cleaner?

One way to do this is to raise the temperature of the liquid. The surface tension of water at 20°C is 72.8 ergs per square centimeter, but at 100°C it is only 58.9. This means that hot water is more effective as a cleansing agent than is cold water.

Another way to increase the cleansing power of water is to add a detergent which can be either a soap or a soapless chemical detergent. Detergents not only act on the lipids, fats that hold the dirt, but also make the water "wetter" by lowering its surface tension. By reducing the attractive forces between the water molecules, wetting agents cause the water to flow evenly over the surface instead of forming droplets.

In an interesting study by the Consumers Union, the results obtained from washing clothes in a washing machine with hot, warm, and cold water were compared.[1] Using various detergents, researchers found that when the results in hard and soft water were averaged, there was no visible difference between the hot and cold water washes. However, a slight difference in favor of the hot water was detected when a reflection meter was used. For all practical purposes, there is little or no difference between using hot or cold water for washing clothes. It may be that the mechanical agitation of the clothes by the washing machine helps to dislodge the dirt particles even in cold water; also, the lowering of the surface tension by the detergent may be more important than the lowering of the surface tension by heating the water. As the study suggests, using cold water may be one way to save energy both for the household and for the nation, since the cost of heating the water is a large part of the energy cost of washing.

Another factor is the use of "builders" in detergents. These are chemicals used to soften the hardness, that is, the minerals, in water. There has been much op-

---

[1]*Consumer Reports.* Oct. 1974.

position to the use of phosphates for this purpose; the states of New York and Indiana and a few other localities have banned phosphates in detergents, and some are limiting the amount permitted. Now, however, it is realized that phosphates are not the only or necessarily the main source of eutrophication (the rapid growth of algae in lakes), but that other chemicals such as carbonates are also involved. The replacement of phosphates with other builders such as carbonates, silicates, and citrates may be causing different problems. As in most situations, the cost/benefit ratio must be evaluated; it is important to secure all facts possible before banning certain compounds.

Disinfectants are usually solutions of low surface tension, allowing the liquid to spread out more effectively over the cell wall of the bacterium. You may have used an antiseptic called ST 37. This is a chemical, hexylresorcinol, having a surface tension of 37 ergs per centimeter square.

An interesting clinical application of surface tension is the Hay diagnostic test for jaundice. Normal urine has a surface tension of 66; but if there is bile in the urine, the surface tension is reduced to about 55. Powdered sulfur sprinkled on the normal urine will float, but it will sink in the solution with less surface tension.

ADHESION AND COHESION

The attraction that the molecules of water (or any other substance) have for each other is called *cohesion*. Molecules of unlike substances (such as water and glass) also exert a mutual attraction known as *adhesion*. You will remember that if you fill a graduated cylinder with water to the very brim, the surface of the water will bulge out. The curved upper surface of a liquid column is called the *meniscus*. The upward bulge of the water surface in this case is a *convex* meniscus because its direction of curvature is outward. It is caused by the cohesion of the water molecules.

If you pour out some of the water from the cylinder so that the water level is below the top of the vessel, you will notice that the meniscus is now *concave;* that is, it sags inwardly. Why? The reason is that there is adhesion between the molecules of the water and those of the glass. This adhesive force is stronger than the cohesive force of the water molecules, and therefore the water molecules try to spread out on the glass surface. Since the dry area is above the water level, the water tends to climb up the sides of the cylinder.

If the cylinder were partially filled with mercury instead of water, the meniscus would be convex instead of concave because the cohesion of the mercury molecules is so much greater than their adhesion to the glass. In fact, this is one reason why mercury is used in thermometers—it will not stick to the glass. Mercury is so cohesive that it does not even feel wet to the touch.

All measurements of both aqueous and alcoholic solutions or suspensions should be made from the lowest point of the meniscus.

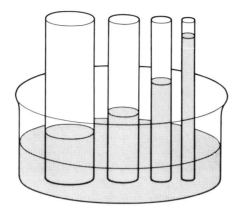

Figure 12-5.
Capillary action.

CAPILLARY ACTION

We noted a moment ago that the molecules of water in a glass cylinder tend to climb up the cylinder walls because of adhesion. If a good-sized cylinder (say, the diameter of a graduated cylinder), open at both ends, is immersed in a pan full of water, the water will not rise to any extent in the cylinder because the weight of the water column is greater than the upward force of adhesion. However, if we substitute a very narrow tube for the cylinder, the weight of the water column will be less than the adhesive force, and will rise in the tube against gravity to a level well above that of the water in the pan. It will stop rising only when the weight of the column and the force of the adhesion balance each other. The narrower the lumen of the tube, the higher the water will rise (Fig. 12-5).

The rise of liquid in a tube due to molecular adhesion is called *capillary action*, and the tube is termed *capillary tube*.

In determining the height to which a liquid will rise in a capillary tube, other factors besides the diameter of the tube must be taken into account. The complete formula for computing the rise is as follows:

$$h = \frac{2T}{rdg}$$

where h is the height in centimeters, r is the radius of the tube in centimeters, d is the density in grams per cc., g is the force of gravity of 980 cm. per second^2, T is the surface energy in ergs per cm^2.

You will notice from the above formula that the more surface tension (or surface

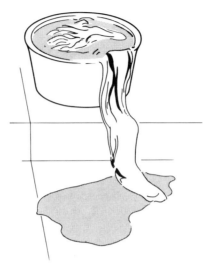

Figure 12-6.
Capillary action in stocking.

energy), the higher the liquid will rise. Also, the smaller the radius of the tube, the greater the height will be.

Use is made of capillary action in pulling blood up in a tube to determine the clotting time. A little piece of the tube is broken off at measured intervals of time until a clot is formed.

Blotters work by capillary action; so do filter paper, surgical packs, and absorbent cotton. The wick of a candle or a kerosene lamp works on the same principle. You may have seen nylon stockings handing over the edge of a sink, dipping into water and making a puddle on the floor (Fig. 12-6).

This same setup has been suggested as a way of feeding plants while one is on vacation.

DIFFUSION

If a crystal of copper sulfate is dropped into a vessel of water, it will go into solution; and after some time has passed, the entire contents of the vessel will have turned a uniform shade of blue. The copper sulfate particles (solute) have dispersed themselves equally throughout the water (solvent). This tendency of the solute particles to move freely throughout the solvent is called *diffusion*. Remember that in diffusion, the particles (molecules or ions) always move from a higher concentration to a lower one.

Diffusion occurs in gases as well as in liquids. If a small bottle of perfume is left opened, the perfume will evaporate and its molecules will tend to disperse themselves equally throughout the room. In the body, oxygen and food substances diffuse into the blood and then into the tissues. Waste materials diffuse in the opposite direction.

OSMOSIS

Cell membranes all have the characteristic of selective permeability; that is, they permit certain favored substances to pass through while barring others. Exchanges of materials are constantly taking place through the membranes of the body cells; and to understand the nature of these, we must first know something of that process called *osmosis*.

Imagine an oblong trough divided into two compartments by a partition made of a semipermeable membrane. A weak sugar solution is poured into one compartment and a strong sugar solution into the other. Now, it happens that a semipermeable membrane will allow the solvent (water) to pass through it, but will block the passage of the solute (sugar). What happens, then? The membrane will try to equalize the concentrations of the solutions on either side of it, and so water molecules will go from the *less* concentrated solution to the *more* concentrated one (Fig. 12-7). Here we have the process of osmosis, defined as *the passage of a solvent through a semipermeable membrane*.

As the water molecules pass through the membrane to equalize the concentration, the volume of the strong solution will increase; that of the weaker will decrease.

The difference in height of the two levels (ΔH) produces a hydrostatic pressure called the *osmotic pressure*. The greater the difference in concentration, the higher the osmotic pressure (provided the membrane does not break) (Fig. 12-8).

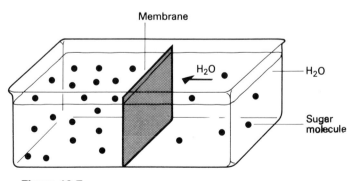

Figure 12-7.
Osmosis of water from a less concentrated solution to a more concentrated solution.

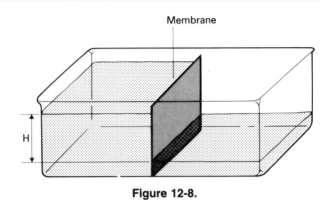

Figure 12-8.
Osmotic pressure.

It is tremendously important that normal osmotic relationships be maintained between body tissues, blood corpuscles, and any other solutions that enter the body by routes other than the mouth (in this case the intestines automatically make these adjustments). What are some typical osmotic relationships?

Consider, for example, the red blood cells. The plasma in which these corpuscles are suspended is slightly salty, with a concentration of about 0.9 percent NaCl. Within the red cells the concentration is the same—that is, "isotonic." Because of the equal concentration on either side of the red cell membrane, there is no osmotic pressure. If we inject a salt solution into the bloodstream, this too must be isotonic or "normal saline," 0.9 percent (incidentally, "normal" as it is used here is not the same term as the "normal" of chemistry, where it refers to an equivalent weight of solute per liter of solution).

Now, what would happen if you were to inject 9.0 percent saline instead of 0.9 percent? By losing the decimal point you stand a very good chance of losing the patient.

Saline of this strength (called *hyper*tonic, since it is above normal percentage) is of a stronger concentration than that within the red cell, and so water would pass from the weaker to the stronger—that is, *out* of the cell. The result would be a shriveling of the cell accompanied by a notched appearance called *crenation* (see Fig. 12-9).

If a *hypo*tonic solution (less than 0.9 percent) is injected, the water would again go from the weaker to the stronger, according to the rule (Fig. 12-10).

The water, then would invade the red cells, causing them to swell and perhaps burst. This is hemo-lysis or *hemolysis*. You can demonstrate this condition to yourself by immersing a dried prune in water and seeing what will happen. A few years ago, a cosmetic company spent a great deal of money trying to develop a cream that would do the same thing, drawing water to the skin so it would be full and firm and wrinkle free. They actually had a cream that would do this, and hopes were high. Unfortunately, the irritation that drew the water to the face lasted only

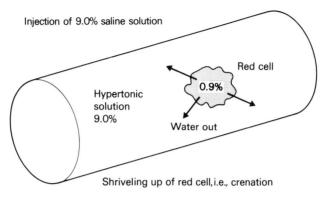

Figure 12-9.
Injection of hypertonic solution, causing crenation.

a short time. Those who tried it looked beautiful and youthful for a short time, but then the wrinkles reappeared.

Wherever salt solutions and semipermeable membranes are brought together, you can usually predict the outcome if you keep this rule in mind: *The water goes where the salt is.*

For instance, suppose you have a sprained wrist that is beginning to swell painfully, indicating that there is an excess accumulation of water in its tissues. If you soak it in a strong solution of Epsom salts, the water goes where the salt is. Through osmosis, the water is drawn out of the irritated area and the swelling is reduced. Saline cathartics work on the same principle. The higher salt concentration will draw water out of the tissues and into the intestines, increasing the fluidity of their contents and aiding elimination. Actually, in the body there is also some diffusion

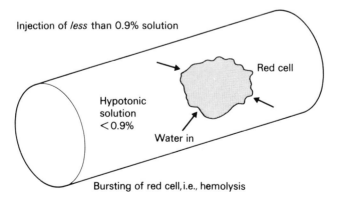

Figure 12-10.
Injection of hypotonic solution, causing hemolysis.

of the salt into the tissues; so prolonged use of salt laxatives may increase the salt concentration in the tissues with its attendant water retention.

One point should be emphasized in connection with osmosis. Osmosis is the passage of the solvent molecules from a lesser to a higher concentration. Not all physiological processes employ osmosis. In many cases a process of diffusion is involved, that is, a passage of material from a higher to a lower concentration. In breathing, for instance, the oxygen diffuses from the lungs into the bloodstream, and from the bloodstream into the cells.

THE ARTIFICIAL KIDNEY (HEMODIALYSIS)

The function of the kidneys is to filter out impurities from the blood. As a result of the catabolism (or breakdown) of proteins, the body must rid itself of urea, uric acid, creatinine, and creatin. In addition, the kidneys excrete about 50 percent of all excess water, and also salts such as sodium chloride and ammonium chloride.

The blood flows very slowly through the glomerular capillaries at a rate of about 18 inches per hour. The number of glomeruli in each kidney has been estimated to be from one to five million, and the surface area for filtration at about 35 square feet per kidney.

In cases of kidney failure the various waste products resulting from catabolism fail to be excreted. Instead, they accumulate in the blood and cause a very serious condition—actually a form of poisoning—called *uremia*.

The artificial kidney is a machine that is designed to perform the work of the kidneys for a period of time. Like the kidney, it works on the principle of *dialysis*, that is, the passage of smaller molecules through a semipermeable membrane while larger molecules are retained. It is, in a sense, filtration.

Various kinds of artificial kidneys are available. The type illustrated in Figure 12-11 employs long cellophane tubing wrapped spirally on a rotating drum that is

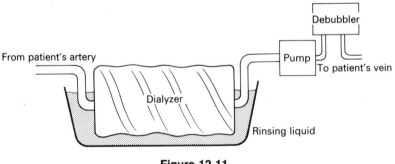

Figure 12-11.
Artificial kidney.

immersed in a tank containing the exchange liquid. The blood flows slowly through the tube, allowing waste products to diffuse through the membrane into the rinse liquid, going from a higher to a lower concentration. The larger protein molecules cannot pass through the membrane. The rinse liquid is kept at about 0.6 percent NaCl, 0.2 percent $NaHCO_3$, 0.04 percent KCl, and 1.5 percent glucose. Why not simply use plain water? This would draw too much water into the blood, causing hemolysis of the red blood cells—the erythrocytes. In the case of edema, more glucose (up to 3 percent) may be used in the bath to draw water out of the system. Hemodialysis is thus not only a means of drawing excess salts out of the patient, but can also be used to add some electrolytes that are in deficient supply. Another way to replace fluids is by intravenous administration.

Among the newer types of hemodialysis units is one in which plastic hollow fibers perform the actual dialysis. Several thousand fibers are bundled together in a container about six inches tall. The blood is passed through these fibers. These newer, lighter units have made it possible for patients to be treated at home under certain conditions.

In *acute* renal failure the artificial kidney is used to remove the waste products so that the natural kidneys may have time to resume normal function again. In cases of *chronic* kidney failure hemodialysis enables some patients to live fairly normal lives. The patient is dialized once or twice a week for a period of 5 to 15 hours each time. A semipermanent cannula is kept in place in the patient's arm. One type is a Teflon U-tube implanted in the radial artery and vein of the forearm, or sometimes in the leg. The dialysis machine can then be easily attached to the two joints. Many patients must report to a dialyzing center once or twice a week; many others can be treated at home with the newer types that utilize disposable plastic dialyzing membranes.

PERITONEAL DIALYSIS

Another form of dialysis is *peritoneal dialysis*. It was found as early as 1923 that the peritoneal membrane could function as an excellent dialyzing membrane. It is estimated that the peritoneum has a filtering surface of 22,000 square centimeters, compared to the kidney, which has 18,000 square centimeters. By injecting a solution into the peritoneal cavity, it is possible to remove catabolites and other unwanted substances and to change the volume or electrolyte content of the extra-cellular fluid. In order to prevent absorption of water, the dialyzing solution must be hypertonic to the plasma of the patient. This is usually brought about by the addition of dextrose to the solution.

Peritoneal dialysis is not suitable when severe uremia or drug intoxication is present and rapid action is needed. Then the artificial kidney is preferable.

Generally about two liters is injected over a period of 10 minutes. This is allowed to remain in the peritoneal cavity for about 10 to 20 minutes to equilibrate with the extracellular fluid. The liquid is then removed by siphoning. Fresh dialyzing fluid is now added, and the cycle repeated for 12 to 36 hours (Fig. 12-12). A number of prepared solutions are available containing sodium, chloride, calcium, magnesium, lactate, and dextrose; potassium is usually not included, as the solutions are often used to treat hyperkalemia.

The aim of both hemodialysis and peritoneal dialysis is to return to the body fluids the normal amount of various electrolytes and to remove the waste products of metabolism.

The usual normal osmotic values for humans is 280 mOsm/liter. The dialyzing solutions sold by Travenol vary in their concentration of dextrose from 1.5%, 4.25%, and 7.0%, giving values of 366, 505 and 644 mOsm/liter. If edema is not present and rapid dialysis is not required, the 1.5% solution is used in order to avoid hypovolemia and reduce the loss of proteins. The 1.5% dextrose solution has an osmotic value or "pull" of 366 mOsm/liter. This is just a little more than the normal value of 280 mOsm/liter, and the salts will be removed gradually. Remember, if the concentration gradient is larger, that is, the solution is much more hypertonic compared with the patient, more water will be attracted from the blood and also more proteins. This would be the case with the 4.25% and 7.0% dextrose solutions. It is therefore advisable to use the lower concentration, unless edema is present and large amounts of salts must be dialyzed.

Another modification of peritoneal dialysis is now being used. It is called *continuous ambulatory peritoneal dialysis (CAPD)*. The patient uses a plastic bag with the proper concentration dialyzing solution, administering it by gravity flow through a tubing inserted in the trocar. After all the solution has run into the peritoneal cavity, the patient rolls up the plastic bag and wears it beneath his or her clothing; this enables the patient to be ambulatory. Later, the fluid is drained back into the same bag by gravity drainage, and the bag is discarded. A bag containing fresh solution is attached, and the procedure is repeated. The advantage is greater freedom for the patient. Figures 12-13A and B illustrate this procedure.

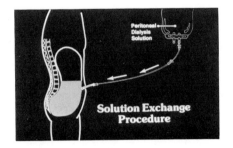

Figure 12-12.
Peritoneal dialysis.
Source: Courtesy © 1980 Travenol Laboratories, Inc.

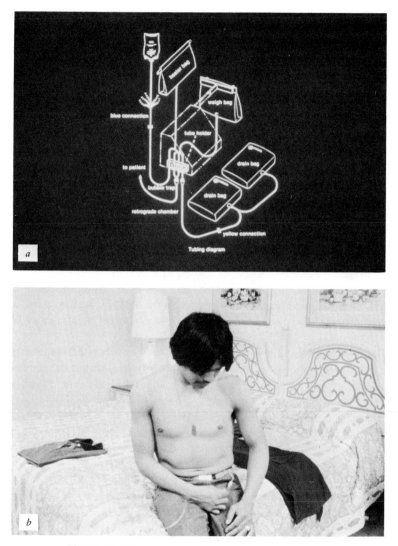

Figure 12-13.
(A,B) Continuous ambulatory peritoneal dialysis (CAPD).

The physiological importance of osmosis can also be illustrated by considering the pressures in the capillaries. The hydrostatic pressure in the capillaries at the arteriolar end is about 35 mm Hg. The pressure decreases from the arteries owing to the frictional losses of energy, as well as the larger *total* cross-sectional area of the capillaries.

Because the capillary walls are so thin and allow water and electrolytes to pass

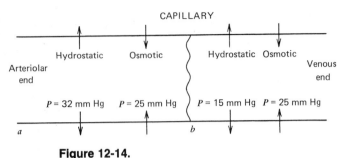

Figure 12-14.
Hydrostatic and osmotic pressures in the alveoli.

through, why don't we all "spring a leak" at the beginning of all the capillaries? Actually, there is some movement of water, electrolytes, and nutrients from the capillaries into tissue spaces, but this is limited by the *osmotic pull*—also called *oncotic pressure*—of the electrolytes and plasma proteins.

We see in Figure 12-14, Part (*a*) that there is a hydrostatic pressure of approximately 32 mm Hg and an osmotic pressure of 25 mm Hg, leaving a net pressure of 7 mm Hg to allow the water, electrolytes, and nutrient molecules to flow out to the tissues.

Further down the capillary, the amount of water is decreased, and the proteins of the blood have not escaped (Fig. 12-14, Part (*b*)). This means the concentration in the capillary will be increased, and the osmotic pull will be increased from 25 mm Hg to about 30 mm Hg.

Meantime, because of frictional losses, the hydrostatic pressure is decreased to about 15 mm Hg. Thus, there is an attraction of 30 minus 15 or 15 mm Hg, pulling water, electrolytes, and metabolic waste products back into the capillaries.

QUESTIONS

1. Using an eyedropper, suppose that you were to produce drops of water, alcohol, and soap solution respectively. Would the drops be the same size for all these liquids? If not, which liquid forms the largest drops, and which the smallest?

2. How does a strong solution of Epsom salts reduce edema? Would a hypotonic solution do the same?

3. What is the principle behind the practice of shaking down a thermometer?

4. Why will a drink of saltwater increase rather than decrease the sensation of thirst?

5. What similarities and differences exist between the artificial kidney and peritoneal dialysis?

6. What is the effect on the red blood cells when a hypotonic solution is injected?

7. Osmosis was defined as the passage of solvent molecules through a semiperme-able membrane from a less concentrated to a more concentrated solution. Some define it as the flow of water from a higher to a lower concentration. Can you reconcile the two statements?

8. If a patient is suffering from hyperkalemia, would you use a dialysis solution with no, a little, or a great deal of potassium?

9. Would a detergent solution rise higher or lower in a capillary tube? Explain using the formula in the text.

10. Explain why water and mercury have different types of meniscus in a glass tube.

CHAPTER **13**

ELECTRICITY

*Electricity in one form or another is constantly
present in the hospital setting. Using it to best
advantage requires knowledge of its basic
characteristics; benefits, exemplified by
development of such equipment as the EEG, ECG,
and cardiac pacemaker; and related hazards, such
as electrical shock and operating-room explosions.*

Electricity is a form of energy resulting from a flow of electrons through a substance
(usually a metal) called a *conductor*.

Electricity can assume two forms. The first is called *static* electricity and is
characterized by the accumulation of a static (nonmoving) charge of electricity upon
the surface of a nonconductor (such as rubber, glass, or nylon). The second form
of electricity is *dynamic* electricity, in which the charge is in motion along a
conductor. Dynamic electricity produces useful energy. Let us investigate the characteristics of both types.

STATIC ELECTRICITY

We are all acquainted with static electricity. On a cold, dry day in the winter a
walk across a rug will produce enough static electricity so that when we touch a
metal object, a spark jumps. What is the origin of static electricity? To understand
it, we must look to the smallest division of matter, the atom.

The basic structure of the atom is a dense nucleus composed of two main types
of particles, protons and neutrons. Revolving about the nucleus, like satellites in
orbit, are particles of a third type: electrons. In an atom of any element, the number

of protons in the nucleus is equal to the number of electrons outside the nucleus. Since protons are said to be *positively* charged, and electrons *negatively* charged, the atom is electrically neutral. This balance of positive and negative particles is shown in Figure 13-1.

By rubbing, it is possible to remove electrons from one substance to another, thus unbalancing the distribution of particles and causing an electrical "charge" to be built up. If one of these substances is an insulator, the charge will be retained. Rubbing a glass or plastic rod with wook, silk, or nylon will produce static electricity. If a hard rubber rod is rubbed with wool, electrons are accumulated on the surface of the rubber. Because electrons are negatively charged, the rod will become negative. It does not matter whether there is an accumulation or a deficiency of electrons; charges will be built up in either case. Of course, if electrons are lost, the surface will become positively charged. As is well known, opposite charges attract one another and similar charges repel.

Everyday examples of static electricity are numerous. Running a comb through the hair will charge the hair, and each strand will tend to be repelled by its neighbor, with the result that the hair fluffs out. The chain dangling from the back of a gasoline truck is for the purpose of grounding the truck, that is, connecting the truck to earth by a conductor so that electrical charges, built up by the friction of the tires, are drained off into the ground. The flexible metal rods usually found sticking up out of the road at the toll stations of bridges and turnpikes also serve to drain off electrostatic charges from the car. Without them, the position of toll collector would be a more exciting occupation than it is.

When blankets or sheets are pulled off the bed, friction will cause much static electricity. Research has shown that large electrical potentials (charges) may be built up. Voltages of 10,000 or more are easily produced. Walking across the floor will produce a fairly heavy charge.

Static electricity will not accumulate so readily when the humidity is high, because water is a good conductor and the charges will be carried away by the moist air. In an operating room the danger of explosion may be minimized by having the

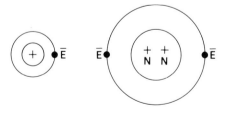

Hydrogen $^1_1$H   Helium $^4_2$He

Figure 13-1.
Structures of the hydrogen and helium atoms.

relative humidity at about 60 percent. However, high humidity is no guarantee that an explosion will not occur.

The problem of static electricity has been made more acute with the advent of synthetic fibers and plastics. Garments may cling to the body because of it, and dust and lint will be attracted to the charged surfaces. Compounders of chemical specialties are now making antistatic sprays, wiping cloths, polishes, and other products to help eliminate static "shocks." The basic ingredient of these antistatic products is a surface-active agent, related to emulsifiers and detergents. One to 2 percent added to the last rinse in the home laundering of synthetic fabrics will render these materials fairly free of static electricity.

DYNAMIC ELECTRICITY

Electricity that flows through some conductor, as water does through a pipe, is dynamic electricity. Actually, if a static charge builds up so as to cause a spark, it is dynamic for a fraction of a second, because the electrical pressure, or potential, is sufficient to cause the air to conduct momentarily. But once the spark jumps, the charge is dissipated; there is no electricity left. Dynamic electricity, on the other hand, will continue to flow through a wire, and to work for us, as long as an energy source (such as a generator or battery) is supplied.

A battery or a generator can be compared with the pump that keeps water circulating through a pipe. Either type of "electron pump" has two connections called *poles,* one of which is negative and the other positive. Electrons flow out of the battery or generator by way of the negative pole. From here they flow through a conductor (usually a copper wire) into the *load* (such as a light bulb or toaster), where they do useful work. From the load the electrons flow through another conductor back to the battery or generator by way of the positive pole. Dynamic electricity, then, flows through a closed circuit just as the blood does within the body (the battery or generator is the "heart").

Note that the electrons always flow from negative to positive. Another way of saying this is that *electrons flow from the higher concentration to the lower.* The greater the difference in electron concentration (also called pressure) between the two poles, the greater the tendency for the electrons to flow. Electrical pressure (or potential) has another name also: *electromotive force,* the unit of which is the *volt.*

No electrical conductor is perfect; as electrons flow through a wire, they will encounter some resistance on the way. The better the conductor, the less its resistance (a piece of copper has less resistance than a piece of iron of like dimensions). Also, the thinner the wire, or the longer the wire, the more electrical resistance it will offer. The unit of electrical resistance is the *ohm.*

The flow of electrons through a conductor is called the *current,* and the unit of measurement of a current is the *ampere* (usually shortened to "amp").

A formula shows the relationships between electromotive force, resistance, and current. It is known as Ohm's law, and is stated

$$E = IR$$

where E = electromotive force (volts), I = intensity of current (amps), and R = resistance (ohms).

We can just as easily state Ohm's law this way:

$$\text{Volts} = \text{amps} \times \text{ohms}$$

A moment's reflection will make these relationships clear. For instance, if the electrical pressure (volts) goes up, it will force electrons through the wire at a greater rate; therefore the current (amps) will increase. If the pressure is kept the same, but a poorer conductor is substituted, the resistance (ohms) is thereby increased; and consequently the current will decrease. People who have traveled in other countries where the household voltage is 220 instead of our usual 110 have often discovered that their electrical equipment burned out because the higher voltage pushed twice as much current through the equipment as it was designed for.

If too much current is allowed to flow through a wire, heat is generated in the wire and a fire hazard may exist. A fuse is a device designed to interrupt the flow of current when it exceeds a predetermined amount, such as 15 or 20 amps. A typical fuse contains a thin ribbon of metal inserted in the circuit. If the current becomes excessive, the ribbon melts ("blows"), breaking the circuit and thus shutting off the current. In some installations an automatic circuit breaker takes the place of a fuse. Excess current will cause the circuit breaker to "kick out" and stop the flow. When the cause of the current surge is removed, the circuit breaker can be reset automatically or manually.

Any device in which electricity is put to work consumes power. The unit of electrical power is the *watt.* The formula for determining the power consumption of an electrical device is

$$\text{Watts} = \text{volts} \times \text{amps}$$

This is a handy equation. If we assume a 110-volt system, we only have to divide the total wattage rating of an electrical device by 110 to find the amps that will flow through it. For example, a machine listed for 1,800 watts will use approximately 16 amps. If the fuse is designed to carry a maximum current of 15 amps, use of this machine will blow the fuse.

A thousand watts equals 1 kilowatt. The consumption of electric power is measured in the number of kilowatts used per hour—that is, kilowatt-hours.

MAGNETISM

A magnet is a piece of iron or steel with its electrons oriented in such a way that it will attract other bits of iron or steel. If a bar magnet is suspended horizontally from a thread, it will swing about until it comes to rest in a North-South direction. That end of the bar that always points North is called the magnet's north pole; the opposite end is the south pole. The tendency of a bar magnet to align itself toward North and South is, of course, the principle of the compass.

A magnet has about it an invisible magnetic field, which can be described as an area in which magnetic forces are present. If a piece of glass is placed on top of a magnet, and iron filings are sprinkled over the glass, the filings will arrange themselves in a pattern called *lines of force* (see Fig. 13-2), which roughly outlines the magnetic field.

In one respect magnetic poles behave like electrostatic charges: Like poles attract one another; unlike poles repel.

Physicians often use small but very powerful magnets to remove bits of iron or steel from the eye.

GENERATORS AND MOTORS

An electric generator is a machine that transforms mechanical energy into electrical energy. An electric motor does exactly the opposite; it consumes electrical power,

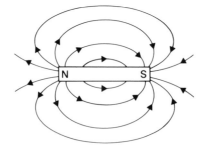

Figure 13-2.
Magnetic field around a bar magnet.

transforming it into a rotary motion that can do work. Motors and generators are very similar in construction, and some machines are designed to serve as either.

There are two elements common to every motor and generator. One is a magnet and the other is a rotating coil of wire. It will be easier to understand the purposes of these elements if we look momentarily to their historical origin.

During the early part of the nineteenth century a Danish scientist, Hans Oersted, discovered that a current-carrying wire had a magnetic field around it. When the current was turned off the magnetic field promptly collapsed.

In 1831, Michael Faraday, an English scientist, thought that the magnetic field of the wire with the electric current in it might interact with the magnetic field of a permanent magnet. This turned out to be the case. It was found that:

- If a wire is *moved* in a magnetic field, an electric current *will be produced* in the wire.
- A *current-carrying wire* in a magnetic field *will move*.

These discoveries did not seem very significant at the time, but the implications were tremendous. In fact, here is an excellent example of the impact of science upon society. The first principle is the basis of all electric generators, and the second is the basis of all electric motors. Imagine life today without electricity—no electric lights, fans, heaters, elevators, movie projectors, radio, TV, and many hundreds of other electrical appliances.

THE GENERATOR

As was mentioned above, by moving a wire in a magnetic field, current will be produced. Basically, a generator consists of a coil of wire made to rotate between the poles of a permanent magnet (Fig. 13-3). As one side of the coil moves up, the current flows in one direction; as the same part of the coil moves downward while it rotates, the current will flow in the opposite direction—that is, it alternates. Household current changes direction 120 times per second, which is the same as making 60 complete cycles per second. Direct electric current (DC) is produced in a similar manner except that the current is kept flowing in one direction only by means of a special device called a commutator. Household current is nearly always alternating, the reason being that AC is easier to transmit over long distances than DC.

The transmission of electric power is accomplished in the following manner. First, there must be the original source of energy, such as a steam or hydroelectric plant, which turns the AC generator. The alternating current from the generator then is fed into a device called a *transformer,* which "steps up" the voltage to a very high level—perhaps as much as 220,000 V. The reason for the very high voltage (also called "high tension") is that it can be transmitted over long distances with less power loss than a low voltage. The power is now sent from the transformer

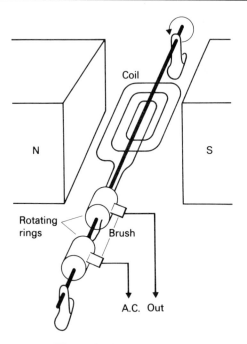

Figure 13-3.
Alternating-current generator.

across country on the familiar spidery steel "high tension" lines to the city. At this point the voltage is "stepped down" by another transformer to about 4,400 or 2,200 volts. Other step-down transformers reduce the voltage to 110 or 220 for use in individual buildings. Very often a dwelling will use two different voltages: 110 for lights and appliances, and 220 for heavy-duty installations, such as ovens and freezers.

Direct current is not used in cross-country power transmission because there is no convenient and efficient way to raise and lower the voltage; a transformer will operate only on alternating current.

THE MOTOR

The electric motor is very similar to the generator in construction because its basic elements also consist of a rotating coil in a magnetic field. Because a wire carrying an electric current will move in a magnetic field, a current is passed through the coil of the motor; this coil (called the *armature*) will rotate to produce usable power. Electric motors have a wide variety of uses in the hospital: electric saws for casts;

refrigeration and suction machines; needle-cleaning machines; centrifuges; passive massage apparatus for the physiotherapy department; and many others.

Electric motors are put to an almost infinite variety of uses in the everyday working world, a good illustration of the transfer of energy. It is interesting to realize that the ultimate source of this energy is the fusion of hydrogen atoms within the sun.

ELECTRIC LIGHTING

We are so accustomed to illumination from electric lights that we tend to take it for granted. It is only when we have a power failure and have to use candles or kerosene lamps that we really appreciate the advantages of electric lighting. Conventional (incandescent) lamps have a tungsten filament that glows white-hot as the electricity forces its way through the high resistance. The metal would burn (oxidize) at the high temperature, so the air is removed from the glass bulb. Nitrogen used to be introduced into the bulb; but now an inert gas is used, prolonging the life of the filament.

Fluorescent lamps operate on a principle different from that of incandescent lamps. They are evacuated tubes containing some mercury vapor. As the current flows through the tube, ionization of the vapor takes place, producing ultraviolet light. This hits the coating on the inside wall of the tube, called a phosphor, which is made of some salt such as zinc sulfide, calcium tungstate, magnesium tungstate, or zinc-beryllium silicate. These salts absorb the short wavelengths of ultraviolet light, and re-emit longer wavelength light in the visible region.

The same type of fluorescence is also involved in the picture tube of a television set and in screens for fluoroscopy. The Patterson Type A screen uses cadmium tungstate which produces visible light with a wavelength of about 5,000 Angstroms, while the Type B screen uses zinc-cadmium sulfide, which produces slightly longer light with a peak radiation at about 5,550 Angstroms. Angstrom is the unit of wavelength of electromagnetic radiation.

The efficiency of fluorescent lamps is much greater than that of incandescent bulbs, about four times as much. In Figure 13-4, it can be seen that only about 10 percent of the energy output of the incandescent lamp is emitted as light and all the rest as heat. (This is the case in many arguments, too—much more heat than light is emitted.) The fluorescent lamp has a higher percentage of its energy output emitted as light, and so produces less heat.

If you read the previous figure carefully, you may have noted what seemed to be a discrepancy. Figure 13-4 shows that the 40-watt fluorescent lamp emits 20.5 percent of its radiant energy as light, whereas the 100-watt incandescent bulb gives off only 10 percent of its energy as light, which would seem to make the fluorescent

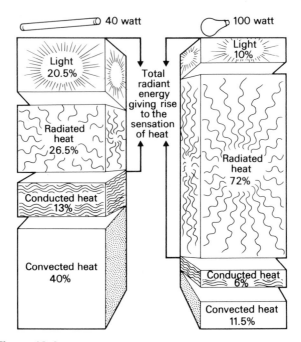

Figure 13-4.

Energy distribution of 40-watt fluorescent and 100-watt incandescent lamps.

Each has about the same lumen output.

Source: *Westinghouse Engineer.* November 1954, pp. 201–205.

twice as efficient, rather than four times as was stated. The reason is that the efficiency increases with the wattage. This can be seen in the table below. The lumen is the standard unit of illumination.

Incandescent		Fluorescent	
Watts	*Lumens per Watt*	*Watts*	*Lumens per Watt*
25	10.6	15	35
40	11.7	20	38
75	14.9	30	44
100	16.3	40	47
1,000	20.7	—	—

If we compare the lumens per watt for the same wattage (let us say 40 watts), we find that the fluorescent lamp gives 47 lumens compared with 11.7 for the tungsten type. This increased efficiency is a good reason for using fluorescent lamps where a heavy electric light bill is involved.

In order to increase the efficiency of illumination of the streets, in 1973 and 1974 New York City spent $30 million to replace many of the mercury lamps with sodium vapor lamps. The 175-watt and 250-watt mercury lamps provide, respectively, 8,000 and 12,500 lumens, whereas the new 150-watt sodium lamps will give 16,000 lumens. This change is expected to save 15,000,000 kilowatts each year for a saving of $500,000.

THE ENERGY SHORTAGE

That the United States is short of energy sources has only recently become quite clear to most citizens, yet the shortage has been in the making for a number of years. Even though the United States is richly endowed with natural resources such as coal, oil, and gas, our comsumption of them has increased so rapidly in the last fifty years, that we have become increasingly dependent upon oil imports; at the same time the possibility that we will deplete our own resources has been predicted. The efforts of environmentalists to improve the quality of our air and water, a very needed and laudable effort, has at the same time made it difficult to use oil and coal containing higher amounts of sulfur.

The use of scrubbers in the exhaust stacks will take out most of the sulfur dioxide and particulates, but this greatly increases the cost.

What is needed is to secure newer sources of energy, such as solar energy, and at the same time reduce our waste of power. Industrial and domestic conservation efforts are having a significant effect on reducing energy consumption. One possible source, and one it is believed we must develop unless our society and mode of living are to be radically altered, is the use of nuclear energy. Full development of nuclear energy has been impeded for the past few years by lawsuits brought by various organizations and individuals concerned about the possibility of radiation damage to both the environment and the people. There are two ways to "harness the atom" to produce nuclear energy: fission—the nucleus of the atom splits into fragments which release energy; and fusion—the nuclei combine and release energy. Nuclear energy is discussed in Chapter 14.

THE X-RAY TUBE

X-rays were discovered by Wilhelm Roentgen in 1895 while he was experimenting with a device called a cathode tube, the ancestor of the modern x-ray tube. The x-

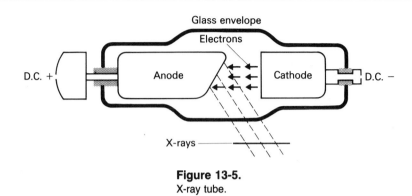

Figure 13-5.
X-ray tube.

ray tube is a glass tube with two electrodes in it and with the air evacuated from it (Fig. 13-5). One electrode is called the *anode*, or "target," and the other electrode is called the *cathode*. When the cathode is heated in the same manner as is a light-bulb filament, electrons are "boiled off" from it (hence the term "cathode rays"). A high-voltage source is connected between the anode and the cathode, causing the free electrons emitted by the cathode to flow between it and the target. When the electrons hit the target, they are absorbed by the target and re-emitted as x-rays. The target becomes very hot when subjected to the energy of the high-velocity electrons; therefore it is made of tungsten, which has a high melting point. For very high voltage x-rays the rotating anode is used. By spinning the target, one part of it is heated while the other part has a chance to cool off. Water is also used to cool the anode for high-voltage applications.

The frequency of the x-rays emitted by the target depends upon the voltage employed, and the quantity of radiation depends upon the current flowing between the anode and the target. You will recall Planck's equation,

$$E = hV$$

Energy equals Planck's constant (6.6×10^{-27}) times the frequency. This means that the higher the energy, the higher the frequency. Now, if the voltage is increased, the electrons have more energy, and the x-rays will have more energy, and their frequency will be greater. These higher frequency x-rays will have a shorter wave-length and more penetration. Or, to put it briefly, *the higher the voltage, the more penetration*. In practice, the x-ray technician measures the thickness of the particular part of the body to be x-rayed and looks up the required voltage on a chart.

The quantity of x-radiation depends on the current and time of exposure. The higher the current, the more radiation.

Although x-rays are invisible, they register on a photographic plate as light rays do. The bones show up well on the plate because they absorb more radiation than do the other body tissues. In order to make x-ray photographs of other structures, such as the stomach and the intestines, a radiopaque substance (such as a solution of barium sulfate) is introduced into them. The barium will absorb the radiation and thereby cause the particular organ to stand out in good contrast.

It is very important that anyone working in the presence of x-rays or similar radiation minimize the exposure of his own body by using proper shielding. Such precautions will be discussed in the next chapter.

ELECTRICITY AND THE HUMAN BODY

The body is, to a large extent, an electrical machine, in addition to being a "wonderfully and fearfully made" chemical and mechanical entity. Nerve impulses are continually flashing back and forth between the brain and the various muscles and organs. A nerve is normally in a *polarized* state; that is, it has an excess of positive charges on the outside of the membrane, and an excess of negative charges on the inside. A voltage difference of about 80 millivolts (thousandths of a volt) results. What actually makes the charges separate in this manner is not completely understood (Fig. 13-6a).

However, during the last decade, biochemists have been able to begin to explain the excitability of bilayer lipid membranes and the "gating" mechanism that allows the flow of sodium and potassium ions across the membrane, thereby changing the potential. During the resting state, there are more potassium ions inside the cell and more sodium ions outside the cell, but the concentration of the K^+ ions are

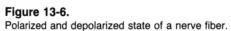

Figure 13-6.
Polarized and depolarized state of a nerve fiber.

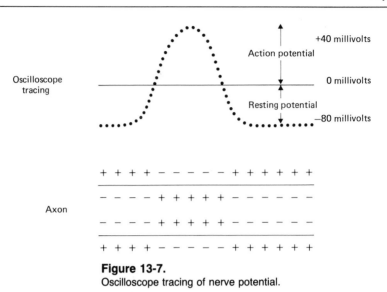

Figure 13-7.
Oscilloscope tracing of nerve potential.

greater, thereby resulting in a positive charge outside and a negative charge inside. "When an axon is stimulated, there is a temporary change in the structure of the membrane making it permeable to sodium. As sodium ions rush into the axon, the negative polarization of the inside with respect to the outside of the membrane is lowered. Sodium continues to pour in until the membrane becomes highly positive—40 millivolts."[1]

When a stimulus excites a nerve ending, there is a change in the permeability (diffusibility) of the nerve membrane. The unequal charges now neutralize each other; that is, *depolarization* occurs (see Fig. 13-6b).

The change in potential can be measured and can be demonstrated on the oscilloscope tracing. As the nerve changes from a position of polarization to one of depolarization, "overshooting" occurs, and for a fraction of a second the polarities are reversed (Fig. 13-7).

A wave of depolarization is passed along the nerve at a velocity of from 1 to 125 meters per second. The frequency of the waves may vary, and will control the amount of muscle contraction that may result.

One thing that is worth pondering is how a depolarization wave can produce such a variety of sensations. Think of all the colors and shades you can detect, the variety

[1]Bauman, G. The cell membrane ready for action, *Chemistry* 51, Dec. 1978. This is part of a three-part series in the November 1978, December 1978, and January 1979 issues of *Chemistry* that gives an excellent biochemical description of cell membrane excitation.

of musical tones, and the odors and tastes you can differentiate. The brain's capacity to interpret is awesome indeed.

AVOIDING ELECTRICAL INJURIES IN THE HOME

Muscle tissue is also a producer of electricity. Each time that a muscle contracts, a measurable voltage is generated. Not only does muscle tissue act as a generator; it also can simulate a motor. If a piece of muscle tissue is touched by two electrodes connected to a battery, it will contract. This is the reason why an appreciable quantity of electricity can cause a violent shock. Death by electrocution is usually attributed to two main causes: ventricular fibrillation resulting from overstimulation of the nerves controlling the heart muscle; and respiratory failure due to paralysis of the nerves controlling the respiratory mechanism. Everyone should remember that household voltage can be lethal, particularly when the part of the body which makes electrical contact is wet. Perspiration and urine are particularly good conductors because of their salt concentration.

One side of the household electrical outlet is always connected to the ground. It can happen, then, that a voltage difference of 110 to 115 may exist between an electrical appliance and some conductor which reaches the ground, such as a water pipe or a radiator. Never touch the two simultaneously. Beware especially of electric light chains and AC–DC appliances. If the metal base of your AC–DC appliance feels "tingly" when you touch it, it is potentially dangerous. Unplug it and then replug it, this time with the prongs reversed. This will keep the appliance at ground, or zero, potential. Light chains now are usually made with an insulated link.

A person's body has a resistance of approximately 1500 ohms when dry. This drops to about 500 ohms when wet. According to Ohm's law, discussed earlier in the chapter, $E = I R$. If we assume a value for E of 120 volts, then the current (I) will be:

$$120 = I\ (500)$$

$$I = \frac{120}{500} = 0.240$$

$$I = 0.240 \text{ amps or } 240 \text{ mA when wet.}$$

$$120 = I\ (1500)$$

$$I = \frac{120}{1500} = 0.080 \text{ amps when dry}$$

Thus it can be seen that water and electricity do not mix, and the danger of water getting on electrical wires is obvious. A person's body will react to even a 1-milliamp current of 60 cycles; with 100 mA there may be ventricular fibrillation.

To avoid accidental damage from electrical "shorts," most heavy electrical equipment has a three-pronged plug, with the third wire leading to ground. Should a "short" occur, the current will travel through the ground wire (and blow a fuse) rather than through the person.

If only a two-pronged receptacle is available, an adapter, into which the three-pronged male plug from the equipment is inserted, can be used; the adapter in turn plugs into the two-pronged wall socket. This however means that there is no ground wire unless the short connecting wire on the adapter is attached to ground. This is often done by loosening a screw on the wall plate, slipping the adapter ground wire underneath, and tightening again.

AVOIDING ELECTRICAL INJURIES IN THE HOSPITAL

Whenever health workers find it necessary to use *any* electrical equipment they must be aware of the danger of electrical shock due to short circuits or leaking of electricity through frayed insulation. When an electric defibrillator is being used, there is especial need to remember that the current can take other pathways through the patient and the bed to ground. The following quotation summarizes the situation well:

> Any piece of equipment may be hazardous if there is an ungrounded line, a ground object, and a low resistance pathway. Such electrical apparatus as heating units, nebulizers, air-conditioners, hypothermia equipment, hemodialysis units, suction machines, and electrical fans [may be dangerous]. The frame of a motor-operated bed or an arterial pressure transducer may be just as lethal a form of shocking the patient as current flowing through the pacemaker electrode. Opportunities for electrical hazards in the hospital are further multiplied by moisture, bodily exposure, and the number of electrical appliances in direct contact with patients.[2]

OPERATING ROOM HAZARDS

The operating room is a possible source of explosions. The use of cyclopropane and ether as anesthetics can be potentially dangerous. When these gases are mixed with air, oxygen, or nitrous oxide in certain proportions, explosive mixtures will result. The explosive range of air-cyclopropane is from 2.4 to 10.3 percent. The explosive range for ether is from 1.8 to 36.5 percent. If the mixture is within the explosive range, a small spark will be sufficient to set off the explosion. Ether is approximately two and one-half times as dense as air, and will tend to settle close to the floor. Where closed breathing circuits are used to administer the anesthetic,

---

[2]Brennan, J., Ervin, D., and Worrell, J. Electrical hazards in patient care situations—A manual for nurses, 1969. Unpublished.

the most dangerous area generally will be within a two-foot radius of the anesthetic machine.

How can the accumulation of static electricity be prevented in the operating room? As we saw, a high relative humidity helps. Also, if all equipment and all personnel are grounded, the static charges will be drained off. The floor, therefore, should be conductive, not insulated. Terrazzo floors may be used with a close-meshed grounded copper grid on which the personnel will be standing at all times. There also are available plastic floors which have been rendered conductive by the addition of carbon or metal particles dispersed throughout their material. These floors should not be waxed, since wax is an insulator. Soap should be thoroughly rinsed off the floor for the same reason.

Whatever the type of floor installation in use, it and other articles employed in the operating room (such as the operating table, pad and pillow covers, and shoes) should be periodically checked by a device called a *conductivity tester*. This type of tester determines whether or not the article in question is capable of conducting sufficiently to drain off any static charges that may accumulate. All materials have a certain *resistance* to the flow of electricity through them. If the resistance of, say, a pair of shoes is too high, they can be considered nonconductive and therefore dangerous in the operating room. Conductivity testers, then, measure the electrical resistance of objects. The usual type of tester has a pair of metal bars called electrodes, which are connected to the device by wires. The electrodes are placed on the object to be tested. When the test button is pressed, an electric current from an outside source is made to flow out of one electrode, through the tested object, into the second electrode, and back to the tester. If the object is sufficiently conductive, enough current will flow through it to cause a small bulb to light up.

As we learned earlier, the unit of electrical resistance is the ohm. The resistance of the conductive floor should be less than 1,000,000 ohms when measured between two five-pound electrodes with a 2½-inch diameter surface placed three feet apart on any point on the floor. The resistance must be more than 25,000 ohms in order to protect against shock from electrical devices.

A conductive floor is not adequate by itself to prevent the accumulation of static electricity. The equipment and personnel must make definite contact with the floor by means of conductive shoes, slip-on sandals, or stick-on tapes. The use of *intercouplers* is also helpful. These are flexible conductive cables that are attached to the operating table, the surgeon, the patient, the anesthetist, and other personnel as needed. The ends of the intercouplers are all connected to the ground. The important principle behind all these devices is to provide a path through which static charges can be drained off to the ground before they can accumulate sufficiently to generate a spark.

Obvious potential causes of explosion are open flames, smoking, the use of live cautery or diathermy, faulty electrical wiring, and sparking electrical equipment. Even an unshielded wall switch may ignite a combustible mixture. If live cauteries are used, a nonflammable type of anesthetic should be administered.

All staff members using the operating room should be educated to the potential danger of explosive anesthetics, and no one should be admitted to the operating room without having been rendered conductive.

ELECTRIC SHOCK THERAPY

The use of chemicals to treat mental disorders began in 1933 with insulin, followed by pentylenetetrazol (Metrazol) in 1935, and later by other chemicals. Electric shock therapy, also called electroconvulsive therapy, was first used in 1938 in Italy, and by 1940 was being used in the United States. Some physicians still prescribe this therapy in the treatment of certain mental disorders. Sometimes a combination of drugs and electric shock is prescribed. The drugs atropine and methohexital are given just before the shock; the patient then has a brief seizure, after which oxygen is administered via mask.

Before the shock is administered, a current of approximately 1 milliamp (1/1,000 amp) is applied to the patient's head to measure resistance through it. The strength of the current ranges from 200 to 1,600 mA. The duration varies from 0.1 to 0.5 second; the voltage is between 70 and 120 volts. A rubber sheet is placed beneath the patient to avoid grounding him.

Recent experiments in which implanted cerebellar stimulators act as "brain pacemakers" may ultimately show us how to reduce the severity of epileptic seizures. Other experiments, in which minute channels are created into specific areas of the brain, may make it possible to deliver chemicals for treatment directly to the part of the brain where it is needed.

MEDICAL ELECTRONIC AIDS

The development of electronic equipment has in some ways revolutionized both hospital practice and medical health care. The "marriage" of medicine and electronics has produced many new types of equipment, some of which are used in the hospital setting and some of which are used in the medical office setting. Let us look at a few of these electronic medical aids.

Some idea of the variety of electronic medical aids can be gained merely by listing the most important ones: monitors that record the heartbeat, the arterial and venous blood pressure, the temperature, the respirations; defibrillators to restore a regular heartbeat; pacemakers to maintain a regular heartbeat; instruments to measure pulmonary function; and instruments to measure the activity of the brain (electroencephalograph), and heart (vectorcardiograph, cardiac catheter, phonocardiograph), and the muscles (electromyograph).

We shall first consider the principle behind these instruments. To convert the input from the patient into electrical energy, a device called a *transducer* is used. (Think back to the discussion of the generator—the generator transforms mechanical energy into electrical energy.) There are many types of transducers, and there are literally hundreds of all sizes and shapes now available. Many are miniaturized so they can be implanted in the patient. The transducer converts mechanical energy (the patient) into electrical energy (the machine), which in turn is amplified and recorded. For instance, the strain gauge transducer records the increase in pressure inside a small cup placed over the tip of a finger. As the blood pulses through the finger, the finger expands or contracts slightly, changing the air pressure inside the cup; this change is picked up by the transducer, which then records the change in electric current identical to the change in pressure. It is important to keep in mind that when physiological activities are to be monitored, some means is required to change the mechanical signal into an electrical signal, which can then be displayed on an oscilloscope or printed on paper tape or magnetic tape. The newest method is to feed the data into a computer. The computer analyzes the data to determine deviations from normal, produces a printout, and sounds an alarm whenever a deviation occurs.

SOLID TRACE CENTRAL STATION (Right Oblique View)
(4-PATIENT, 2-PARAMETER ECG AND PRESSURE)
Consisting of two instrument cabinets:
  (1) CB-12A cabinet (left), containing:
    1 WR-4A Central Station Writer
    2 HR-2 Heart Rate Display Modules
    2 PR- 5 Pressure Display Modules
    1 ST-412 Central Station Display

  (2) CB-12B cabinet (right), containing:
    1 ST-412 Central Station Display
    2 HR-2 Heart Rate Display Modules
    2 PR- 5 Pressure Display Modules
    1 WR-4A Central Station Writer

Figure 13-8.
Computaview patient monitoring system.

Source: B-D Electrodyne, Division of Becton, Dickinson & Co., Sharon, Mass. 02067.

In comparison with industry, hospitals have been slow to adopt electrical and mechanical aids to efficiency. Nevertheless, the enormous increase in numbers of hospital patients is forcing change, and there is a strong trend toward increased automation in hospitals. While most such innovations are welcomed by both staff and patients, automation does raise one unpleasant possibility. Imagine rows of patients all wired with transducers, with the nurse taking readings of temperature, cardiac activity, pulse rate, blood pressure, and respiratory function simply by flipping various switches at a central control panel, while inspecting the patient through closed-circuit TV. Everyone who deals with patients should always remember that patients are not robots. No machine, however efficient, should ever be allowed to interfere with or replace the personal relationship which is at the very heart of the therapeutic process.

THERMISTOR THERMOMETERS

There are a number of devices for measuring temperature besides the common mercury thermometer. One of these is called a *thermistor thermometer*. The main component of a thermistor thermometer is a piece of crystal which conducts an electric current. As the temperature of the crystal changes, its electrical resistance will change also. In this unit a constant voltage is supplied by a small battery, and the current flowing through the crystal is read on a scale calibrated in degrees. As the temperature of the crystal changes the crystal's resistance, the current varies accordingly, and so with it the reading on the scale.

The thermistor thermometer comes with a number of probes for both internal and dermal temperature determinations. They are available in both single and multiple channel units. The probe may be located up to 1,000 feet from the unit, and the readings may be made in 5 to 7 seconds with greater accuracy than with a mercury thermometer.

THERMOCOUPLES

A *thermocouple* is another device used to measure temperature. It consists of two wires made of different metals, joined at one end. When the temperature at the junction is higher than that at the other end of the wires, a small electric current is generated. This can be measured, and a temperature scale can be calibrated. Various pairs of metals used are copper-iron, antimony-bismuth, and platinum-iridium. A thermocouple used to measure high temperatures is called a pyrometer. Advantages of a thermocouple apparatus are that it can be made very sensitive and is easily read on a meter.

ELECTROENCEPHALOGRAPHY

There are an estimated 15,000,000,000 cells in the brain; each is capable of generating an independent electrical signal, and each is linked to all the other brain cells. It is said that 100,000,000 bits of information are received by the brain every second, analyzed, and stored; and each time a group of cells responds to a stimulus, some very minute activity is generated.

The *encephalograph* (EEG) records the total output of brain cells; the record is called the *electroencephalogram*. The EEG produces very weak electrical signals (approximately 1/100 as strong as does the electrocardiograph); the signals must be amplified over 10,000,000 times in order to be recorded. The large current resulting from the amplification varies exactly as the original; these current variations are recorded by a pen on a calibrated paper strip. The electroencephalogram, or encephalogram, as it is often referred to, records four major brain wave forms. The alpha waves (8 to 13 Hz) originate in the occipital region, and the beta waves (13 to 22 Hz) originate in the cerebral area of the brain (concerned with thinking). These are prominent while the person is awake. The theta (4 to 8 Hz) and delta (0.5 to 4.0 Hz) are more prominent during sleep, and originate in the central and parietal regions.

To produce an electroencephalogram a number of electrodes are applied to the scalp, and the resulting activity is recorded; each reading requires two electrodes, one on the scalp and one attached to a neutral area such as the ear. The tracing records the *difference* in electrical activity between the two electrodes. Recordings are made of both sides of the brain (frontal, central, parietal, temporal, and occipital areas).

Two types of electrodes are used. One is a flat disk-shaped electrode which is fastened to the scalp with conducting electrode clay. The other type is a fine No. 27 needle connected to the copper lead wire. This needle is inserted just below the superficial layer of the scalp. The needle type may give fewer false waves than the disk type.

Patients who are having an EEG taken must be relaxed, and should keep their eyes closed to reduce external stimuli. They must remain awake and should refrain from any muscular movement.

The EEG has proved valuable for diagnostic purposes. For example, tumor tissue is electrically inactive, but the immediately surrounding tissue may give abnormal brain wave readings. In epilepsy the characteristic wave is a slow three- or four-cycle wave of moderately high amplitude, occurring in a random fashion.

Subdural hematoma and some types of drug poisoning may be diagnosed by electroencephalography. A recent application has been an effort to diagnose mental retardation in infants, and to diagnose migraine headache, by this means.

The electroencephalograph is used in sleep clinics to study patterns of sleep. For this purpose, it is used in conjuction with the electrooculograph (EOG), which measures eye movements, and the electromyograph (EMG), which measures muscle tone.

Investigation of sleep affords us an interesting example of electronics applied to health care; let us take a closer look at this relationship. It has been learned that there are five stages of sleep. One stage is called REM (rapid eye movement). During REM sleep, the eyes move very rapidly under the eyelids; a person awakened from REM sleep always reports that he has been dreaming. It is characterized by heightened EEG activity, fluctuations in pulse, respiration, and blood pressure, as well as increased gastric secretion. Then come stages 1, 2, 3, and 4—quiet or non-REM sleep— with stage 4 being the deepest sleep. Figure 13-9 depicts a typical sleep pattern for a young adult.

It will be noted that there are several short periods of wakefulness, and this is true even of persons who say they slept soundly through the whole night. An actual sleep tracing for one night in a single subject would require about 2,000 feet of tracings. There is a marked decrease in stages 3 and 4 with advancing age, and also a shortening of the total sleep time. You probably know people who wake up at 5 or 6:00 AM and who cannot get back to sleep. This is normal, and it is valuable to know from these studies that such is the case. Research into sleep is important also to reveal the effects of medication upon sleep. It has been found, for instance, that people frequently sleep well after taking a placebo, or "dummy" medication. On the other hand, sleep studies demonstrate that certain medications do definitely improve sleep.

The encephalograph is also used to study patients who manifest various psychic disturbances. You probably have known a few persons who are given to violent "temper tantrums" bordering on rage. Their encephalogram shows a "brainstorm," with exaggerated peaks or spikes. The part of the brain that seems to be involved is the amygdala, an almond-shaped structure in the inner part of the brain. By passing a weak current through wire into the amygdala, the neurosurgeon can stimulate the brain electrically to reproduce such violent anger and rage. One result

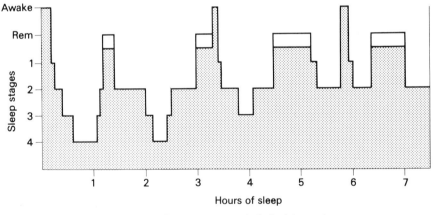

Compact portrayal of all-night tracings

Figure 13-9.
Encephalographic tracings during sleep.

Source: Reprinted with permission of Roche Laboratories, a Division of Hoffman-LaRoche, Inc., Nutley, N.J. 07110.

of these investigations has been a surgical procedure called prefrontal lobotomy, in which a portion of the amygdala is removed with consequent reduction or even complete elimination of violet behavior patterns. It is said that some 600 lobotomies were performed in 1973. This sort of neurosurgery has come under heavy criticism, because it can change the person into a "vegetable," and there is a great deal of discussion about whether it should ever be performed.

ELECTROCARDIOGRAPHY

The heart muscle, like any other muscle, is accompanied by electrical changes in the course of its action. Much can be learned of the general condition of the heart by recording these electrical impulses and analyzing them. A device which "picks up" these currents, amplifies them, and records them is called an *electrocardiograph*. The actual record produced by this device is an *electrocardiogram*.

The action current generated by the heart muscle was first observed in experiments on frog hearts in 1855 by Koelliker and Muller in Germany. Then, in 1877, Waller in England made the first electrocardiogram on a frog, and on a human in 1889.

Very weak electric currents can be measured by a device called a *galvanometer*. In 1903 Einthoven developed a string galvanometer that was sensitive enough to react to the small current produced by the heart. The string is made of a very thin quartz fiber, coated with platinum, silver, or gold to conduct the electric current, and is placed between the two poles of a magnet. You will recall, from the discussion of a motor, that a current-carrying wire moves in a magnetic field. Thus the thin wire would move sideways as the current flowed; as the direction of the current changed, the wire would move in the other direction. You will find it easier to understand the principle of the galvanometer if you hold your thumb and first two fingers in the positions indicated in Figure 13-10.

The magnetic lines of force extend in the direction at which your middle finger points (at right angles to the palm). Imagine your index finger to be the wire suspended between the poles of the magnet. If the current is made to pass through the wire in the direction at which the index finger points, the wire will move in the direction indicated by the thumb (the thumb actually is at right angles to the index finger). This manual analogy is called Fleming's rule, and it illustrates the operation not only of galvanometers but also of electric motors.

The Einthoven galvanometer is the main component of one form of electrocardiograph. The movements of the string, which follow the electrical fluctuations of the heart muscle, are recorded on a paper strip.

A second type of electrocardiograph is known as the amplifier-oscillograph apparatus. The heart-muscle currents are amplified by vacuum tubes or transistors and then are fed into the oscillograph portion, which consists of a pair of wires placed between the poles of a magnet. Attached to the wires is a small mirror. As the wires move, the light beam reflected by the mirror traces a photographic pattern

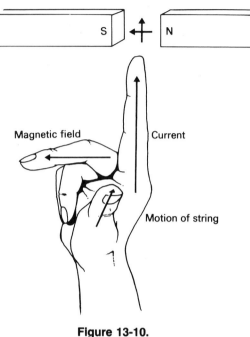

Figure 13-10.
String galvanometer.

on sensitized paper. A more recent variation of this apparatus has a heated stylus attached to the moving wire. The stylus traces the pattern on a piece of plastic-coated paper.

The recordings have very little meaning by themselves; but by comparing tracings of normal hearts with tracings of abnormal ones, these waves can be analyzed. The waves are divided into various segments, each with its identifying letter. Each segment represents a particular phase in the function of the heart (see Fig. 13-11, with the interpretation of the electrocardiogram).

A number of wire connections, also called leads, are used with the electrocardiograph. Lead 1 measures the current between the right and left arm. Lead 2 has the electrodes attached to the right arm and the left leg. Lead 3 reads the left arm and left leg current. These three leads constitute the standard leads. There are also nine other leads which may be used in special cases (Fig. 13-12).

Although the ECG is a very important diagnostic tool, it is not infallible. For example, one study showed that 17 out of 106 patients with triple-vessel coronary artery disease had completely normal resting cardiograms.[3] The investigators therefore stated that the ECG is not a reliable predictor of the extent of coronary disease.

---

[3] A. Benchimol, C. Harris, K. Desser, B. Kwee, and S. Promisloff: Resting electrocardiogram in major coronary artery disease, *JAMA*, June 11, 1973, p. 1489.

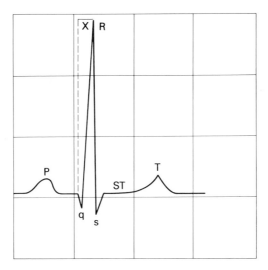

Figure 13-11.

One complete normal cardiac cycle as recorded by the electrocardiograph.

The vertical lines represent time intervals of 0.04 second. The horizontal lines represent voltage differences of 0.1 millivolt and indicate the magnitude of positive and negative deflections. The P wave shows activity of the atria, the QRS complex that of the ventricles. The ST segment is an interval of zero voltage following maximum activation of the heart muscle. As the muscle relaxes, another voltage difference is produced and is represented by the T wave. The time interval between the beginning of the P wave and the R wave is the time of conduction of the beat from atria to ventricles.

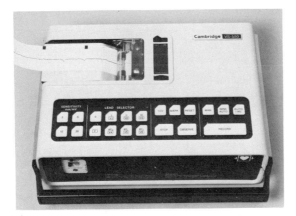

Figure 13-12.

Portable electrocardiograph.

Source: Photograph courtesy of Picker International, Spring Street, Ossining, N. Y. 10562.

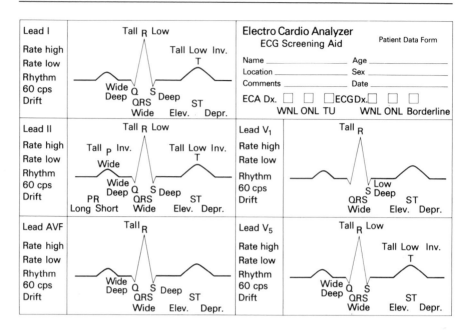

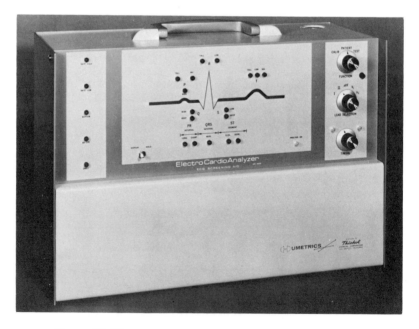

Figure 13-13.

Electrocardioanalyzer.

Source: Courtesy of Humetrics Corporation, 353 North Oak Street, Inglewood, Calif. 90302.

The stress ECG is helpful in detecting latent and stress-dependent cardiac insufficiency using an ergometer as described in an earlier chapter. The use of radioisotopes in determining cardiac function is becoming an important procedure, and is discussed in the next chapter.

To achieve optimum protection of both patient and operator, newer ECG machines have no direct voltage connection between the power line and the instrument. All internal supply voltages are taken from the secondary winding of the safety transformer. Thus, by using two coils of wire that do not touch, necessary current can be induced in the secondary winding without a direct connection to the first. The importance of this improvement is that in the event the insulation wears out, the patient will not be in any danger of injury or death due to electrical shock. The high degree of electrical insulation also keeps the current leakage below 10 microamps, so the operator and the patient should have no fear of any short circuit.

A recent advance combines an ECG machine with a small, portable digital-analog computer. This unit, known as the ElectroCardioAnalyzer, has been developed for mass screening purposes (Fig. 13-13). It is automated, so a person without medical training can operate it. The instrument examines 20 elements of the ECG, using five leads; appropriate lights flash when a reading is abnormal. The operator notes

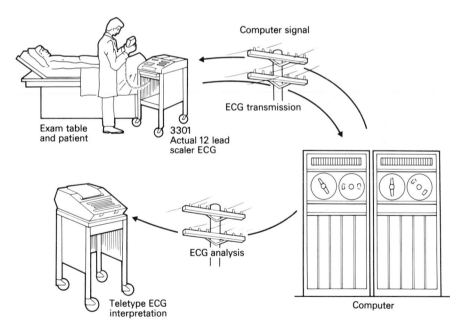

Figure 13-14.
Telephone transmission of ECG data.

Source: Courtesy of Picker International, Spring Street, Ossinging, N.Y. 10562.

which buttons light up, and a cardiologist can then evaluate the data. Such equipment seems to hold promise for improved mass screening, although precise testing with standard equipment must follow.

Another promising development is telemetry, which is the making of measurements at a distance from the subject, with the measurable evidence being transmitted by radio signals. The telemeter will receive up to 20 ECG data from ambulatory patients. A miniature radio transmitter located inside the telemeter sends the signals to a centrally located nurses' station. It was pointed out previously that resting ECGs do not always demonstrate heart abnormalities, so this equipment is helpful only in some cases.

Still another modification involves the use of telephone lines to transmit the ECG from the terminal to a computerized ECG interpretation center. An example is shown in Figure 13-14.

VECTORCARDIOGRAPHY

Vectorcardiography is an extension of electrocardiography. You may recall from our discussion of vectors in Chapter 3 that a scalar quantity gives the magnitude of a force, and a vector also indicates direction. The same type of electrodes are used but in somewhat different positions. The overall electric potential from the heart is measured, and the direction of the vector from point to point is recorded on an oscilloscope that can then be photographed.

The advantage of the VCG is that it gives a better spatial representation and is more reliable for diagnosing difficulties such as right ventricular hypertrophy and right bundle branch block. The disadvantage is that it is more difficult to understand; also it requires additional equipment, including two parallel amplifier channels, a cathode ray tube, and a special camera.

HIGH-FREQUENCY CURRENTS

As was noted earlier, ordinary house current alternates, or reverses its direction of flow, at the rate of 60 cycles per second. If the frequency were increased to, say, 20,000 cycles (i.e., 20 kilohertz) or higher, some of the electrical energy would actually leave the wire and travel through space in the form of radiation known as radio waves.

House current is produced by a mechanical generator of the type previously discussed. Radio waves, also known as high-frequency alternating currents, could conceivably be produced by the same type of generator; but such a method would not be very practical. The generator would have to be rotated at fantastic speeds.

Instead, another device, called an *oscillator,* is used. An oscillator is a vacuum tube or transistor circuit that can produce high-frequency currents up to many thousands of megahertz (a megahertz is 1,000,000 cycles per second).

Besides radio communication, oscillators have a number of other uses. Two medical applications of oscillators are electrocautery and diathermy.

ELECTROSURGERY

The usual electrosurgical apparatus contains an oscillator that produces a high-frequency current of frequencies above 500,000 cycles per second (500 kHz). Two electrodes are connected to the device. The large electrode is placed beneath the patient, and the "active" second electrode is sterilized before use. The purpose of the current is to raise the temperature of a small part of tissue so as to coagulate it, boil the tissue fluids, or char the tissue. The machine may be used in different ways.

Electrocoagulation.   Electrodes are in direct contact, and the heat is generated in the tissue to cause coagulation.

Electrodesiccation.   The active electrode is held at a very short distance from the tissue. This causes a spark to jump from the electrode to the tissue, which in turn causes evaporation of moisture with coagulation, penetrating less deeply than when in contact.

Electrohemostasis.   The bleeders are clamped with a hemostat in the usual manner. The electrode is touched to the clamp, and the dehydration and coagulation take place under the clamp.

Electric Cutting.   The fine tungsten wire makes a minute arc (a continuous spark or tiny flame) just before touching the tissue. This cuts the tissue and seals off the blood vessels. The arc separates the wire from the tissue so that the wire does not actually touch it. That is, the arc does the cutting rather than the wire.

Electrocautery is especially valuable in surgery of the brain, the prostate gland, and the liver.

HIGH-FREQUENCY DIATHERMY

Diathermy is a method of producing heat in the superficial or deep tissues by means of high-frequency electric currents. Heat is generated in the tissues through the process known as *induction;* that is, the proximity of the electrodes of the diathermy apparatus can produce heat in the tissues without actual contact.

An oscillator is used to produce an electromagnetic field fluctuating at a frequency between 500 and 1,000 kHz (although some recent machines use currents up to 2,500 megahertz, called "microwaves"). The part of the body to be treated is placed within the electromagnetic field. As the field changes, the electrons in the tissue cells are pushed first in one direction and then in the other. The continuous "shaking" of the electrons produces heat by increasing the average kinetic energy of the molecules.

Diathermy can be applied to a large area or be localized by wrapping an insulated cable about the part requiring treatment. It is useful in the treatment of such conditions as arthritis, muscle strains, respiratory diseases, various inflammations, and other disorders in which deep heat is of value.

Other modern devices using the principle of induced heat are induction furnaces and "electronic ovens" in which food is cooked very quickly by high-frequency currents. These microwave ovens are being increasingly used in restaurants and in private homes.

BILATERAL IMPEDANCE RHEOGRAPHY

It has been estimated that 50,000 deaths occur annually as a result of pulmonary emboli, with limb-deep venous thrombosis as the major cause. It was pointed out in an earlier chapter that venous thrombosis may occur in as many as 30 percent of patients undergoing major orthopedic or abdominal surgery. Venography or fibrinogen labeled with radioactive iodine (^{125}I) will detect obstructions, even though clinical signs are lacking. However, the pain and/or cost associated with these procedures makes them unsuitable for routine use in all operations.

The word "impedance" indicates a measurement of resistance, and "rheography" is the science of the flow of matter. In this case, then, the flow of blood in the veins is determined by the increase in resistance to a high-frequency current of 100 kHz. Using Ohm's law, $E = IR$, by measuring the current I, and the voltage E, the change in resistance can be found. With occlusion and shutting off of the venous return, pooling takes place in the obstructed portion. The greater volume of blood will produce an increase in conductivity, that is, a decrease in impedance. The change in impedance (ΔR) is divided by the original impedance R_0. In a normal leg, there is a marked change in impedance; if there is a blockage, blood in the leg has pooled, so the change will be less. It has been found that a normal leg will have a change of 0.19 percent, whereas one with a venous occlusion will have a change of *less* than 0.19 percent.

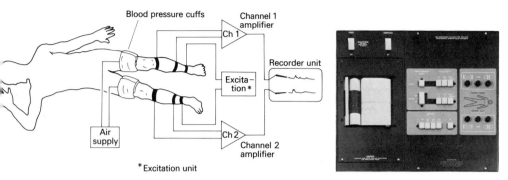

Figure 13-15.
Bilateral impedance rheograph.

Source: Courtesy of Beckman Instruments, Inc., Schiller Park, Ill.
60176.

As seen in Figure 13-15, a pressure cuff is inflated just above the knee, and the impedance is measured bilaterally. The shutting off of the venous return can also be accomplished by the patient's holding the breath with maximum inspiration, or by a Valsalva maneuver, in which inspiration is accompanied by straining of the thoracic and abdominal muscles. It has been found that a mechanical method gives more easily reproducible readings.

Figure 13-15 shows the arrangement. The bands serve as the electrodes; an electrolyte gel is applied to provide good contact.

Here is a description of a case illustrating the use of bilateral impedance rheography. A 56-year-old man developed swelling of the right ankle 10 days following homograft aortic valve replacement. Impedance rheography with the Valsalva maneuver indicated a right limb-deep venous thrombosis. Ascending venography demonstrated a thrombosis of the deep femoral vein. The patient was treated successfully with anticoagulants.[4]

THE ELECTRIC CARDIAC PACEMAKER

The control of the contractions of the heart is found in the sinoauricular node—the pacemaker. Impulses are sent out at the rate of 70 to 80 per minute. Then these impulses are transmitted from the auricular wall to the ventricles. It sometimes

[4]For a good discussion of bilateral impedance rheography see A. Gazzaniga et al. *Arch. Surg.* 104:515–519, 1972.

Rheography

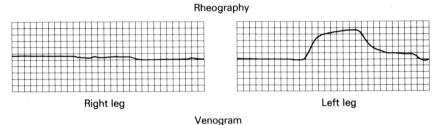

Right leg   Left leg

Venogram

Figure 13-16.
Rheography venogram.
Source: Courtesy of Beckman Instruments, Inc., Schiller Park, Ill.
60176.

happens that there is a disturbance of the normal heartbeat, which results in either arrhythmia or complete heart block. The electric cardiac pacemaker has successfully maintained a normal heartbeat under such conditions.

The electric pacemaker consists of an inlying cardiac catheter inserted through the jugular vein. Electric shocks of 6 milliamps at 2 or 3 volts, each of about 3 millisecond duration, are sent at the rate of 70 per minute into the heart. Normal heartbeats override the electric pulses; but if the heartbeat should falter, the battery signals are sufficient to keep the heart beating.

The first pacemaker was installed in a 38-year-old Swedish engineer, Arne Larsson, in 1958. After more than 20 years he was still alive and very active. The pacemaker has maintained his heartbeat about 400 million times since implantation. In the early models, the stimulator was placed externally, on the patient's chest. Now most are implanted beneath the skin of the chest or abdomen. The Cardiac Pacemakers, Inc., report that in October 11, 1977, 2½-hour-old Leslie Jane Nelson became the youngest person to receive a pacemaker implant and do well. One problem is that children have a much faster pulse rate, so that a pacemaker set for an adult rate of 70 beats per minute would be too slow, and if set at 119 beats per minute, it would be too fast as the child matured. The new Microlith—P can be programmed noninvasively, and the rate can be adjusted periodically to closely duplicate the rate changes of a normal child.

In 1980 it is reported that 60,000 implants were performed, half of them in the United States. Earlier batteries only lasted for about 4 months, and the batteries would have to be changed involving a minor operation. Now these batteries have an expected lifetime of 8 years.

There are two main types of pacemakers. One is called the *fixed rate* type, which produces a constant rate of signals, about 70 to 72 per minute. The other is called the *demand* type, where the unit does not go on when the patient's own heartbeat is sufficiently strong, but takes over when the natural pacemaker of the heart fails.

The nuclear units are powered by 400 mg of the radioactive isotope, plutonium-

238, in the form of an oxide. Radiation from this source produces heat, which in the American models is converted to electricity by means of a thermocouple. These units are designed to last at least 20 years.

A third type is also being developed. It has a rechargeable cadmium battery. A patient can recharge the battery at home by putting on a small vest containing a recharging coil utilizing household current, which recharges the battery through the skin by induction in a few hours once a week. This type should last about 40 years.

Some concern has been expressed as to the danger of electromagnetic radiation in the environment and its effects on pacemaker functioning. The fixed rate type is less susceptible to outside interference, and so is the implanted type as compared to external models. In general, there seems to be only minimal interference from electric shavers and microwave ovens, although a certain amount of care would certainly seem in order. The wearer must always avoid standing close to a microwave oven.

How does one test a pacemaker battery? It is one thing to have your car battery "go dead"—but a patient could not risk having his pacemaker battery "go dead." It is possible to measure the signal emitted by the pacemaker on an oscilloscope both as to form of the wave and also as to frequency. In addition, an x-ray film of the chest can be done to determine whether there is any visible deterioration in the pacemaker. Can a patient at home have his pacemaker checked? Yes, today

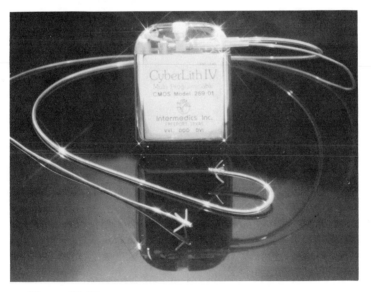

Figure 13-17.
259-01 A-V sequential pacemaker.

Photograph courtesy of Intermedics Inc., P.O. Box 617, Freeport, Tex. 77541.

this is possible. By means of a magnet, the pacemaker is converted to asynchronous mode. The patient then sends the information over a special sensing device by telephone to a hospital where a computer analyzes the functioning of the unit. For example, by 1973 the Beth Israel Medical Center in Newark had performed at least 20,000 such remote checks of pacemakers. Thus we see how electronics has revolutionized medicine.

NEW CLINICAL USES OF ELECTRICITY

The uses of electricity in the health professions seem to be endless, so just a few examples may suggest the scope of these applications. An especially interesting application is bone healing. There is the case of a 14-year-old boy who suffered from congenital pseudarthrosis, which involves bone loss in the weight-bearing bones. Despite his being immobilized for very long periods, the fracture of the tibia did not heal. The decision was made to use electric current. Two small holes were drilled into the tibia, and platinum electrodes were inserted. A current of 3.9 microamps was applied for 125 days. Two D cells (regular flashlight batteries) were the source of current. After four months the bone was healed.[5]

In some instances electrical stimulation can relieve chronic pain. Two types of stimulation are involved: dorsal column and transcutaneous stimulation. The idea behind them is what is called "gate control" theory, that is, by stimulating the nerve fibers it should be possible to disrupt the transmission of pain perception. Several thousand patients have been given this therapy, with reported beneficial results in bursitis, toothache, headache, contusions, acute back and neck strains, broken bones, and ruptured intravertebral disks.

The laser beam is being utilized in electrosurgery, in cases where delicate surgery is required, and where preservation of function is important. The infrared beam can be focused to a spot less than 1 mm in diameter, and will cut or vaporize the tissue without damage to adjacent tissue. It can even cut bone.

QUESTIONS

1. Would eight 25-watt incandescent lamps give more, less, or the same amount of light as one 200-watt lamp? Explain your answer.
2. Would one 40-watt incandescent bulb or one 40-watt fluorescent bulb give more light?

[5]Lavine, Lustrin, Shamos, Rinaldi, and Liboff, *Science,* Vol. 175, 1972.

3. Why is a 10,000-volt static charge usually harmless to the body, while contact with the 120-volt household current may be very damaging?

4. A motor is rated at 2 kilowatts. Assuming a 120-volt line, what is its wattage rating? Suppose the motor operated on 220 volts; what is its amperage?

5. In buying an air-conditioner, why is the BTU/amp ratio important?

6. Would you expect more or less static electricity to be present on a humid day?

7. Using Planck's equation, explain why you would use higher voltage for a chest x-ray film compared with one of a hand.

8. What is the reason for using three-pronged plugs for appliances?

9. What is a transducer? To what form of energy are all the parameters such as temperature and pressure converted?

10. How can the electric encephalograph be used in studying sleep?

11. How does the vectorcardiograph differ from the scalar type of electrocardiograph?

12. Define the words "impedance" and "rheography," and explain how equipment based upon these can be used to detect deep venous thrombosis.

13. The nuclear-powered pacemaker has one important advantage over the usual cadmium battery type. Explain.

14. How is it possible to evaluate the function of a pacemaker while the patient is at home?

15. Assuming a cost of 10 cents per kilowatt-hour, how much would it cost to run two 100-W electric bulbs for 8 hours daily for 30 days?

16. An energy efficiency rating (EER) number is now required to be displayed on a number of appliances. These numbers generally range from 7 to 12, with the higher value more efficient. Explain why you might decide to pay more for a unit with a higher EER rating.

17. How much would it cost per year to use a coffeemaker rated at 800 watts for 1 hour daily? Assume a cost of 12¢/kwh.

CHAPTER 14

NUCLEAR ENERGY
AND NUCLEAR MEDICINE

*The mushroom-shaped cloud is not the sole legacy
of the atomic age. Knowledge that built atomic
bombs can also be used to serve medicine
diagnostically and therapeutically to trace and
destroy malignancies. The field of nuclear medicine
is dynamic; recent procedures—such as neutron
activation analysis and radioimmunoassay—are
constantly developed, tested, and refined in ongoing
research.*

What is meant by "splitting the atom," "radioactivity," "pile," "chain reaction," and "isotope"? Many interested people try to acquire some understanding of these terms by consulting a scientist. Suppose a layman were to ask his scientist friend what "isotope" meant. He would probably be told that an isotope is a different form of the same element in which the atoms have the same atomic number, but a different atomic weight. And to make it clearer, he might be told that the two types of atoms have the same number of protons, but a different number of neutrons. Our intrepid seeker of knowledge would probably say "Oh" and resign himself to an atomic future filled with unintelligible technical terms. Why this block to understanding? One reason is the lack of a technical vocabulary. In the foregoing explanation, one abstract and unknown term after another was used in trying to make clear what "isotope" means. Science is not the only field having its specialized language. Anyone who has tried to explain a baseball game to a foreigner has to a milder degree experienced the same problem. As in all learning of abstract material, we have to build up an understanding of the basic concepts, using as much concrete material as possible.

Let us begin with the most fundamental discussion possible in physics—the nature of matter.

THE STRUCTURE OF MATTER

Through the centuries scholars and scientists have been trying to find the answer to the question: Of what "stuff" is matter composed? As we take any sample of "solid" material and study it, we find that it is not really solid at all; it is made up of many small particles separated from one another. We find that all the matter in the world is made up of only 92 different elements (with the synthetic elements above uranium the number is 105). This is already quite a simplification, when we consider the hundreds of thousands of different substances in the world.

The smallest unit of matter is called the atom. All atoms are made up of the same kinds of particles. The three basic particles of the atom are the proton, the neutron, and the electron.[1] The elements differ from each other only in the *number* of these particles within their atoms.

What are protons, neutrons, and electrons? The proton is a positively charged particle with a mass or weight of one atomic mass unit (amu). The neutron is an electrically neutral particle, also with a mass of one atomic mass unit. The electron is negatively charged, with a mass only 1/1,800th of that of the proton or the neutron.

Particle	Charge	Mass
Proton - p	+	1
Neutron - n	0	1
Electron - e	−	Negligible (1/1,800)

ATOMIC CONFIGURATIONS

How are these atomic particles combined to form the various elements? Almost all the mass is concentrated in a small compact *nucleus* where we find the protons and the neutrons. The electrons are located outside the nucleus, revolving at high velocities. *The number of protons determines the type of element, and is called the atomic number.*

Thus, the hydrogen atom, which is the simplest of the elements, has an atomic number of 1. This means that it has 1 proton. Because the hydrogen atom is electrically neutral, the number of electrons is the same as the number of protons,

---

[1]About 100 different "particles" have been found in the atomic nucleus. Many of these exist for less than one-billionth of a second. They include the meson, neutrino, antineutrino, and positron.

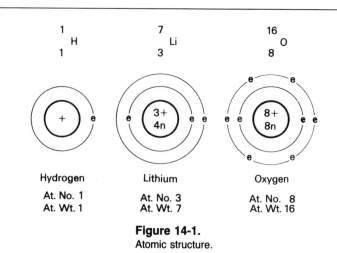

Figure 14-1.
Atomic structure.

that is, the atomic number. Another term used to describe the atom is the *atomic weight*. This gives the weight of the atom; and since the weight is almost completely that of the protons and the neutrons, the atomic weight is the sum of these particles. Diagrams of a few atoms with their configurations are shown in Figure 14-1.

Now we can go back to the original question as to the structure of matter. A seemingly solid piece of iron is found to be composed of atoms of iron with a great deal of space between the atoms. Furthermore, the atoms themselves are not solid spheres but mostly space; there is a small, dense nucleus with some negative particles revolving around it. "But," you might say, "if the atoms can't be seen, how can we now that they are made up mostly of space?" The British scientist Rutherford ran a relatively simple experiment in the late nineteenth century which led to this conclusion. He shot a beam of alpha rays, which are positively charged helium nuclei, at a thin gold foil. The number of particles which came through deflected and undeflected were counted, and the surprising result was that most of the particles were undeflected. This was interpreted to mean that the atom was made up mostly of space, and the deflected particles indicated a positively charged nucleus.

The normal hydrogen atom $^1_1$H has an atomic number of 1 and an atomic weight of 1. This means that is has just 1 proton and 1 electron, but no neutrons. All other elements have an atomic weight greater than the atomic number, and have neutrons in addition to protons and electrons. To repeat, all elements are composed of the same fundamental particles differing only in the number of them.

ISOTOPES

In 1913 the English chemist Soddy, with the Polish chemist Fajans, discovered that the atomic weight of certain samples of lead differed from the figure in the periodic table. For such deviations Soddy coined the name "isotope" from the Greek words *iso* (the same) and *topos* (place). Isotopes, then, are atoms of the same element having *different* weights owing to a different number of neutrons in their nucleus, but occupying the same place in the periodic table because they have an *equal* number of protons.

Now let us take as an illustration an isotope that is of current interest. The terms ^{235}U and ^{238}U are familiar ones, used whenever atomic energy is discussed. What do they mean and what is the difference between them? First of all, they are both uranium atoms; therefore, they must have the same atomic number, that is, the same number of protons; and they are isotopes of each other. The atomic number

I A	II A	III A	IV A	V A	VI A	VII A	VIII A			I B	II B	III B	IV B	V B	VI B	VII B	NOBLE GASES	
1 H 1.0079																1 H 1.0079	2 He 4.00260	2
3 Li 6.941	4 Be 9.01218											5 B 10.81	6 C 12.011	7 N 14.0067	8 O 15.9994†	9 F 18.998403	10 Ne 20.179	2 8
11 Na 22.98977	12 Mg 24.305											13 Al 26.98154	14 Si 28.0855†	15 P 30.97376	16 S 32.06	17 Cl 35.453	18 A 39.948	2 8 8
19 K 39.0983	20 Ca 40.08	21 Sc 44.9559	22 Ti 47.88†	23 V 50.9415	24 Cr 51.996	25 Mn 54.9380	26 Fe 54.847†	27 Co 58.9332	28 Ni 58.69	29 Cu 63.546†	30 Zn 65.38	31 Ga 69.72	32 Ge 72.59†	33 As 74.9216	34 Se 78.96†	35 Br 79.904	36 Kr 83.80	2 8 18 8
37 Rb 85.4678†	38 Sr 87.62	39 Y 88.9059	40 Zr 91.22	41 Nb 92.9064	42 Mo 95.94	43 Tc (98)	44 Ru 101.07†	45 Rh 102.9055	46 Pd 106.42	47 Ag 107.868	48 Cd 112.41	49 In 114.82	50 Sn 118.69†	51 Sb 121.75†	52 Te 127.60†	53 I 126.9045	54 Xe 131.29†	2 8 18 18 8
55 Cs 132.9054	56 Ba 137.33	57 *La 138.9055†	72 Hf 178.49†	73 Ta 180.9479	74 W 183.85†	75 Re 186.207	76 Os 190.2	77 Ir 192.22†	78 Pt 195.08†	79 Au 196.9665	80 Hg 200.59†	81 Tl 204.383	82 Pb 207.2	83 Bi 208.9804	84 Po (209)	85 At (210)	86 Rn (222)	2 8 18 32 18 8
87 Fr (223)	88 Ra 226.0254	89 †Ac 227.0278	104 Unq§ (261)	105 Unp§ (262)	106 Unh§ (263)													

*	58 Ce 140.12	59 Pr 140.9077	60 Nd 144.24†	61 Pm (145)	62 Sm 150.36†	63 Eu 151.96	64 Gd 157.25†	65 Tb 158.9254	66 Dy 162.50†	67 Ho 164.9304	68 Er 167.26†	69 Tm 168.9342	70 Yb 173.04†	71 Lu 174.967†	2 8 18 32 9 2
†	90 Th 232.0381	91 Pa 231.0359	92 U 238.0289	93 Np 237.0482	94 Pu (244)	95 Am (243)	96 Cm (247)	97 Bk (247)	98 Cf (251)	99 Es (252)	100 Fm (257)	101 Md (258)	102 No (259)	103 Lr (260)	2 8 18 32 ? 9 2

Figure 14-2.
Periodic Chart of the Elements.

Source: Reproduced from Periodic Chart of the Elements Fisher Scientific Co.

of uranium is 92; that is, 92 protons in the nucleus. The 235 and 238 are the atomic weights of the two atoms. If the total weight of one of these atoms is 238, and we can account for 92 of that by the weight of the protons, the difference ($238 - 92 = 146$) must be the number of neutrons. Similarly, for ^{235}U, the number of neutrons will be 235 minus 92, or 143. This doesn't seem to be much of a difference; but as it turned out only ^{235}U was usable in the initial atomic bomb as the explosive material. The ^{235}U was derived from ^{238}U at the Oak Ridge, Tennessee, federal facility.

At present there are 81 stable elements, seven radioactive ones found in nature, and 17 prepared artificially, making a total of 105. But isotopes offer a different story. We now have about 280 *stable* isotopes and about 1140 *unstable* isotopes, a total of well over 1,400 atomic species. Unstable isotopes decompose by giving off radiation. These will be discussed later. Thus, there are an average of about 14 isotopes for each element, though isotopes are more numerous among the elements with even atomic numbers, and a few elements do not have any isotopes. Figure 14-2 shows the periodic table of the elements.

RADIOACTIVITY

Until 1896, it was taken for granted by scientists that the elements were composed of atoms that were as solid, immutable, and indestructible as Democritus had said they were more than 2,000 years ago. Roentgen had discovered the year before that cathode rays produced another type of ray when they hit another substance, and that these secondary rays, which he called x-rays, were very penetrating. They were able to pass through material and darken or produce an image on a photographic plate. Within a short time they were employed by doctors in setting fractures and in making diagnoses. In 1896 a French physicist, Becquerel, discovered that a piece of pitchblende, a uranium salt, gave off a ray that penetrated the metal film cover, and fogged a film. A substance that spontaneously emits some kind of radiation is *radioactive*. The discovery of radioactivity destroyed the idea of the unchanging, immutable characteristic of the atom.

The phenomenon of radioactivity excited tremendous interest, and many scientists began to investigate this mysterious occurrence. The well-known story of Pierre and Marie Curie's search for the source of radioactivity exemplifies the long, patient, and arduous labor that usually must precede a great discovery. The Curies found that pitchblende contained small quantities of two elements that were very radioactive. They called these polonium and radium.

The discovery of natural radioactivity raised two important questions: What are the rays composed of, and what happens to an element when it emits these rays?

It was found that there were three types of radiation from a radioactive substance, not one, as previously suggested; these could be separated by passing the rays

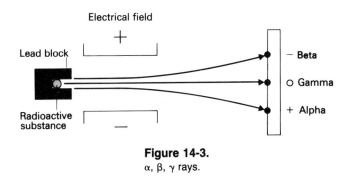

Figure 14-3.
α, β, γ rays.

through an electric or a magnetic field. In an electrical field, one was attracted to the negative pole and another toward the positive pole; the third was not affected at all. These were called respectively, A, B, and C rays. However, because the custom in science is often to use Greek symbols, the Greek letters α *(alpha)*, β *(beta)*, and γ *(gamma)* were used to designate these powerful rays.

Figure 14-3 shows the separation of the three types of radiation.

The main characteristics of alpha, beta, and gamma rays are summarized in the following table.

Radiation	*Charge*	*Description*
alpha	+	helium nuclei (2 protons, 2 neutrons) or ^4_2He
beta	−	electrons (similar to cathode rays)
gamma	0	similar to x-rays—electromagnetic radiation

This leads us to the second question: What takes place in the atom when radiation occurs? It seemed probable that *some* change should occur in the atom if it shoots off particles. It was found that radium, which has an atomic weight of 226, produced a gas called radon. The atomic weight of radon was determined, and found to be 222, four less than radium. The assumption, then, was that radium emits a particle with a number between radium and radon; that is, something having a mass of four units. Now, the helium nucleus has a weight of 4. Therefore, it was necessary to prove whether or not alpha particles were helium nuclei.

Rutherford devised a clever experiment for this purpose (see Fig. 14-4). He made a tube with very thin glass walls so that the alpha rays could penetrate it, and enclosed this in a heavy-walled glass container. The inside tube was filled with radon and then was closed at both ends. All the air was removed from the outside container, which was sealed off by mercury. After several days the mercury level was raised to force any gas present into the top of the outer tube, into which two platinum wires projected.

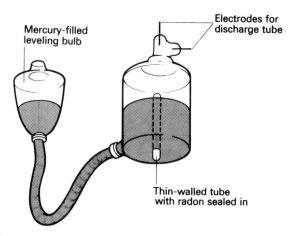

Figure 14-4.
Rutherford's appparatus for demonstrating that alpha particles are
charged helium nuclei.

Source: Drawn from J. Sacks. *The Atom at Work.* Ronald Press, New
York, 1951, p. 39.

When an electric current was made to pass through the gas between the wires,
a discharge occurred; that is, the gas glowed. The light was passed through a
spectroscope, which as you will recall is an instrument that breaks light up into its
various components. Every element has its own characteristic wavelength or color,
and in this experiment it was found that the gas in the outer tube showed the same
color as the inert gas helium. The impact of this experiment was tremendous. It
proved that instead of there being unchanging, immutable, indivisible atoms, here
was an example of a radioactive element that "has the intrinsic property of disintegrating into other elements, and that nothing we can do can in any way affect that
disintegration."[2]

The next discovery was that not only did radium change into radon and helium,
but that radon in turn emitted an alpha particle and became another element, polonium. In fact, there is a whole disintegration series beginning with uranium and
culminating in lead, with radium and radon as only two of the 15 elements involved.

The table below shows the various stages involved in the disintegration of ^{238}U.
Note the term "half-life." This is the time required for *one half* of a given number
of nuclei to change by giving off some type of radiation or particle. The half-life
is constant for a particular nuclear species.

This table looks imposing, but it is relatively easy to figure out what to expect
from a particular radiation. The gamma ray has no weight (unless one considers

[2]J. Sacks: *The Atom at Work,* New York, Ronald, 1951, p. 38.

the energy-mass equivalency, explained later in this chapter) and gamma emission affects neither the atomic number nor the weight. On the other hand, the alpha particle is a helium nucleus, which means it is composed of two protons and two neutrons. As a result, *an alpha emission will lower the atomic number by two units and the atomic weight by four units*.

THE URANIUM DISINTEGRATION SERIES

Element	Atomic Weight	Atomic Number	Half-life	Radiation
Uranium	238	92	4.5 billion years	alpha
Thorium	234	90	24.1 days	beta, gamma
Protactinium	234	91	1.18 min	beta, gamma
Uranium	234	92	250,000 years	alpha
Thorium	230	90	80,000 years	alpha, gamma
Radium	226	88	1622 years	alpha, gamma
Radon	222	86	3.82 days	alpha
Polonium	218	84	3.05 min	alpha, beta
Lead	214	82	26.8 min	beta, gamma
Bismuth	214	83	19.7 min	beta, gamma
Polonium	214	84	0.00016 sec	alpha
Lead	210	82	21 years	beta, gamma
Bismuth	210	83	5.0 days	beta
Polonium	210	84	138 days	alpha
Lead	206	82		

The beta ray is composed of electrons. Here it may be somewhat more difficult to see what effect beta emission will have. In the first place, one might ask, "How can an electron come from the nucleus?" the accepted theory being that the nucleus of any atom (except hydrogen) is composed only of protons and neutrons. One way out of the problem is to consider a neutron to be made up of a proton and an electron. If the neutron then emits (throws off) the electron, the proton would remain. The net result would be that in *beta emission* a neutron would be changed to a proton, which would mean that the *atomic number* (the number of protons) will *increase* by one unit; and since the weight of an electron is negligible, there will be *no change in the atomic weight*.

Let us see whether this checks with the table. Uranium-238 loses an alpha particle. That means that the weight will decrease from 238 to 234. The atomic number of uranium is 92, which should decrease by 2 units to 90. A check on the table shows that this is right. Because the atomic number has changed, it indicates that the uranium has changed to some other element. The element on the periodic chart having an atomic number of 90 is thorium. However, the usual thorium has a weight of 232, so our element is actually an isotope of thorium.

Note also in the table that as the thorium in turn disintegrates, it emits beta rays. Therefore, the atomic number will *increase* by one unit; and, as you can see, the next element in the series, protactinium, has an atomic number of 91. But because at this stage no alpha particles are emitted, the atomic weight remains the same (234).

The reason why an element is radioactive is that it has an unstable nucleus. The half-life is a good measure of its instability. Obviously, an element such as uranium with a half-life measured in billions of years is much more stable than one like polonium, in which the half-life is measured in seconds.

NUCLEAR FISSION AND THE ATOMIC BOMB

Having briefly reviewed the basic concepts in atomic structure, we can consider the phenomenon of nuclear fission. We have seen that some elements spontaneously undergo disintegration of the nucleus and emit radiation. Disintegration also can be produced artificially. In 1919 Rutherford showed that the nucleus of a *stable* atom could be disturbed. He succeeded in changing a few atoms of nitrogen into atoms of oxygen by bombarding them with alpha particles (note also that a hydrogen nucleus is emitted).

$$\frac{4}{2}He + \frac{14}{7}N \rightarrow \frac{17}{8}O + \frac{1}{1}H$$

For centuries the alchemists labored in vain to change a base metal into gold. Here is actual transmutation occurring; one element is changed into another. This has since been done many times to many different elements. Even gold has been made. Unfortuntely, the cost involved is usually high.

In 1932 Chadwick in England found that in some bombardments, another particle was emitted. This was assumed to be a neutron. It has a charge of zero, but a weight of 1, as discussed earlier. In the following reaction

$$\frac{9}{4}Be + \frac{4}{2}He \rightarrow \frac{12}{6}C + \frac{1}{0}N$$

an alpha particle ($\frac{4}{2}He$) plus beryllium produces carbon-12 plus a neutron.

Fermi, an Italian, reasoned that since neutrons are electrically neutral, they should be effective in penetrating nuclei because they would not be diverted by either the positively charged protons or the negatively charged electrons. This proved to be correct. Since 1934 many experiments have been conducted in which radioactive isotopes were produced through neutron bombardment.

It was not until 1939 that the nuclear reactions involving neutron bombardment showed any great difference from other nuclear reactions. The usual neutron-bombardment reactions consisted in shooting a particle into the nucleus and having another knocked out of it. This type of "atom smashing" was very important theoretically, and did produce transmutation, but did not arouse much interest in the general population. However, when nuclear *fission* was put to use in the atomic bomb, the work of the physicists became a matter of interest to the general public. The absorption of a neutron by a uranium nucleus caused that nucleus to split into approximately equal parts with the release of enormous quantities of energy. Furthermore, as a nucleus was split, it emitted more neutrons. This discovery raised the startling possibility of a chain reaction.

THE CHAIN REACTION

It was found that ^{235}U, when hit by a neutron, splits into two nearly equal parts (with atomic weights of approximately 130 and 100). Not only is a great deal of energy liberated during this process, but also up to three neutrons are emitted as well (Fig. 14-5).

It is most important to understand that when one neutron hits the nucleus, two or three neutrons are emitted. Each of these three neutrons may in turn hit a ^{235}U nucleus, releasing three more neutrons. Thus, nine neutrons are emitted, producing nine fissions and releasing 27 more neutrons, and so on (Fig. 14-6).

This is the principle of the chain reaction. Such a reaction might be easier to visualize if you imagine a number of mousetraps, all of them set and each of them arranged so as to hurl three Ping-Pong balls when sprung. If you drop a Ping-Pong ball (a neutron) on a trap (the nucleus), the trap snaps (energy release) and flings three balls into the air. Each of these balls lands on the trigger of another trap. This

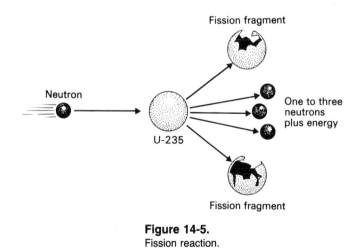

Figure 14-5.
Fission reaction.

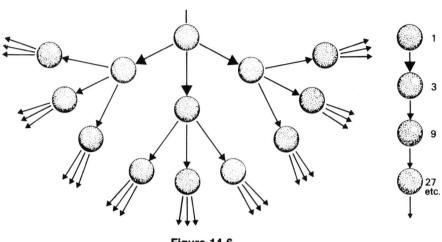

Figure 14-6.
Fission chain reaction.

very lively performance will gather speed and continue until all the traps have been sprung.

It immediately became apparent that, were it possible to achieve a chain reaction, an enormous quantity of energy would be released in the course of the rapidly accelerating reaction within a tiny fraction of a second. In short, a bomb of hitherto unimaginable destructive power became a practical possibility.

ENERGY-MASS EQUIVALENCY

By this time you might be asking a very fundamental question, "Where does this energy come from in the nuclear fission?" First of all, it should be stated that atomic energy is different from all our usual forms of energy production. Burning and other exothermal chemical reactions are the transformation of atoms or compounds into new combinations having a lower energy content, with the differenece in energy being emitted (in the form of heat, light, etc.). Chemical reactions involve changes only in electronic configurations and do *not* change the structure of the nucleus, as is the case in nuclear reactions.

Before we attempt to answer the question of where the energy comes from in nuclear fission, we have to take a look at Einstein's equation, $E = MC^2$.

$$E = \text{energy}$$

$$M = \text{mass}$$

$$C = \text{the velocity of light}$$

Einstein's equation actually means that all matter (i.e., anything having mass) is capable of being transformed into energy, and vice versa. The greater the mass, as the formula shows, the greater the energy. This energy should be considered as inherent within the atom; do not confuse it with the energy figures derived from Newton's laws. For instance, if you had a ball of uranium—say, the size of a tennis ball—and catapulted it at a house, it would acquire a certain amount of kinetic energy and do about as much damage as a heavy stone. But if your ball could be made to fission, releasing the energy contained within its atoms, presumably it could wreck a large town or a small city. Refer back to Einstein's formula and recall the speed of light. This is a huge figure; when squared, it is astronomical. When you multiply the mass of your ball by this figure, it is obvious that inherent within the ball is a fantastically large amount of energy.

When Einstein presented his special theory of relativity in 1905, two other "laws" were believed very reliable. The first was that matter can be transformed, but neither created nor destroyed. The second was that energy can be transformed, but can be neither created nor destroyed. Here Einstein challenged both laws, and said that mass and energy were interchangeable. As Sacks says in his excellent book, *The Atom at Work,* "When Einstein published that equation, it did as much violence to the thinking of physicists and chemists as the atomic bomb did to Hiroshima."[3] It took some time before experiments could be devised to prove or disprove his thesis.

The final acceptance came in 1919 when Einstein calculated the degree to which a star would be seemingly displaced from its true position in the sky when its light traveled past the sun during a total eclipse of the sun. If his theory were correct, since light has energy it should have some mass, and therefore the light beam should be attracted toward the sun by gravitation; and its path thus would be bent. This was the case (see Fig. 14-7).

Thus, by 1920, it was accepted by most physicists that light had mass, and that energy and mass were related. Now we have the answer to the question that was raised earlier, "Where does the energy come from in nuclear fission?" It is the difference in the energy content between the original nucleus and the resulting nuclear fragments. And, since $E = MC^2$, if the products have less energy than the original material, we should expect the product to have less mass also. This is what is found in the fission (splitting) of either ^{235}U or ^{239}Pu. In the fission of ^{235}U, for example, the difference in the weight of the neutron plus ^{235}U minus the sum of the masses of fission fragments is equal to the energy that is given off; that is, the products weigh a little less than the sum of the starting material. It should be noted here that since the mass loss is multiplied by a tremendously large constant, the speed of light, squared $(3 \times 10^{10})^2$, even a very small weight loss will mean a tremendous amount of energy liberated. In order to get a better understanding of

[3]*Ibid.* p. 52.

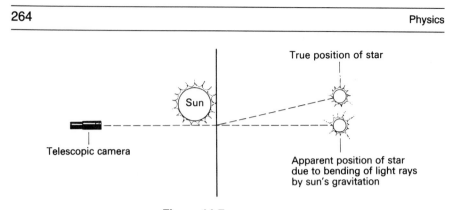

Figure 14-7.
Gravitational bending of light.

Einstein's equation, let us make a calculaton. In the metric system, E is expressed in ergs, M (mass) in grams, and C (the speed of light) as 30,000,000,000 cm/sec. If we assume that only 1 gram (about 1/30th of an ounce) of material is lost, we have:

$$E = 1 \times 30{,}000{,}000{,}000 \times 30{,}000{,}000{,}000$$

$$E = 1 \times 9 \times 10^{20} \text{ ergs (9 with 20 zeros)}$$

An erg is a small unit; but even in kilowatt-hours, it amounts to 25,000,000 KWH. And since one kilowatt-hour will keep a 100-watt electric light bulb burning for 10 hours, this means that 25 million 100-watt bulbs could be operated for 10 hours by the "destruction" of only 1 gram of matter. Atomic—or better, nuclear—energy not only differs in its derivation from familiar energy-yielding chemical reactions, but also is of a different order of magnitude.

THE HYDROGEN BOMB

As we have seen, the A-bomb is based on the principle of fission, the splitting of the uranium nucleus. The hydrogen bomb, on the other hand, employs the principle of *fusion,* which is the combining of hydrogen nuclei to produce helium nuclei with a tremendous release of energy.

The energy of our sun and the other stars is believed to be produced by the same process—fusion of hydrogen atoms to produce helium. Fusion is called a *thermonuclear* reaction because a temperature of millions of degrees is necessary in order that it may take place. To attain a sufficiently high temperature to initiate the reaction on earth, an atom bomb is used as a "match." This is one reason why the hydrogen bomb was developed after the A-bomb.

It was found that normal hydrogen ($_1^1$H) would be too slow, but that heavy hydrogen isotopes, either deuterium ($^2_1$H) or tritium ($^3_1$H) could be used. These isotopes are made in an atomic reactor by neutron bombardment.

Basically, the fusion reaction is

$$^2_1H + {}^2_1H \rightarrow {}^4_2He$$

In this reaction, as in the atom bomb, the product weighs a little less than the starting materials. Thus some mass has been destroyed by being converted into energy.

EFFECTS OF NUCLEAR EXPLOSIONS

The effects of a nuclear explosion can be listed as follows: (1) blast, (2) heat, and (3) radiation.

The radius of destruction of the early A-bombs was only about two miles. The newer H-bombs have a destructive radius of more than 50 miles. This means that in a large metropolitan area possibly 10 to 20 million people might be killed in an attack.

The blast and heat effects of nuclear explosions are similar to those from more conventional bombs, but are of an unprecedented order of magnitude. The radiation effect is twofold: the initial and the secondary.

Initial Radiation. This is the radiation given off by the bomb during the explosion. The gamma rays and the neutrons are the most dangerous due to their great penetration. The danger from the initial radiation is over in a few minutes. A one-shot exposure to 400 roentgens would be fatal to about 50 percent of those who were exposed, while 600 or more roentgens would be fatal to almost 100 percent. (The roentgen is one unit in which radiation is measured. It is the unit of quantity of radiation and is based upon the amount of ionization that it will produce.)

The effect of a one-shot exposure to radiation (say from a bomb) varies directly in severity with the amount of radiation. This can be seen in the following table[4]:

Acute Radiation (Roentgens)	Probable Early Effects
0 to 25	No obvious injury
25 to 50	Possible blood changes; but no serious injury
50 to 100	Blood cell changes; some injury; no disability
100 to 200	Injury, possible disability
200 to 400	Injury and disability certain; death possible
400 plus	Fatal to 50 percent of those exposed
600 or more	Fatal

[4]U.S. Atomic Energy Commission: *The Effects of Atomic Weapons,* Washington, D.C., 1950, p. 342.

Another unit of measurement is the *curie*. It was originally defined as the radiation equal to that from 1 gram of radium. It is now defined as the amount of radioactive material that decays at a rate of 3.7×10^{10} (37,000,000,000) disintegrations per second. These units are discussed more fully in a later section.

When tissues are exposed to an overdose of radiation, the energy of the high-velocity particles or x-rays or both is imparted to the electrons within the irradiated matter. Some electrons are thereby displaced from orbit, causing the atoms to become ions—that is, they acquire an electrical charge. Ionization of the tissues causes extreme disruption of their chemical structure, hence of their normal metabolic processes. Peroxides are formed, resulting in local necrosis as well as general systemic effects. The widespread destruction of lymphoid elements may bring about an unfavorable immune response. Enzymes become deactivated, and the selective permeability of the cell membrane may be lost. Finally, because of the reduction in size of the molecules, osmotic pressure of the tissue fluid surrounding the irradiated cells may increase, causing the cells to burst.

Some types of cells are more susceptible than others to the destructive effects of radiation. The lymphoid elements and the bone marrow are most sensitive in this respect; in fact, damage to the bone marrow is probably the most immediately dangerous radiation hazard. Platelet production falls to the point at which a hemorrhage can be uncontrollable. Production of granulocytes is also inhibited, heightening the victim's susceptiblity to infection.

The tissues which are least vulnerable to radiation effects are the central nervous system and skeletal muscle. Technically, the skin lies somewhere between the two extremes of sensitivity; but because of its exposed state, it usually bears the brunt of the radiation damage. Healing is usually a lengthy and uncertain process, with excessive growth of scar tissue.

According to the law of Bergonié and Tribondeau, the vulnerability of a cell to radiation varies directly with its reproductive capacity. This is the basis of radiation therapy, since malignant cells reproduce more rapidly than normal ones. Unfortunately the difference between the sensitivity of cancer cells and normal tissues is not so great as might be desired. It is often impossible to give sufficient radiation to kill all malignant cells without damaging an excessive number of nonmalignant cells in the surrounding area.

Secondary Radiation.   This is due to the fallout: radioactive fission fragments and the material that has been made radioactive by neutron bombardment from the blast. Here again we have to distinguish between two types of fallout. The *immediate* fallout occurs within a few days, and may extend for several hundred miles downwind. This movement would scatter fairly high concentrations of radioactive debris over the countryside, which would make it necessary to evacuate the area for some weeks until the radiation decreased. The *global* fallout contains the radioactive material that is blown up into the stratosphere. This material circles the globe and takes years to settle out. It is eventually dispersed throughout the world.

The furor some years ago over the fallout from test bombs was due to differences in evaluation of the situation in moral terms, rather than to any great disagreement over the scientific facts. It is important that we attempt to establish a proper perspective on this problem. The two most significant elements in the debris are strontium-90 and cesium-137. Strontium-90 is deposited in the bones and other tissues with a high calcium content. There is a possibility that it might be a carcinogenic substance. Its half-life is 28 years. Cesium-137 may produce mutation of the genes. It has a half-life of 30 years.

THE ATOMIC REACTOR

The atomic reactor, also called the atomic pile, employs controlled fission of ^{235}U or ^{239}Pu for the purpose of generating power. One type of reactor, of graphite, is a large block made up of smaller graphite blocks with many holes through them. The fuel, uranium, is placed in aluminum tubes within the reactor. Here it fissions; and at this point the obvious question to ask is, "Why doesn't it explode?" You will recall from the earlier discussion of the chain reaction that when a nucleus absorbs a neutron, three more neutrons are emitted. Now, if two of these liberated neutrons can be "caught" or absorbed by some other substance before they can smash another atom, the reaction will be steady instead of explosive. One atom will yield one "working" neutron, which in turn splits one atom. In the atomic pile, control rods (usually of cadmium) are inserted through the top to absorb the extra neutrons so that a constant reaction rate can be maintained (Fig. 14-8). Another

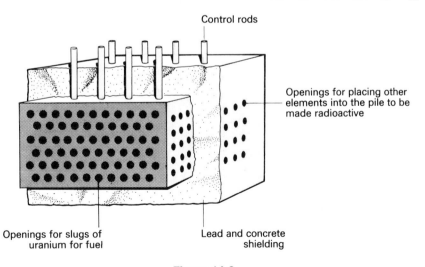

Control rods

Openings for placing other elements into the pile to be made radioactive

Openings for slugs of uranium for fuel

Lead and concrete shielding

Figure 14-8.
Atomic reactor.

reason reactors will not undergo a nuclear explosion is that the ^{235}U percentage is kept low in the range of a few percent compared to that of a bomb.

The heat generated by the atomic reactor can be used to produce power. About 70 stationary nuclear power plants are in service, producing 13% of the country's electricity. Some submarines and surface vessels are also employing atomic power to drive their propulsion machinery.

Another very important use of the atomic reactor is the production of radioactive isotopes. If other elements are placed in the reactor, their nuclei will be hit by neutrons, and they will become radiaoactive. For example,

$$n + {}^{59}_{27}Co \rightarrow {}^{60}_{27}Co \qquad \text{Half-life} = 5.26 \text{ years}$$

Since the neutron is absorbed, there is no change in the atomic number, and the element is still cobalt, but it is one unit heavier. This isotope is radioactive, emitting strong gamma radiation.

The use of the atomic reactor to make radioactive isotopes has made possible the use of radioactive tracers in medicine, science, and industry. A radioactive tracer is simply any chemical element that has been made radioactive. The radiation from such isotopes makes it possible to trace or follow the course of the element through any medium—for instance, the human body, a chemical reaction, or even a washing machine. Some of these applications will be discussed in a moment.

Another very important type of nuclear reactor is the "swimming pool" reactor. This is a much smaller reactor, about 20 feet in diameter and about 20 feet deep. It is filled with deionized water to serve as the moderator, instead of graphite. Many hospitals and research centers now have this type of reactor. It is used to make radioactive isotopes for both diagnostic and therapeutic use. Because the isotopes can be made at the time they are to be used, it is possible to use isotopes of short half-lives, thus reducing the exposure of the patients to excessive radiation. These isotopes are also being made by commercial companies using cyclotrons.

CONTROLLED FUSION

A great deal of research is currently being conducted on the issue of controlling nuclear fusion. As noted earlier, the fusion of deuterium and tritium isotopes of hydrogen requires a temperature of about 10 million degrees Celsius in order to begin the reaction. Once started, the reaction is greatly exothermic, and the energy liberated is much greater than that from the fission reaction.

One of the methods being tested is the use of high-energy lasers to compress glass beads containing the deuterium and tritium gases so as to reach a temperature where the reaction will be initiated. The question is, "How do you contain the hot plasma at 10 million degrees Celsius?" The plan is to use a wall-less container, a "magnetic bottle," using a strong magnetic field to contain the plasma.

If the research is successful scientifically and turns out to be economically feasible, this controlled fusion source should provide a tremendous source of energy without the problem of fission fragments. This prospect, however, seems to be at least 20 to 40 years in the future.

RADIATION UNITS

There are several units used to describe ionizing radiation. One is the *curie,* defined as the amount of radioactive material that decays at a rate of 3.7×10^{10} disintegrations per second (dps). A millicurie (mCi) is therefore 1/1000 of a curie, or 3.7×10^7 dps. and a microcurie (μCi) is 3.7×10^4 dps. Most tracer doses are in the microcurie or low millicurie range.

The SI system has added a new unit, the *Becquerel,* which is defined as one nuclear disintegration per second, that is, 1 dps. Because 1 curie is equal to 3.7×10^{10} dps, and a Bequerel is 1 dps, 1 Ci $= 3.7 \times 10^{10}$ Bq.

The curie, millicurie, and microcurie are units of activity, that is, how much radiation is emitted. It is not a direct measurement of how much energy is absorbed by some material.

Another unit is the *roentgen* (r). This is defined as the amount of x or gamma radiation that will produce an ionization charge of 0.000258 coulombs per kilogram of dry air at standard temperature and pressure. This is also a unit indicating the amount of radiation being emitted. A milliroentgen is, of course one-thousandth of a roentgen.

The *rad* is a unit of absorbed radiation. As a mnemonic device, one often uses the phrase, roentgen absorbed dose, but the term is not an abbreviation. It is defined as the deposition of 0.01 joules of energy per kilogram of absorbing material. The SI has also come up with a new unit here, the *Gray* (Gy), which is equal to 1 joule of absorbed energy per kilogram of tissue. Because 1 rad would deposit 0.01 joules of energy per kilogram of tissue, and the Gray will deposit 1.00 joule per kilogram of tissue, one Gray is equal to 100 rads.

The *rem* is also a unit of absorbed dose of any radiation that has the same effect as a rad of "standard" x-rays. It is often remembered as "radiation effect, man" or "radiation equivalent man."

Relative Biological Effectiveness (RBE).   This is a factor that compares, for the same energy absorbed per unit mass, the biological effect of various types of radiation.

Radiation	RBE
χ, γ, or β	1
Thermal neutrons	5
Fast neutrons	10
α	20

A term very similar to RBE is the quality factor, or QF. It will be seen above that the heavier particles cause more ionization. The *rem* is therefore equal to the number of rads times the RBE.

$$\text{rem} = \text{rads} \times \text{RBE}$$

COMPARISON OF RADIATION SOURCES

In order to evaluate the dangers from various sources of radiation, it is important to get some "feel" for the relative order of magnitude of these sources. The *background* radiation from external radiation such as cosmic and gamma rays is about 50 mrads per year. From sources within the body, the K-40 (radioactive potassium) supplies about 40 mrads, the radium about 20 mrads, and C-14 and radon about 4 mrads, a total of 64 mrads internally. The total background radiation is thus in the order of 110 to 150 mrads per year.

By contrast, a luminous watch may give 20 to 40 mr. Mr is the abbreviation for milliroentgen (1/1,000 roentgen). Living in a brick house subjects a person to about 40 mr, compared with 10 mr in a wooden house. Living at an altitude of 6,000 feet gives us 30 mr above the amount at sea level.

Figure 14-9 summarizes some of the radiation sources. Note that living outside of an operating nuclear reactor exposes a person to an average less than 1 millirem or millirad per year.

Radiation exposure for medical purposes is about 60 mrems per year for diagnostic work and 10 mrems per year for therapy. These figures seem to be going up steadily in this country, and are much higher than those of other countries. Medical exposures could be reduced with better equipment and care in its use. For example, the use of proper cones for x-ray machines will reduce the scattering to other parts of the patient's body and also to the technician.

The National Committee on Radiation Protection recommends the maximum dose for a person not involved in work connected with radiation to be $5 \times (\text{age} - 18)$ rems. This formula is used to calculate the *cumulative life dose*, which should not exceed three rems per 13-week period. A person 50 years of age could then receive $5 \times (50 - 18) = 5 \times 32 = 160$ rems. For a 20-year-old person, the total dose

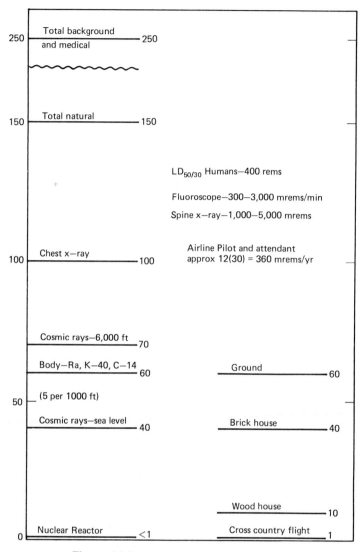

Figure 14-9.
Approximate radiation levels per year (mrems).

would be $5 \times (20 - 18)$ or 10 rems. Lower doses are recommended for young people because genetic damage in this age group has more serious consequences.

The above figures apply to whole-body radiations. For individual organs, the dose may be higher. For example, a spine x-ray may produce 1,000 to 5,000 mr or 1 to 5r, while a fluoroscopic examination may give 300 to 3,000 mr per minute. The time and intensity of x-rays should be reduced to the minimum required.

MANAGING RADIOACTIVE MATERIALS
IN THE HOSPITAL

The dangers to which the nurse, technician, and physician are exposed in using radioactive material come from two sources. The first is the radiation emanating from some radioactive isotope, in the room or within the patient. Since alpha and weak beta radiation have very low penetrating ability, usually less than 3 millimeters (a layer of rubber or several layers of cloth will stop them entirely), they usually will not be a danger as far as direct radiation from the patient is concerned. On the other hand, gamma rays are very penetrating and are dangerous.

The second source of radiation danger is less direct and obvious than the first: contaminated articles and radioactive debris. Radioactive dust may be ingested or contaminate the skin. In either case, both alpha and beta radiations are a menace. Always remember that a radioactive substance within the body is infinitely more dangerous than an equal amount of the substance outside the body. Also, never forget that the effects of exposure to radiation are cumulative.

When any radioactive substance is handled, the following precautions should be closely followed:

1. Remember the Inverse Square Law; that is, the radiation intensity varies inversely as the square of the distance. This means that if the distance to the source is doubled, the amount of radiation will be only one quarter as much. In order to keep the distance as great as possible, tongs should be used when "hot" material is handled.
2. The "dose" of radiation received varies directly with the time of exposure. Therefore, keep the total time of close contact with the patient to a minimum.
3. Shielding reduces the amount of radiation received. Appropriate shields should be employed when radon seeds, x-rays, and certain other sources of radiation are used. Such shields are usually made of lead.
4. Avoid spilling any radioactive liquid. The body should be washed thoroughly to remove any radioactive contamination. Maintain careful cleanliness to help to keep the amount of radioactive dust at a miniumum.
5. Anything used by the patient who is undergoing radioisotope therapy should be touched only with rubber gloves. The radiotherapist will have special instructions for the deposition of used linens, the contents of bedpans, and so forth. It is important to know which isotope is being used, the type of radiation it emits, the route by which it is eliminated, and its half-life, because these all influence the specific precautionary measures.

6. Use film badges, dosimeters, or Geiger counters to check the radiation level of the working areas and of the personnel involved.

7. A pregnant woman should never expose herself to any type of radiation unless conditions are such that the gains will outweigh the risks.

These precautions do not have to be followed in the case of a single patient undergoing radioactive tracer tests; in this case, the radiation is so slight as to be negligible. However, anyone who works with radioactive tracers in large quantities, or routinely encounters these substances, should remember the cumulative effects of radioactivity and take appropriate precautionary measures.

NUCLEAR DETECTION DEVICES

Because alpha, beta, and gamma radiations are invisible, some type of device is required to detect them. The earliest was the spinthariscope, a small container holding a chemical that would give off a flash of light when hit by a particle. This required that the eye be adapted to the dark, and was limited by how fast the person could count these flashes.

The spinthariscope was followed by the Wilson cloud chamber, a container filled with a supercooled vapor, usually alcohol. As the α or β particle goes through the chamber, it may hit a molecule of the vapor and cause ionization. This ionized particle will in turn serve as a center for condensation, and a droplet will form. Thus we do not see the particle itself, but the path it followed. Newer equipment uses liquid helium to produce the paths, which can then be photographed.

Then came the Geiger-Müller tube, popularly known as the Geiger counter, which is a tube usually about 2 inches in diameter and eight inches long. It contains a positive center wire and a negative foil inside the wall. A direct current with a potential of 1,000 to 1,500 volts is applied. The voltage can be adjusted to give the best counts on a plateau where a change in voltage will produce the least effect on the count rate. A thin mica window, through which the radiation passes, is found at either the end or the side of the tube. As radiation enters, the gas inside the tube will become ionized, producing positive and negative ions. This causes a current to flow through the tube, and the scaler or rate meter records the count. The current can also be made to produce a "click" in a pair of headphones or a loudspeaker (Fig. 14-10).

The Geiger-Müller tube is best suited for counting betas, since the heavy alphas will generally be stopped by the window, and most of the gammas will go through the tube without colliding with the gas; thus the equipment has a low efficiency. These machines are capable of counts up to 20,000 per minute without appreciable losses. This is quite a step from counting by eye.

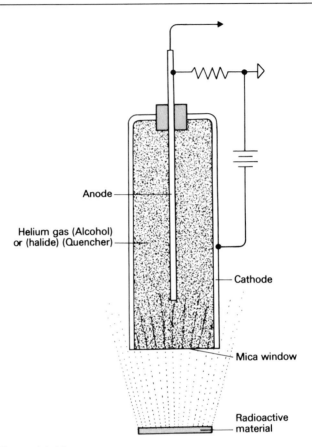

Figure 14-10.
Geiger-Müller tube.

Source Adapted from J. T. Jensen and W. P. Ferren. *College General Chemistry*. Charles E. Merrill Publishing Company, Columbus, Ohio, 1971, p. 89.

To count gammas, a solid scintillation counter is used. This is a crystal of sodium iodide activated by a very small amount of thallium. Since the crystal is much denser than the gas in the Geiger-Müller tube, it has a higher probability of stopping the gammas. The radiation produces a flash of light that is picked up by an attached photomultiplier tube. This light causes an electron to be released from the photo-cathode. This electron in turn hits another cathode, emitting several electrons. This is repeated with about 10 cathodes, producing about 1,000,000 electrons. This current can then be counted, and the energy of the gamma determined (Fig. 14-11).

Another important detection device is the liquid scintillation counter. This is similar to the solid scintillation counter, but it uses a liquid that will scintillate, that is, will produce light. Because the radioactive material is dissolved in the solvent and the scintillation liquid (also called a fluor), the probability of catching the light is very high. This counter is used for weak beta emitters such as tritium and carbon-14, both very often used in biological work as tracers.

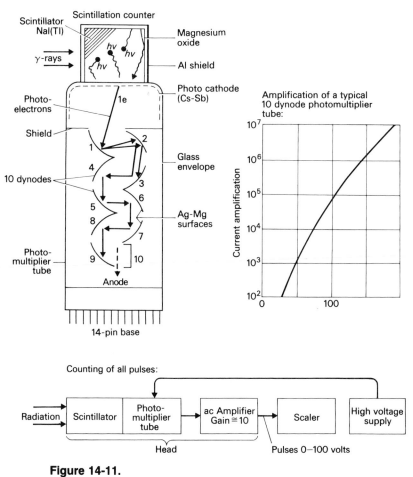

Figure 14-11.
Solid scintillation counter.

Source: Adapted from J. T. Jensen and W. P. Ferren, *College General Chemistry*. Charles E. Merrill Publishing Company, Columbus, Ohio, 1971, p. 90.

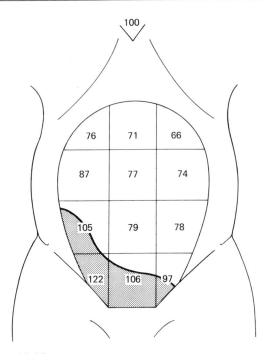

Figure 14-12.
Manual isotope scanning.

Isotope localization of the placenta at 34 weeks gestation in a patient
with third trimester bleeding. The average figure was 94; therefore
the significant counts were 105, 122, 106, and 97. Interpretation: The
placenta was localized on the right lateral and anterior wall of the
lower segment of the uterus (shaded area). Patient was delivered
by cesarean section at 38 weeks and total placenta previa was found,
as depicted in the sketch. This study was made using a dosage of
10 microcuries of radioiodinated ^{131}I serum albumin (human). The
recommended dosage for this purpose is 3 to 5 microcuries, and the
investigators report a subsequent reduction in dosage to 5 microcuries without any appreciable change in the accuracy of the method.

Source: Courtesy of E. R. Squibb & Sons, Inc., P.O. Box 4000,
Princeton, N.J. 08540

SCANNING CAMERAS

To count the gamma radiation emanating from a person who has been given a
radionuclide, solid scintillation crystals are used. If it is desired not only to measure
the total count, but to "map" the counts, to give an outline of the "hot" and "cold"
areas, then a number of counts must be made. Before looking at two methods of
doing this instrumentally, we shall take a look at a simpler procedure that will
illustrate the method.

To determine the location of the placenta in cases of antepartum bleeding, either ^{131}I or ^{51}Cr can be used as a tracer, incorporated into human serum albumin. In general, isotopic methods of determining the location of the placenta are based on the concept that the placenta contains a large pool of maternal blood; when the maternal circulation is tagged with a suitable radioactive isotope the location of the placenta will make itself evident as a local accumulation of the tracer, and can be outlined by means of radioisotope techniques. Figure 14-12 shows how even 12 readings can be significant.

The procedure just described for "mapping" the radioisotope activity involved a single crystal and was repeated 12 times. In order to obtain much greater detail and increase the speed, two more sophisticated devices were developed. One is the *rectilinear scanning camera*. This apparatus moves back and forth over the organ being tested, recording the counts per minute in each position. This is in a sense a mechanized form of the earlier manual method, that was described above. The next step was to couple this type of scanner with a computer using magnetic tape, which prints out the counts as dots, so that a picture of the organ is available in either black and white or color (Fig. 14-13).

Rectilinear scanners still take a fairly long time to scan the entire object, about 30 minutes for a complete organ scan. To increase the speed, another type of camera called the *Anger camera* was developed. This uses a thin sodium iodide crystal about 15 inches in diameter, one-half inch thick, backed by about 37 photomultiplier tubes. The gamma radiation from the organ passes through a number of openings in a lead collimator and hits the scintillation crystal; the light is picked up by one of the photomultiplier tubes. Thus in one exposure it is possible to record the

Figure 14-13.
Computerized scanning gamma camera system.
Source: Photograph courtesy of the Baird Corporation.

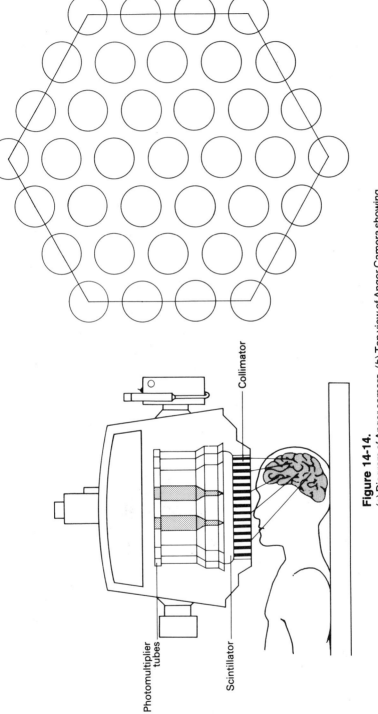

Figure 14-14.

(a) Diagram of Anger camera. (b) Top view of Anger Camera showing arrangement of 37 photomultiplier tubes.

The gamma rays from the injected radioisotope pass through collimators immediately above the patient.

Source: Reprinted by permission from D. E. Tilley, and W. Thumm. *Physics for College Students.* Cummings Publishing Company, Menlo Park, Calif., 1974, p. 33 and 34.

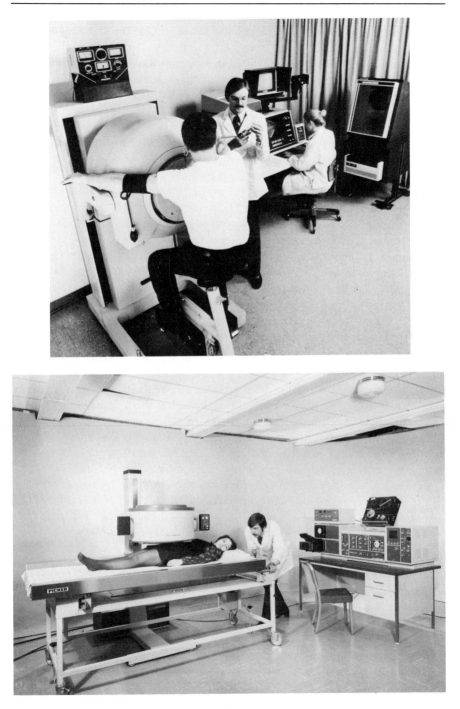

Figure 14-15 (a) and (b).
Anger camera in use.

Source: Photograph courtesy of Picker International.

radiation coming from various positions on the organ, and the time required can be reduced to a minute or two. Figure 14-14(a) shows the diagram of the Anger camera and Figure 14-14(b) shows the arrangement of the photomultiplier tubes.

Figures 14-15 (a) and (b) show an Anger camera in use. This is the Picker Dyna™ Camera 2 C coupled to a computer.

In addition to single organ counters, *whole body counters* have also been developed. Some of these use liquid scintillation counting. The person is placed in a large container filled with the scintillating solution, and the light produced by the radiation from the body is detected by 108 photomultiplier tubes. While the person is being tested, the whole unit slides into a 20-ton shield of 5-inch lead.

The body contains about 70 grams potassium. Radioactive ^{40}K makes up only 0.0119 percent of the potassium. Only 11 percent of the ^{40}K gives off high-energy gammas that can be detected with a whole-body counter, but this amount can be detected quite well. It has been found that fatty tissues have a low potassium concentration and muscles a higher level. Thus it is possible to measure indirectly the amount of fat in a person. Significant decreases in potassium content have been found in patients with diabetic coma, muscular dystrophy, and myotonia atrophica.

HALF-LIFE

Before discussing various types of radioactive tracers, we should mention the various types of half-lives. The *physical* half-life is defined as the time required for one-half of any number of nuclei of a radioactive species to undergo decay. Thus, if it takes one hour for 500 disintegrations out of 1,000 atoms, it will also take one hour for 250 of the 500 to decay, and one hour for 125 out of the 250. This occurs because the rate of disintegration is proportional to the number of nuclei present.

Another half-life important in biological tracer studies is the *biological* half-life. This is the approximate time required for one-half of the radioactive material to be eliminated from the body. For example, the biological half-life for deuterium ($^{2}_{1}$H) is about 10 days, while that of the heavier tritium ($^{3}_{1}$H) is about 19 days.

The *effective* half-life is a combination of the physical (T_p) and the biological (T_b), according to the equation

$$\frac{1}{T_e} = \frac{1}{T_p} + \frac{1}{T_b} \quad \text{or} \quad T_e = \frac{T_p \times T_b}{T_p + T_b}$$

NUCLEAR MEDICINE

In Chapter 13 we described how the rapid development of electronics applied to medicine was associated with new instruments of great value to medical practice.

Similarly, the last two decades have seen a quiet revolution in nuclear medicine—the application of radioactive materials to diagnosis, therapy, and research. The availability of sensitive electronic equipment for measuring radioactivity, coupled with the production of short-lived isotopes, has made possible new tests and treatments that afford maximum information with minimum risk and discomfort to the patient.

The radioactive isotope, or radionuclide, acts chemically like the stable isotope, but emits some type of radiation such as alpha, beta, or gamma rays or a combination thereof. Isotopes have been called "miniature radio stations," and serve as "tags" of chemical substances that are injected or ingested. Thus it is possible not only to follow the *path* of a certain chemical through the body, but also to determine the *rate* at which it is absorbed.

There are two main uses for radionuclides: diagnosis and therapy.

THERAPEUTIC NUCLEAR MEDICINE

Therapeutic nuclear medicine uses the tissue-destroying or activity-suppressing property of specific doses of implanted or injected radioisotopes. Although their effect is similar to that of x-ray therapy, radioisotopes can sometimes be made to irradiate more selectively.[5] For example, sodium iodide—^{131}I—allows the destruction of thyroid tissue by emanating radiation from within the thyroid gland, because of the selective concentration of the iodine. The radiation effect is primarily that of the beta particles, which have a maximum path of 2 to 3 mm in tissue, thereby limiting the damage to surrounding tissue. This radioisotope has been used since the early 1940s to treat hyperthyroidism. It is also used after thyroidectomy to destroy any residual thyroid tissue and to treat any metastases.

Another isotope used therapeutically is gold-198, which is produced from the naturally occurring stable isotope of gold (^{197}Au) by neutron capture. The ^{198}Au isotope has a half-life of 2.7 days and emits primarily a 0.959 MeV beta and a 0.412 MeV gamma. Most of the radiation dose to the body is from the beta, which penetrates an average distance of 0.4-mm tissue and a maximum distance of less than 4 mm. The gamma rays contribute 6 to 10 percent of the total dose. Gold-198 is used in the treatment of soft tissue tumors and of carcinoma of the prostate and bladder. Its major therapeutic action is in controlling effusions in the pleural and peritoneal cavities caused by cancer. The gold is in the metallic form, as a red colloidal solution, and is insoluble and chemically inert. The radioactive gold is absorbed on the cavity surfaces, and there the short-range beta emissions irritate the lining. The usual dose is 35 to 150 mCi for intraperitoneal administration and 25 to 100 mCi for intrapleural administration. This treatment is not a cure, but rather a palliative measure to reduce symptoms (Fig. 14-16).

Phosphorus-32 is also used in therapy. It has a half-life of 14.3 days and emits

[5]Moore, L. L.: Nuclear medicine. *Medical Electronics & Equipment News* p. 5, Sept. 1972.

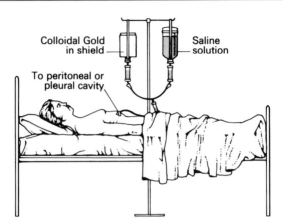

Figure 14-16.
Radioactive gold-198 for intracavity use in metastasized cancer.

Source: B. Duffy. Atomic energy in the diagnosis and treatment of malignant diseases. *Am. J. Nurs.* 55:434, and U.S. Atomic Energy Commission.

beta rays. It is deposited in the bone marrow and in the nuclei of rapidly multiplying cells. It is effective in the treatment of polycythemia vera (overproduction of red blood cells) and also in some cases of chronic leukemia. Phosphorus-32 is also used to locate cerebral and ocular tumors, and in the study of the metabolism of bones and teeth. It is supplied as an aqueous solution of sodium phosphate with sodium chloride for isotonicity and can be administered either orally or by injection.

DIAGNOSTIC NUCLEAR MEDICINE

There has been an explosive growth in the use of radioisotopes in medicine during the last decade. Practically every hospital in the country now uses some radioisotope procedures, and one-third of the patients in a general hospital undergo some type of nuclear test. One reason is that the in vivo procedures are often able to show cases in which there is malfunction of an organ, and the in vitro procedures such as in radioimmunoassays make it possible to detect substances down to part per billion. RIA procedures are discussed in the next chapter.

The radioisotope imaging devices can be contrasted with conventional x-ray procedures. In the latter, the x-ray beam is shot through the patient's organ, producing an image on the film showing the various densities of the bones and tissues. The radioisotope procedures, on the other hand, usually involve the injection of a small dose of a radionuclide into the blood stream, and the gamma rays that are *emitted* will be detected by the camera.

The radioactive element can be used by itself, such as iodine-131, -125, and -123, but the radionuclides are usually bonded to a compound that will be selectively absorbed by a particular organ. This makes it possible for a nuclide such as technetium-99m to be used in so many procedures such as brain, lung, liver, kidney, and bone scans and also in blood-pool imaging. This element meets the three requirements for a good radionuclide for medical purposes: (1) short half-life, (2) gamma energies below 200 keV, and (3) preferably no particle emission.

Technetium-99m, with its 6-hour half-life, would be too short-lived to be delivered from a supplier. However, using a technique called the "radioactive cow," or column generator, this is easily done in the hospital laboratory.

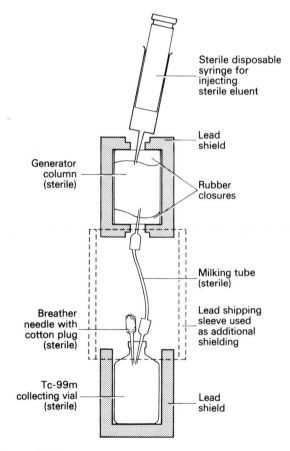

Figure 14-17.
Diagram of technetium cow.

Source: Reprinted by permission from D. E. Tilley and W. Thumm. *Physics for College Students.* Cummings Publishing Company, Menlo Park, Calif., 1974, p. 34.

In this procedure, a longer-lived isotope is eluted, or separated, with a solvent in a chromatographic column; the offspring, which is short-lived, is then drained off for use. The parent isotope will remain on the column, and as it decays, it will continue to produce more of the short-lived isotope. Thus, by successive elutions it is possible to "milk" the generator, or "cow." Figure 14-17 shows how the eluting liquid is passed through the column containing molybdenum-99 adsorbed in alumina. The ^{99m}Tc will collect at the bottom of the column. Figure 14-18 shows a generator with the lead shielding assembly.

Using the concepts described earlier in this chapter, let us try to follow the decay of the molybdenum-99. The $^{99}_{42}$Mo is found in fission fragment or is made by neutron bombardment of Mo-98. The $^{99}_{42}$Mo emits beta particles of various energies. The loss of a beta will *increase* the atomic number by one, thus producing $^{99m}_{43}$Tc.

Molybdenum-99 has a half-life of 67 hours; the hospital will have to secure a new "cow" every week or so since in 2.8 days the activity will be reduced by one-half. The buildup of technetium-99m reaches a maximum after 23 hours, so the cow will give maximum activity if milked every 23 to 24 hours. Figure 14-19 illustrates this activity graphically.

It would be impossible to obtain such short-lived isotopes from a commercial source—they would have decayed almost completely by the time they were delivered to the laboratory. Using a "cow" with a long-lived parent assures a steady supply of the short-lived isotope. Were it not for this type of generator, only hospitals that owned nuclear reactors or generators would be able to use such short-lived isotopes.

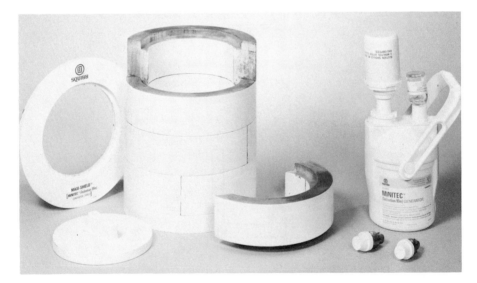

Figure 14-18.
Technotope II technetium-99m generator.

Source: Photograph courtesy of E. R. Squibb & Sons, Inc., P.O. Box 4000, Princeton, N.J. 08540.

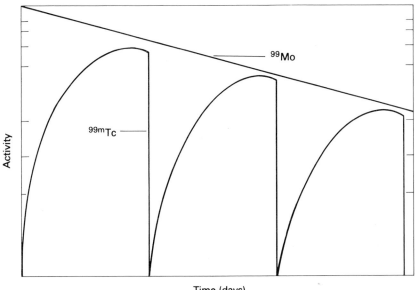

Figure 14-19.
^{99}Mo decay and ^{99m}Tc growth after daily elutions.
Source: Courtesy of E. R. Squibb & Sons, Inc., P.O. Box 4000, Princeton, N.J. 08540.

Now that we have described the production of ^{99m}Tc, let us see how this isotope is put to the service of medicine. The solution is in the form of sodium pertechnetate, $Na^{99m}TcO_4$, and can be administered either intravenously or orally. Its most important application is in scintillation scanning of the brain, scanning of the thyroid and salivary glands, localization of the placenta prior to delivery, and scanning of other blood pools. Brain scans of excellent quality demonstrating midline as well as lateral lesions have been obtained within 10 to 15 minutes after intravenous injection. Localization in the tumor appears to be almost immediate. A dose of 10 mCi is often used, and since it may be taken up by the fetus and excreted in human milk, it should not be administered to pregnant or lactating women.

DECAY SCHEMES

Each radioactive element decays in a specific manner, and a diagram showing the mechanism is called the *decay scheme* for that radionuclide. For example, iodine-131 emits a B⁻ 93% of the time. Losing a B⁻ will *increase* the atomic number Z by 1. By convention, this is indicated by an arrow slanted to the right. If a positron, B⁺, is emitted, the Z value will *decrease* by 1, and the arrow will slant to the left.

Similarly, if the nucleus captures an electron from the K shell of the extranuclear electrons, the negative electron will neutralize a proton in the nucleus, and the Z value will *decrease,* and the arrow will slant to the left in cases of EC.

Let us draw the decay scheme for iodine-131, a B⁻ emitter.

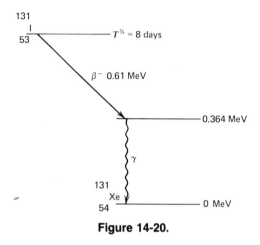

Figure 14-20.

Actually, this is a great oversimplification, since ¹³¹I emits six different betas of different energies and percentages and 14 different gammas. However, Figure 14-20 indicates the most abundant transition.

Gallium-68 is a positron emitter.

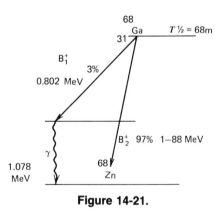

Figure 14-21.

You will notice that 97% of the nuclei emit positrons (B⁺) of 1.880 MeV energy, and 3% emit B⁺ of 0.802 MeV followed by a 1.078 gamma, in both cases producing stable ⁶⁸Zn.

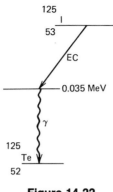

Figure 14-22.

Iodine-125 exhibits electron capture (EC).

The energy of the emitted gamma is 0.035 MeV or 35 keV. In nuclear medicine, the keV units are usually used.

For thyroid uptake studies, the activity of the injected solution is in the 50- to 200-microcurie range.

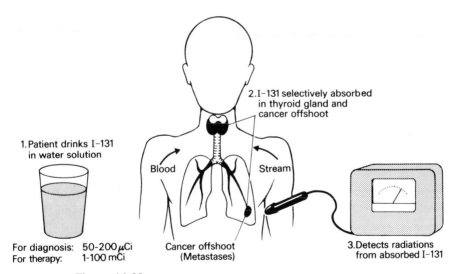

Figure 14-23.
Radioactive iodine-131 for diagnosing and treating thyroid gland disorders.

Source: B. Duffy. Atomic energy in the diagnosis and treatment of malignant diseases. *Am. J. Nurs.* 55:434, and U.S. Atomic Energy Commission.

The normal thyroid picks up about 30 percent of the tracer dose, 60 percent is excreted in the urine, and 10 percent is distributed throughout the body. In hyperthyroidism, the gland takes up on an average 70 percent, and the urine excretes as little as 10 percent of the tracer dose. Iodine-131 is also used to detect thyroid metastases. As noted above, some use is also being made of it in treating cancer of the thyroid. For therapy, dosages in the millicurie range are employed.

BRAIN SCANS

Iodine-131 or technetium-99m-labeled human serum albumin and technetium-99m-labeled glucoheptonate can be used in locating brain tumor. Figure 14-24 shows the use of a rate meter to measure the radiation. This is now done by using the Anger camera to show areas of increased activity.

BONE SCANS

Bone scans have used strontium-85, strontium-87m, and fluorine-18. The most commonly used radionuclide is currently technetium-99m, joined to polyphosphate,

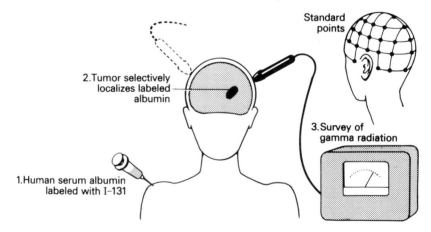

Figure 14-24.
Locating brain tumors with iodine-131-tagged human serum albumin.

Source: B. Duffy. Atomic energy in the diagnosis and treatment of malignant diseases. *Am. J. Nurs.* 55:434, and U.S. Atomic Energy Commission.

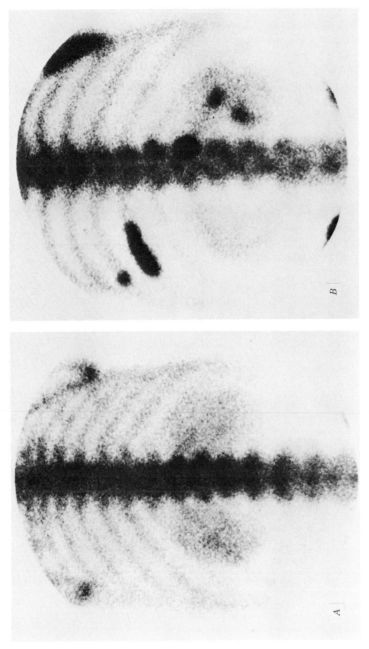

Figure 14-25.
Source: Photographs courtesy of Orlando Manfredi, M.D., Director of Radiology and Nuclear Medicine, St. Vincent's Medical Center of Richmond, N.Y.

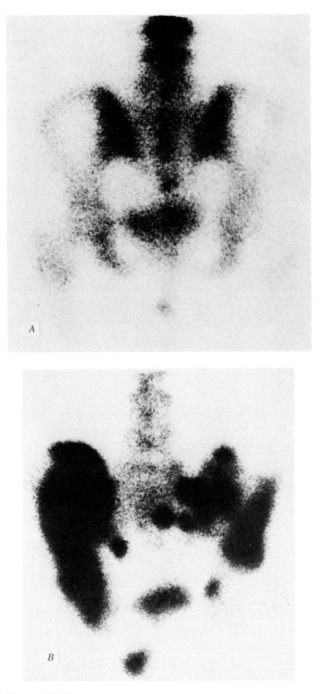

Figure 14-26.
Source: Photographs courtesy of Orlando Manfredi, M.D., Director
of Radiology and Nuclear Medicine, St. Vincent's Medical Center of
Richmond, N.Y.

pyrophosphate, ethylenehydroxyphosphonate (EHDP), and methylenediphosphonate (MDP).

Figure 14-25a shows a scan using the ^{99m}Tc-MDP. Figure 14-25b shows the presence of metastatic disease.

Figure 14-26a shows a normal pelvic scan using ^{99m}Tc-MDP and Figure 14-26b shows a case of metastatic disease.

LIVER SCAN

Technetium-99m-labeled sulfur colloid scans are based on the phagocytosis of the colloidal particles by the Kupffer cells, and as can be seen in Figure 14-27a, a good static image is produced in a normal liver. Figure 14-27b shows a liver with metastatic disease where activity is missing. The spleen also takes up the colloid and is frequently also shown on the image.

The function of the liver can also be evaluated using Rose Bengal dye (tetraiodotetrachlorofluorescein labeled with ^{131}I). The dye is absorbed by the hepatic polygonal cells. It is thus possible to determine both the blood circulation to the various parts of the liver, as well as the functioning of the polygonal cells.

LUNG SCANS

There are two different procedures to evaluate the lungs. To determine the blood supply to the lungs, the *vascularity,* technetium-99m-labeled microaggregated albumin (MAA) is used. The particle size is in the 5- to 100-μm range. These particles will be caught in the capillaries in the lung tissue and will give a uniform scan, as is seen in Figure 14-28a. If there is a pulmonary embolism present, a "cold spot" will be seen. Figure 14-28b shows this in the left lung, indicating a lack of blood flow.

To evaluate the *ventilation,* xenon-133 and xenon-127 gases are used. These are inert gases, with a physical half-life of 5.3 and 36 days, respectively. The biological half-life is short, however, as a person simply breathes the gas, holds his breath for 30 seconds or so, and then exhales the gas into a container. The Anger camera records the radiation during that time, and if there are certain parts of the lungs that are blocked, these will show up as areas of low counts.

OTHER USEFUL RADIOISOTOPES

Iron-59 as ferrous citrate is used to study erythropoiesis and the mechanism of iron metabolism. By measuring the activity of the blood, it is possible to determine the

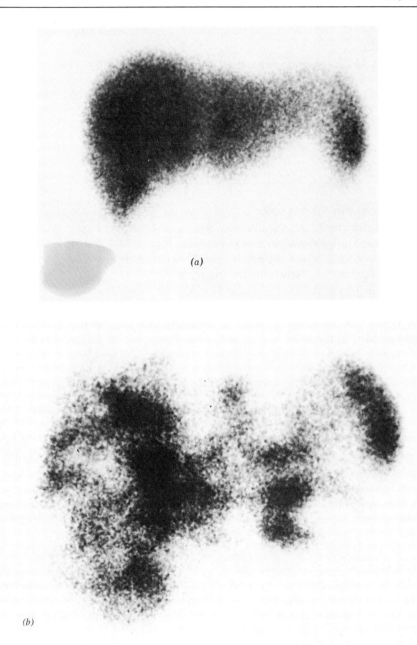

Figure 14-27.
Source: Photographs courtesy of Orlando Manfredi, M.D., Director
of Radiology and Nuclear Medicine, St. Vincent's Medical Center of
Richmond, N.Y.

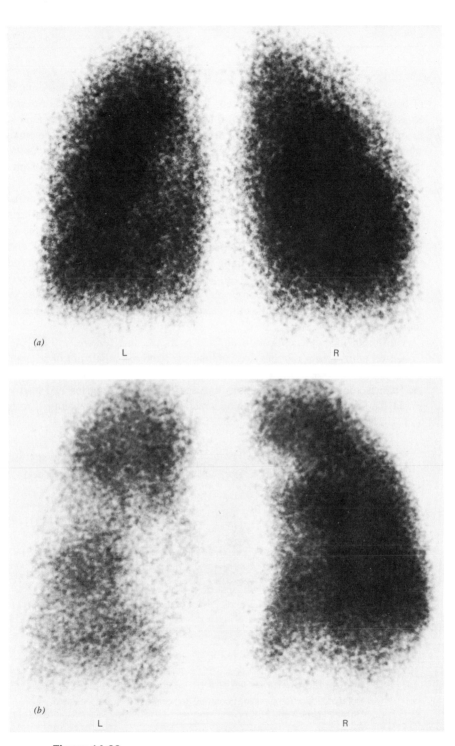

(a)

L   R

(b)

L   R

Figure 14-28.
Source: Photographs courtesy of Orlando Manfredi, M.D., Director
of Radiology and Nuclear Medicine, St. Vincent's Medical Center of
Richmond, N.Y.

rate of red cell production and destruction. Iron-59 appears in the red cells in three hours, and is maximal in 5 to 10 days. *Strontium-87m* is used in the early diagnosis of bone and joint infection in children. It has a half-life of 2.8 hours. *Strontium-85* is used for bone scanning. It has a half-life of 65 days. Use of *selenium-75* is the only direct means of determining the size, shape, and position of the pancreas short of exploratory laparotomy. It has a half-life of 120 days. *Chromium-51* is used as a complex with human serum albumin to detect gastrointestinal protein losses associated with hypoproteinemia. It has a half-life of 27.8 days.

Cobalt-58 and *cobalt-60* as radiocobalamin are administered to secure information about the absorption, fate, and excretion of vitamin B_{12}. Fecal excretion, urinary excretion, and hepatic uptake are measured. The vitamin is stored in the liver, which can be scanned. It is also used in the differentiation of pernicious anemia from other clinical conditions resulting in disturbances of vitamin B_{12} absorption. Pernicious anemia is the only B_{12} deficiency state that is related to an absence of *instrinsic factor*—the substance produced within the body by cells lining the stomach. Vitamin B_{12} is referred to as the *extrinsic factor,* because it is ingested in the food one eats.

A different medical application of ^{60}Co is its use in *teletherapy* (Fig. 14-29). Cobalt-60 emits a beta ray and two gamma rays, and has a half-life of 5.27 years. By irradiation of pieces of cobalt metal the size of a nickel in an atomic reactor, the ^{60}Co isotope is produced. This is encased in a heavy tungsten and lead alloy shield. The beta rays are filtered out by a metal filter, allowing the gammas to come

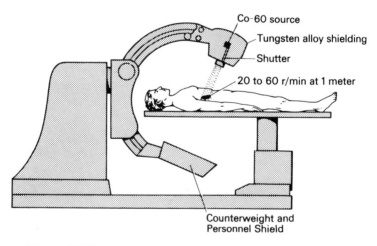

Figure 14-29.
Rotational teletherapy using cobalt-60 gamma rays (Francis Delafield Hospital, New York, N.Y.)

Source: B. Duffy, Atomic energy in the diagnosis and treatment of malignant diseases. *Am. J. Nurs.,* 55:434, and U.S. Atomic Energy Commission.

through. These are similar to high-energy x-rays, except that they are of shorter wavelength, and are used for radiation treatment of cancer. By rotating the head of the apparatus, it is possible to aim a high dose at the malignancy without excessive damage to the skin and tissue in between. The point made earlier regarding the sensitivity of tissues to radiation applies also to x-ray or gamma treatment. It is difficult to kill all malignant cells without causing excessive damage to normal cells. Thus the radiation dose is limited by the radiation sickness that results from damage to normal cells.

NEUTRON ACTIVATION ANALYSIS

Neutron activation analysis is a new analytical method by which very low concentrations of elements, in the range of parts per billion, can be measured. This has important implications in medicine. As is true of many milestones in medicine the principle of neutron activation analysis is quite simple (although very complicated equipment is needed to perform it). A stream of neutrons is produced either by a particle accelerator or by an atomic reactor. If the impurities in the sample are sensitive to neutron capture, they will absorb the neutrons and may become a radioactive isotope. By using a sodium iodide solid scintillation crystal and multichannel analyzer, it is possible to detect the gamma radiation given off by the newly formed radioactive nucleus. Thus a "fingerprint" of the isotope is produced, showing both the energy of the gamma and also its intensity. This fingerprint can both be seen on the oscilloscope and printed out on either paper or magnetic tape, and the counts per minute can be used to calculate the amount of radiation. By running standards of the same elements at the same time, this method can be made quantitative.

Neutron activation analysis is useful in measuring trace elements in biological material and in foodstuffs. Elements such as bromine, arsenic, and mercury may enter goods through fertilizers, pesticides, nematocides, and fungicides. In biomedical research, it has been found that the human body contains 12 elements in substantial amounts. These elements are oxygen, carbon, hydrogen, nitrogen, calcium, phosphorus, potassium, sulfur, sodium, chlorine, magnesium, and iron. There are about 30 other elements found regularly or occasionally in the tissues of animals and higher plants in much lower levels—hence the term trace element. Of these trace elements only seven are known to be essential to life: zinc, copper, manganese, molybdenum, iodine, cobalt, and selenium.[6]

Thus researchers are confronted with an interesting puzzle: a number of elements have been identified, but their biological role is not known. It may be that the

[6]Lukens, L., Guinn, V.: *Notes for Activation Analysis Course*. Gulf General Atomic Co., San Diego, Calif., Feb. 1970.

extremely sensitive neutron activation analysis will shed light on the matter. Many enzyme systems in the body are very sensitive to the absence of trace elements.

By analyzing a single strand of hair, it is possible to measure some 20 elements. Each person's hair is somewhat different, and it can thus serve as a "hairprint." Napoleon, who was very vain, gave out locks of hair to all his favorites, so there were ample samples around for analysis. His hair was found to contain about 13 times as much arsenic as normal. But by cutting the hair into seven sections, researchers found that each section had the same amount, so he must have taken the arsenic over a period of at least a year. King Erick XIV of Sweden was believed murdered, and when his hair was analyzed after four centuries, traces of mercury were found, thus supporting the rumor.

Analysis of human hair is illustrated in Fig. 14-30.

The medical uses of radioisotopes as tracers and for radiation therapy are multiplying as new isotopes are becoming available and as trained personnel learn to utilize this new medical tool. There is a quotation from Shakespeare's *The Tempest* on a government building in Washington, D.C., "The past is but the prologue." A man asked his cab driver what that meant, and the driver replied, "You ain't

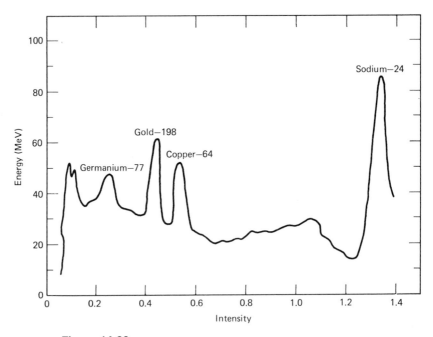

Figure 14-30.
Gamma ray spectrum of neutron-activated human hair.

Source: Adapted from J. T. Jensen and W. B. Ferren, *College General Chemistry*. Charles E. Merrill Publishing Company, Columbus, Ohio, 1971, p. 100.

seen nothing yet, brother." I believe that this can be said also about radioactive tracers. The uses of this tool may be more valuable in the long run than the use of atomic energy for power or for bombs.

QUESTIONS

1. If radium is stored in a lead container, can the disintegration of the radium be stopped?

2. Distinguish between the somatic and genetic effects of radiation.

3. Explain how a rectilinear scanning camera operates. How does this differ from the Anger camera?

4. Explain how a radioactive "cow" functions. What is the value of the technetium cow for a hospital?

5. How many disintegrations per second occur in a 10-millicurie source? How many in a 5-microcurie source?

6. Explain the meanings of physical and biological half-life. Would the half-life be the same for a small and a large amount of radioactive material?

7. "Distance is the best defense against an atomic bomb." What law does this refer to?

8. Explain in terms of half-life why radon seeds can be left permanently in the tissues, whereas radium cannot.

9. When ^{131}I was administered to a patient, high concentrations of radiation were found in the thyroid and in the groin. Explain the significance of each concentration.

10. What would be the effective half-life of a radionuclide that has a physical half-life of 72 hours and a biological half-life of 120 hours?

11. Would you use the formula to calculate the effective half-life if the physical half-life is long and the biological half-life is very short? Explain.

12. Technetium-99m has a half-life of 6 hours. A sample is checked at 8:00 A.M. to be 80 millicuries per milliliter. After 24 hours, how many milliliters would be needed to inject if the patient should have 10 millicuries of activity?

13. $^{232}_{90}$Th emits two B$^-$ particles and two alpha particles in succession. What isotope would result?

14. $n + ^{23}_{11}$Na $= X + \alpha$ What isotope is X?

15. How does the ^{99m}Tc$^-$ sulfur colloid procedure give information about the condition of the lungs? Does this show ventilation or vascularity effects?

CHAPTER 15

CAT SCANS,
ULTRASOUND SCANS, AND
RADIOIMMUNOASSAY PROCEDURES

*The development of scientific detection equipment,
including medical diagnostic machines, is the result
of rapid progress in electronics, sodium iodide
crystals and photomultiplier tubes for the detection
of x-rays and gamma rays, and computers using
microprocessing chips. The integration of the basic
sciences and technologies has resulted in
equipment for clinical diagnosis that has
revolutionized the practice of medicine in many
areas.*

COMPUTED AXIAL TOMOGRAPHY

X-RAY COMPUTED AXIAL TOMOGRAPHY (CAT)

The use of regular x-ray pictures in clinical diagnosis has two principal limitations:
(1) it is difficult to distinguish between various types of tissue since they do not
differ much in absorption of the x-rays; and (2) when an x-ray picture is made of
an organ, one gets an *average* absorption of the x-rays; hence, the picture is not
very clear and can give very little quantitative information.

To avoid this problem, the computed tomograph measures the attenuation of the
x-rays through a section or "slice" through the body. The slice or plane can be
either *longitudinally,* that is, parallel to the head-to-toe axis of the body, or *transverse,* that is, at right angle to the head-totoe axis.

A CAT scanner attempts to determine the attenuation in a single plane by focusing a narrow x-ray beam on a spot in the plane of interest. The emerging x-rays are then detected and measured by sodium iodide crystals coupled to photomultiplier tubes, the same as the gamma detectors described in Chapter 14. The x-rays can also be measured by xenon–krypton gas-filled ionization tubes.

The x-ray beam is then rotated a few degrees, and the intensity is recorded again. By rotating the x-ray tube through 180°, the attenuation of the beams passing through the section of the body from a large number of different angles will enable a computer to average out the attenuation of other tissue layers, leaving a much clearer reconstructed picture of the attenuation of the desired cross section (Fig. 15-1). Since the sensitivities of the current detectors and photomultiplier tubes are very high, the exposure time can be be very brief and still produce pictures that can detect clinical changes in the tissue.

The first CAT scanner was developed in England by the EMI Medical Company, and the first brain scans were done in 1973 and whole-body scans in 1975. Since that time, tremendous improvements have been made in these machines due to the competition of 20 companies that began to make them. By 1981, the number of companies had been reduced to about 10, and EMI was recently bought by General Electric. The cost of a scanner is currently $300,000 to $400,000, and the government is attempting to hold back the proliferation of scanners by having local hospitals share the use of a scanner instead of having one in each hospital.

A brain scan can now be made in a few seconds, and a full-body scan in 6 seconds. The short time required is helpful in eliminating problems due to motion or breathing of the patient. This feature is made possible by the number of detectors used. General Electric has 320 xenon detectors, and Varian has 301 xenon tubes. The x-ray tube and the detectors are on a ring that rotates as a unit.

After the information is stored in a computer, either on tape or disk, it is fed to a cathode-ray tube to make a picture on the scope that can be photographed by a Polaroid camera so as to get "hard" copy. The information can also be retrieved from the computer at a later date for comparison purposes.

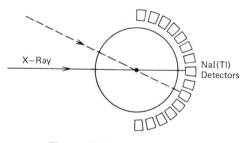

Figure 15-1.
CAT scanner detector assembly.

EMISSION COMPUTED AXIAL TOMOGRAPHY—ECAT

The next step in the development of tomographic scanners was to join the x-ray CAT scanner with gamma-emitting radionuclides. The same principle of averaging out radiation from the gamma source located on a spot in a plane applies here, except the radiation is *emitted* from the tissue, whereas the x-ray scanners shoot radiation from the outside of the body. In this method, the sodium iodide crystals are on a ring surrounding the patient in 360°. By using focused collimators it is possible to get sharp scintillation pictures at the selected focal plane. Transverse emission computed axial tomography (ECAT) thus gives a cross-sectional view of the distribution of the radioisotope. By repeating this procedure at different focal lengths, a number of cross-sectional pictures are possible.

POSITRON EMISSION COMPUTED AXIAL TOMOGRAPHY—PETT

This is very similar to the ECAT, except that instead of using a radionuclide that emits gamma rays, a *positron* emitter is used. By injecting a positron-emitting isotope with a short half-life, an additional benefit results. When a positron (also called β^+) collides with an electron (β^-), they "annihilate" each other. The particles actually cease to exist (self-destruct) and are changed into energy and will appear as two gammas coming off at 180° from each other. According to Einstein's equation, $E = MC^2$, the total energy resulting from the conversion of mass to energy in this case will be 1.02 million electron volts (MeV), and this will mean that each gamma will have 0.51 MeV energy. Please note that they come off at 180° from each other. By circling the patient with NaI(T1) crystals, the gammas will be detected that are coincidence. That is, if they are emitted simultaneously or within about a nanosecond (10^{-9} sec), the machine will record the event. In Figure 15-2, only detectors 3 and 12 will receive the two gammas from a particular annihilation at the same time, and will record the direction and activity in the computer.

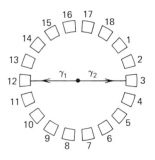

Figure 15-2.
Detection of two gamma rays at 180°.

By rotating the detector ring, it is possible to measure thousands of paths and average out the attenuation of the tissues in other planes, giving a good picture of the distribution of the positron-emitting radionuclide in the plane selected.

The latter two scanners, the ECAT and the PETT, are still not widely used but are being tested in large research hospitals where there are particle accelerators such as cyclotrons that can produce short-lived isotopes.

Positron-emitting nuclides that are used include gallium-68, carbon-11, nitrogen-13, oxygen-15, rubidium-84, and fluorine-18. Fluorine-18 has a half-life of 110 minutes, enabling the scientists to remove the sample from the cyclotron, purify it, dissolve it, inject it into the patient, and secure the scan. Other radionuclides have half-lives of only a few minutes, so it takes great speed and teamwork to use them. On the other hand, the short half-lives make it possible to use high activity samples that yield high counts.

The positrons travel only a few millimeters in the surrounding tissue before colliding with an electron and undergoing annihilation. The resolution of PETT is currently about 1 to 2 cm, and this makes it possible to detect small lesions and areas of uptake in an organ.

One recent experiment used fluorine-18-labeled deoxyglucose to trace regional brain metabolism. Both labeled and unlabeled deoxyglucose are absorbed by the brain, and the higher the energy requirement of the cell, the higher the absorption. This compound is not easily metabolized, and accumulates in the cell, where it emits positrons, undergoes annihilation, and produces two gammas that are detected. It is thus possible to get some idea of the activity of that section of the brain.

Similar experiments have shown that stimulating the left side of a person's visual field with light produced increased activity in the right side of the visual cortex as indicated by increased scintillation in that section. Schizophrenics tend to use less glucose in the forebrain than do normal subjects. This type of evaluation of the *function* of sections of an organ may have great potential for learning more about the biochemistry of the body.

ULTRASONOGRAPHY

Ultrasonography is a very new diagnostic modality that is a supplement to x-ray and nuclear medicine procedures. In 1943 in Austria, Karl Dussek attempted to *transmit* ultrasound through the head, but was not successful because of the interference by the skull. Soon after World War II, a method of sending ultrasound pulses into metal to detect flaws by measuring the echos was tried on living tissues and succeeded in detecting internal structures. This procedure has now developed in the last 10 years into an important diagnostic tool. The important point to note about this procedure is that it measures the *echo* of the ultrasound pulse from the interfaces between various tissues.

The audible range of hearing for humans is about 20 to 20,000 Hz. Higher-frequency sounds are called *ultrasound*. The frequencies employed in medical ultrasonography are in the range of 1 to 20 MHz (million cycles per second). We noted in the chapter on sound that as the frequency of a wave increases, the wavelength decreases. Therefore, a frequency of 7.5 MHz will have a wavelength of about 0.2 millimeters, but a 1.0-MHz sound wave will have a much longer wavelength. Because the wavelength of the sound must be smaller than the object studied, it would seem that one would use a very high frequency for all applications in order to have good resolution. However, as it is often the case in life, there are tradeoffs necessary, or as former President Truman said, "There is no free breakfast." As the frequency increases and Hertz value goes up, the attenuation, or the absorption of the sound, increases. Put another way, the attenuation varies *directly* with the frequency, and the penetration through tissues varies *inversely* with the frequency. So we have to pick the frequency that will give the best resolution yet will have sufficient penetration.

It is for that reason that in the use of ultrasonography for the eye, where less penetration is required, one can use 18 to 20 MHz, whereas for cardiac and obstetrical scans requiring greater penetration, the frequency will be in the 1- to 5-MHz region.

The velocity of sound usually used in medicine is about 1,450 meters per second in fat, 1,600 m/sec in muscle, and 4,000 m/sec in bones.

The ultrasound waves are generated by a device called a transducer that converts electrical energy into high-frequency sound waves, but can also pick up the sound waves and convert them back into electricity. The transducer has a crystal that will change dimensions when a current is applied. The rapid movement of the crystal as the current changes will produce vibrations, thereby causing sound waves. This is called the *piezoelectric effect,* and the piezoelectric crystal can be made of quartz, lithium sulfate, barium titanate, or lead zirconium titanate.

In order to have good coupling of the sound wave with the skin, a liquid such as mineral oil or water is applied to the skin, and the probe is held in contact with it. When the probe is moved over the area of interest, an echo pattern can be seen on the oscilloscope, and can be stored in the microcomputer. It is thus possible to detect the outline of organs, find tumors, and, with the newer procedures, determine the motion of certain organs.

The uses of ultrasound all depend on the echos from the interfaces of various tissues. A soft mass tissue would have a low-amplitude echo, liquid very little, and a wall a high-amplitude echo. When the speed of the wave and the duration in microseconds of the pulse emitted are known, it is possible to calculate the distance the sound has traveled before the echo is returned, and thus have some measurement of the location of the object.

The most common use of ultrasound is in obstetrics, where it is estimated that it is employed in about half the pregnancies and deliveries. It is possible to predict multiple births, localize the placenta, and detect abnormalities. By measuring the biparietal diameter of the fetal head, the weight and age can be estimated.

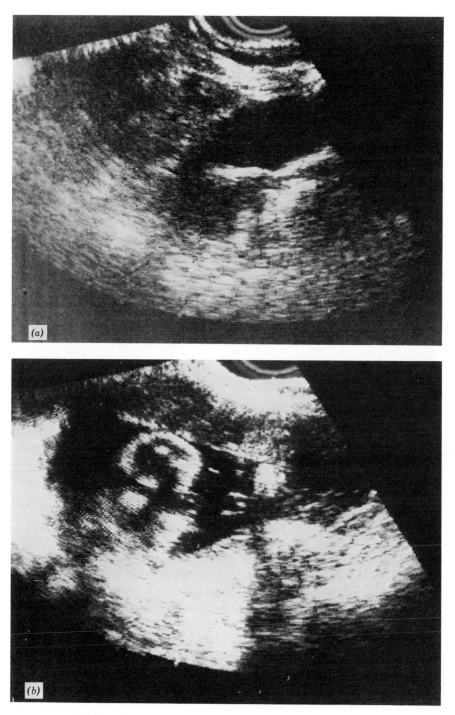

Figure 15-3.

(a) Normal pelvis. (b) Fetus at 30 weeks gestation.

Source: Photograph courtesy of Orlando Manfredi, M.D., Director of Radiology and Nuclear Medicine, St. Vincent's Medical Center of Richmond, N.Y.

Another application of ultrasound techniques is in brain scanning. Echoencephalography is used to study the ventricals, tumors, and midline displacement. This would use a frequency of 2.5 MHz and 0.6 mm wavelength. Abdominal ultrasound can also be used to study the kidneys, liver, and spleen. For ophthalmological studies, the interocular structure can be evaluated in cases of opaque lenses.

The ultrasound machines can operate in a number of different modes. One mode is unconverted echo mode that gives high black and white contrast. In order to have more detail, a computer can produce gray scale images.

In order to be able to see movement, multiple transducers are used. This type is called "real-time" devices. By taking a series of pictures by the rapid movement of the probe, it is possible to see movement. This is used in echocardiography where the movement of the walls of the heart and the valves can be visualized. In this connection, another mode, called Doppler scanning, is also used. If a sound of a certain frequency is reflected by a tissue wall moving toward the probe, the waves will be compressed, and there will be an increase in the frequency. Conversely, if the wall is moving away from the probe, tbe waves will be stretched, and the frequency will decrease. The machine can detect these changes in frequency and can measure the motion of the wall as well as the rate of flow of blood.

RADIOIMMUNOASSAY PROCEDURES

Radioimmunoassay (RIA) procedures are in vitro procedures, that is, they are done in test tubes, and are to be distinguished from the in vivo nuclear imaging procedures described earlier. RIA procedures do not involve injection of a radionuclide into the patient. Instead, a blood sample is taken and it is then analyzed by the use of radioactive substances.

As the name implies, RIA involves immune-response reactions, the binding of an antigen to an antibody, using radionuclides. Rosalyn Yalow and Solomon Berson discovered the method in 1955, in attempting to develop a highly sensitive test for insulin. Since then, especially the last decade, this principle has been extended to a large number of substances of interest in medicine. Commercial companies are continually developing new kits to detect concentrations in the nanogram (10^{-9} g) range.

Radioimmunoassay procedures are based on the principle of competitive binding. If we have an *antigen (Ag)* in the presence of a specific *antibody (Ab),* they will join to form the *antigen–antibody complex.* Where does the competition come in? If one can synthesize or label an antigen molecule with a radionuclide, we have a labeled antigen, Ag*. By using a limited amount of antibody, there will be a competition between the unlabeled antigen (Ag) and the labeled, radioactive antigen (Ag*) for sites on the antibody (Ab), and the number of Ag–Ab and Ag*–Ab complexes formed will follow the Law of Mass Action, which states that the rate of a chemical reaction varies directly with the concentration of the reactants.

A simple example will illustrate this. Let us assume

$$
\begin{array}{c}
6 \text{ Ag} \\
+ \quad 8 \text{ Ab} \\
6 \text{ Ag*}
\end{array}
$$

How many antigen–antibody complexes will be formed? Because there are 8 Ab, we can expect eight complexes if the reaction goes to completion. Also, because there are an equal number of Ag and Ag* antigens (6 + 6), we would expect an equal probability of them hooking up with the Ab molecules.

$$
\begin{array}{ccccc}
6 \text{ Ag} & & 4 \text{ Ag–Ab} & + & 2 \text{ Ag} \\
& + 8 \text{ Ab} = & & & \\
6 \text{ Ag*} & & 4 \text{ Ag*–Ab} & + & 2 \text{ Ag*}
\end{array}
$$

$$
\text{bound} \qquad \text{free}
$$

What would happen if we had the same number of Ag* and Ab molecules, as in the previous example, but had 18 Ag molecules?

$$
\begin{array}{ccccc}
18 \text{ Ag} & & 6 \text{ Ag–Ab} & + & 12 \text{ Ag} \\
& + 8 \text{ Ab} = & & & \\
6 \text{ Ag*} & & 2 \text{ Ag*–Ab} & + & 4 \text{ Ag*}
\end{array}
$$

Because the concentration of the nonradioactive antigen molecules (Ag) is three times that of the radioactive antigen (Ag*), there is a 3 : 1 chance of the Ag forming the complex.

Please note that if we have a fixed, known, limited amount of Ag* and Ab, the amount of the normal antigen can be determined by the amount of complex being formed.

The next step in the process is to separate the complex, the "bound" portion from the "free." This is done by a number of methods. The free antigens can be adsorbed on dextran-coated charcoal, resin strips, or silicates, or by precipitating with alcohols, polyethylene glycol, or some inorganic salts. A double antibody procedure is also used at times. The most convenient method is to have the antibody attached to either glass beads or to the inside of plastic test tubes. The antigens, both Ag and Ag*, will bond to the Ab on the test tube. After the proper reaction time, the material is decanted and washed, and the tube is counted in a gamma scintillation counter. This gives the amount of radioactivity in the complex.

How can this process be quantified? A commercial kit will have a number of standard tubes containing known amounts of antigen. These standards are run at the same time and under the same conditions of temperature, time, and mixing as the unknown samples. If we measure the activity of the *bound* portion, that is, the complex, the *less* of the antigen (Ag) there is in the patient's blood sample, the *more* the radioactive antigen (Ag*) will have an opportunity to bind to the antibody. Therefore, a high count of the complex means a lower amount of the patient's antigen.

Similarly, if the patient has a great deal of Ag, less Ag* will be bound, and the count rate will be decreased. By running about five standards in duplicate and plotting a curve such as that shown in Figure 15-4, one can then measure the activity of a number of unknown samples and find the concentration of the antigen by locating the counts per minute on the *y*-axis, and reading the concentration on the *x*-axis.

The *free* antigen, both Ag and Ag*, can also be counted. In this case, the higher the Ag value of the patient, the less Ag* was bound in the complex, and more Ag* will be in the free state. Therefore, the curve for this case, where one is counting the activity of the free portion, will be as seen in Figure 15-5.

One of the requirements for a sensitive RIA method is to have an antibody that is *specific* for the particular antigen being measured, one that has a high *affinity* for the antigen, that is, it will form a stable complex. These antibodies are normally produced in an organism in order to fight off substances that are foreign to it. In determining the amount of T-3 (triiodothyronine), T-4 (thyroxine or tetraiodothyronine), or digoxin natural to humans, we could not produce our own antibodies to these. The usual method is to inject one of these substances into an animal such as a rabbit. The rabbit will produce a specific immunoglobin or antibody to that antigen. After a month or two, blood can be withdrawn and the antibody diluted and standardized for use in the RIA procedure.

It is also necessary to produce a radioactive antigen to be used in the competitive

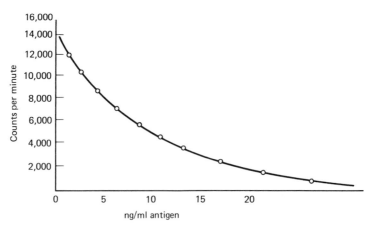

Figure 15-4.
Generalized plot of bound portion.

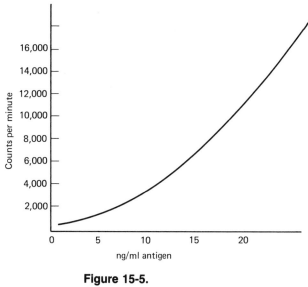

Figure 15-5.
Generalized plot of free portion.

binding procedure. What is needed is a gamma emitter that is easily counted by the solid scintillation method using the NaI(T1) crystal. In addition, it should have a fairly long half-life and must be able to be incorporated into the antigen molecule. Iodine-125 is one such a radionuclide, having a half-life of 60 days, and is used in making T-3* and T-4* molecules, as well as a number of other radioactive antigens.

RIA tehniques have been extended to include assays in which reactors other than antibodies may be used. Names of such assays include radioassay, saturation analysis, displacement analysis, radioligand binding assay, and competitive protein binding (CPB).

At present a number of companies are distributing kits for analyzing more than 200 substances found in body fluids. It is expected that RIA will become an increasingly important clinical tool. A current use of RIA is in determining transmission of one hepatitis virus, called hepatitis B, from carrier mothers to their newborn babies (a carrier is one who has no symptoms of a disease yet transmits it to others). A second recent application is in the detection of certain hormones thought to be influential in the treatment of breast cancer.

QUESTIONS

1. Why are ordinary x-ray pictures unable to discriminate well with various organs as distinct from bones?

2. Why do tomographic x-ray pictures give much better details?

3. Why is a computer required in obtaining a CAT scan?

4. What are the basic differences among CAT, ECAT, and PETT?

5. What advantage is gained by using a positron-emitting radionuclide in PETT?

6. Define annihilation and show connection with Einstein's equation,

7. Explain why ultrasonography of the heart is called echocardiography.

8. Explain how a compromise has to be made between resolution and penetration in using ultrasound.

9. Explain what is meant by the piezoelectric effect and how it is used in an ultrasound transducer.

10. Explain the competitive binding aspect of radioimmunoassay procedures.

11. If the activity of the "bound" portion in an RIA test is measured, would a high level of antigen in the patient give a high or low count rate?

12. Why is it necessary to run a set of RIA tubes of known antigen concentrations at the same time as running the samples from the patients?

GLOSSARY

accommodation   The ability of the lens of the eye to focus on both near and far objects through adjustment of its curvature.

alpha particle (α)   A nuclear particle composed of 2 protons and 2 neutrons. The nucleus of the helium atom is an α particle.

ampere   A unit of flow of electricity; one ampere is equal to 6.24×10^{18} electrons per second.

Anger camera   A device that uses a large sodium iodide crystal to detect gamma radiation.

annihilation   The reaction of a positron with an electron, destroying the mass of the particles, and producing two 0.51-MeV gamma rays at 180° from each other.

antigen–antibody complex   The structure formed when an antigen and antibody molecule combine.

astigmatism   Inability of the eye to focus vertical and horizontal light equally due to deformation of the cornea.

atomic number   The number of protons in the nucleus of an atom. Every element has a specific atomic number, which is different from that of any other element.

atomic weight   The atomic weight or mass of an element is equal to the sum of the protons and neutrons in the nucleus; usually stated as atomic mass units (AMU).

audiometer   An electronic instrument used to test hearing ability by generating tones of known frequencies and intensities.

Becquerel (Bq)   One nuclear disintegration per second: 3.7×10^{10} Bq equals 1 Curie.

Bernoulli's principle   The pressure of a flowing liquid is least at the point of greatest linear velocity. The Venturi tube, which creates suction in an atomizer or an aspirator, is an application of Bernoulli's principle.

Beta particle   (β⁻) An electron emitted from a nucleus. β⁺ is a positron emitted from a nucleus.

biological half-life   The time required for one-half of the substance of a radioactive isotope to be cleared from the body.

Boyle's law   The volume of a gas varies inversely with the pressure at constant temperature: $PV = k$.

calibration   The process of marking a scale on a measuring instrument, such as lines on a ruler or a thermometer, or units of pressure on a gas gauge.

cardiac pacemaker   An instrument that stimulates, by electrical impulses, contraction of heart muscle at a certain rate; used in arrhythmias and in heart block.

CAT scanner Computed axial tomography scanner using x-rays.

center of gravity A point at which the total mass of an object is considered to be located. The object is balanced in every direction around this point.

centripetal force An inward force on a body moving in a circle, producing an acceleration toward the center.

change of state The conversion of a solid, liquid, or gas to either one of the other physical states.

condensation The conversion of a gas into a liquid state.

conduction The transmission of energy in the form of electricity, heat, or sound through matter.

convection The transfer of heat in a gas or a liquid by eddy motion.

curie (Ci) The amount of activity of a radioactive element emitting 3.7×10^{10} particles per second or 3.7×10^{10} Becquerels: 1 mCi = 1/1,000 of a curie, 1 μCi = 1/1,000,000 of a curie.

deductive reasoning A method of reasoning whereby a principle or a law is employed to predict what might happen in a specific case.

density Mass per unit volume, that is, the amount of a substance in a given volume, such as grams per milliliter: $D = m/V$.

determinate errors Errors that are constant because of incorrect calibration of an instrument or some other error that is repeated in each measurement.

dialysis A type of filtration through a membrane in which small particles can pass through the membrane while larger molecules such as proteins cannot.

diffusion The movement of matter (solid, liquid, or gas) from a more concentrated region to one of lesser concentration.

doppler effect The change in the frequency of sound waves when the source is either traveling away from or toward the hearer.

ECAT scanner A computed axial tomography apparatus using emission of gamma rays instead of x-rays.

electrical resistance The opposition by a conductor to the passage of an electrical current. The unit of electrical resistance is the ohm. $E = IR$.

electromagnetic spectrum A broad band of electromagnetic radiation ranging from the long wavelength electric and radio waves to the very short wavelength x-rays and gamma rays.

electromyography (EMG) The evaluation of a condition of a muscle by measuring its electrical properties.

electron A negatively charged particle having -1 charge and a mass equal to approximately 1/1,860th of a proton.

energy The ability to do work.

equilibrium A condition where the sum of the forces acting upon an object cancel out; balance.

fission Splitting into parts. In physics, a nuclear reaction in which an atomic nucleus splits into fragments (usually two fragments of comparable mass) with release of tremendous quantities of energy.

fulcrum The point about which a lever moves.

frequency   The number of cycles per unit time, with reference to sound waves and electromagnetic radiation.

friction   In physics, resistance to the relative motion of one body sliding, rolling, or flowing over another with which it is in contact.

fusion   The union of atomic nuclei to form heavier nuclei resulting in release of enormous quantities of energy. The fusion of two deuterium or tritium atoms to produce helium is the basis for the H-bomb.

galvanometer   An instrument that measures the rate of flow of electricity.

gamma rays   Very penetrating, short wavelength electromagnetic radiations similar to x-rays, emitted by radioactive nuclei.

generator   A machine that converts mechanical energy into electrical energy.

Hertz (Hz)   Vibrations per second.

hydrometer   An instrument for measuring the density of a liquid, usually a long glass float marked with a scale indicating the depth of immersion.

hydrostatic pressure   The pressure exerted by a column of liquid. The difference in height between the pressure source and the pressure outlet is called the head of fluid.

hyperopia   Farsightedness; ability to focus on distant objects but not on near objects due to insufficient curvature of the lens.

illumination   The luminous flow per unit area at any point on a surface exposed to light.

indeterminate errors   Random or accidental errors that are not consistent.

inductive reasoning   A method of reasoning whereby consideration of specific cases leads to formulation of a general hypothesis.

inertia   A property of matter by which it remains at rest and resists acceleration.

intensity of heat   A measure of the average kinetic energy of the molecules of a substance; the temperature of the substance.

ionizing radiation   A stream of particles or photons that causes the molecules to dissociate into ion pairs.

isotope   One of two or more atoms whose nuclei have the same number of protons but different numbers of neutrons.

laminar flow   Smooth, nonturbulent flow of a viscous fluid in layers. It is faster than turbulent flow.

mass   A quantity of matter; the measure of a body's resistance to acceleration.

mechanical advantage   The ratio of the force performing the work of a machine to the force applied to the machine.

metric system   A decimal system of weights and measures based on the meter as a unit length and the kilogram as a unit mass.

motor   A rotating machine that transforms electrical energy into mechanical energy.

myopia   Nearsightedness; ability to focus on near objects but not on distant objects due to excess curvature of the lens.

neutron   An electrically neutral or uncharged particle existing along with protons in the atoms of all elements except the lightest hydrogen nucleus.

nucleus   The positively charged dense center of an atom composed of protons and neutrons.

orifice   An opening; in physics, an opening through which a gas or a fluid can flow.

parallax   An apparent change in direction of an object caused by a change in the observer's position that affords a different line of sight.

Pascal's law   Pressure applied to a liquid at any point is transmitted equally in all directions.

PETT   Position emission transverse tomograph detector using the gamma rays resulting from positron–electron annihilation.

Piezoelectric effect   The changing in dimensions of some crystals when an electric current is applied, or the reverse—is used to produce ultrasound waves for ultrasonography.

phonocardiography   The conversion of heart sounds into electrical impulses that are recorded on a graph.

physical half-life   The time required for half of the atoms of a radioactive substance present at the beginning to undergo decay.

physical resistance   Any force that acts to oppose or retard motion.

pitch   The quality of sound dependent on the frequency of vibration of the waves producing it.

placebo   An inactive substance, usually a pill, used in controlled studies to determine the effectiveness of medicinal substances.

presbyopia   A decrease in accommodation of the eye due to loss of elasticity of the crystalline lens; the eye cannot focus sharply on nearby objects.

pressure   Force applied over a surface, measured as a force per unit area, e.g., pounds per square inch.

pressure gradient   The difference in pressure between two points in a fluid.

proton   A nuclear particle having a mass number of 1 and $+1$ charge.

quantity of heat   A measure of the total energy of a substance, including the total quantity of the substance.

rad (radiation absorbed dose)   A unit of measurement of the absorbed dose of radiation, corresponding to energy transfer of 100 ergs per gram of absorbing material.

radiation   The emission and movement of waves or particles; also the waves or particles, such as light, sound, or particles emitted by radioactivity.

radioactivity   The spontaneous emission of radiation, including particles, gamma rays, and electrons, as a result of disintegration of a nucleus.

rbe (relative biological effectiveness)   An expression of the effectiveness of other types of radiation compared to that of gamma or roentgen rays.

refraction   The bending of a light ray in passing obliquely from one medium to another of different density.

rem (roentgen equivalent man)   The amount of ionizing radiation required to produce a biological effect equal to one roentgen of x-ray or gamma ray dosage.

resultant   The sum of two or more vector forces.

RIA   Radioimmunoassay procedure using gamma-emitting radionuclides in antigen–antibody reactions.

roentgen   The international unit of x-radiation or gamma radiation.

scanning camera   A device that uses a smaller NaI(T1) crystal to detect gamma radiation. The machine moves back and forth over the patient recording the radiation at each spot.

siphon   A bent tube with two arms of unequal length used to transfer liquids from a higher to a lower level.

specific gravity   The ratio of the density of a liquid to the density of water at the same temperature.

speed   The rate or a measure of the rate of motion, especially distance traveled divided by the time of travel, for example, miles per hour.

standard conditions   Temperature of 0°C and pressure of 760 mm Hg used in comparisons of gas volumes.

surface tension   A property of liquids due to the unbalanced molecular cohesive forces at or near the surface causing the surface to contract and to have properties resembling a stretched elastic membrane; measured in dynes per centimeter squared.

thermal expansion   Increase in linear dimensions of a solid or in volume of a fluid due to a rise in temperature.

thermostat   A device placed in a heating system by which temperature is automatically maintained at a certain level; based on the difference in thermal expansion of two dissimilar metals.

timbre   A quality of sound dependent chiefly on the presence or absence of overtones and their relative intensity.

transducer   A device that converts one form of energy to another such as electricity into sound or pressure into electricity.

turbulent flow   Motion of a fluid subject to random fluctuation of speed and pressure.

ultrasonography   The use of high-frequency sound above the normal human hearing level of 20,000 Hz.

velocity   The time rate of linear motion in a given direction.

viscosity   The degree to which a liquid resists flow under an applied force.

visible spectrum   The part of the electromagnetic spectrum to which the human eye is sensitive; it extends from a wavelength of 400 nm to 800 nm. (A nanometer [nm] is equal to one-billionth of a meter.) Earlier terminology used the angstrom as the unit of measurement.

wavelength   The distance between two points of corresponding phase in consecutive cycles in a periodic wave.

weight   The force with which a body is attracted toward the earth by gravitation.

BIBLIOGRAPHY

GENERAL TEXTS

Aird, E.A. *Introduction to Medical Physics*. Philadelphia: International Ideas, 1975.

Benedek, G.B., and Villars, F.M. *Physics with Illustrative Examples for Medicine and Biology*. Reading, Mass.: Addison-Wesley, 1979.

Cameron, J.R., and Skofronick, J.G. *Medical Physics*. New York: Wiley, 1978.

Cromwell, A., Weibell, P., Pfeiffer, F., and Usselman, B. *Biomedical Instrumentation and Measurements*. Englewood Cliffs, N.J.: Prentice-Hall, 1973.

Damask, A.C. *Medical Physics*. New York: Academic Press, 1978.

Flitter, H.H. *An Introduction to Physics in Nursing*, 7th ed. St. Louis: Mosby, 1976.

Fung, Yuan Cheng, et al. (eds.). *Biomechanics: Its Foundations and Objectives*. Englewood Cliffs, N.J.: Prentice-Hall, 1972.

Geddes, L.A., and Baker, L.E. *Principles of Applied Biomedical Instrumentation*, 2nd ed. New York: Wiley, 1975.

Haga, E. (ed.). *Computer Techniques in Biomedicine and Medicine*. Philadelphia: Auerbach, 1973.

Index Medicus and Cumulative Index Medicus. An excellent reference source for articles in all areas of medicine. References are published monthly and also in a cumulative annual edition, and are cross-listed under several headings.

Jensen, J.T., and Ferren, W.P. *College General Chemistry*. Columbus: Charles E. Merrill, 1971.

Karselis, T. *Descriptive Medical Electronics and Instrumentation*. Thorofare, N.J.: Charles B. Slack, 1973.

Lawrence, J.H., and Hamilton, J.G., (eds.). *Advances in Biomedical Engineering and Medical Physics*, vols. 1–14. New York: Academic Press, 1948–1973.

Nave, C., and Nave, B. *Physics for the Health Professions*, 2nd ed. Philadelphia: Saunders 1980.

Tilley, D.E., and Thumm, W. *Physics for College Students*. Menlo Park, Calif.: Cummings, 1974.

Wiggins, A.W. *Physical Science with Environmental Applications*. Boston: Houghton Mifflin, 1974.

MECHANICS AND BODY MECHANICS

Drake, S. Newton's apple and Galileo's dialogue. *Sci. Am.* 243:150, 1980.

Steindler, A. *Kinesiology of the Human Body Under Normal and Pathological Conditions.* Springfield, Ill.: Thomas, 1970.

Tricker, R.A., and Tricker, B.J. *The Science of Movement.* New York: American Elsevier, 1967.

Wells, F. (ed.). *Kinesiology: The Scientific Basis of Human Motion,* 5th ed. Philadelphia: Saunders, 1971.

Williams, M., and Lissner, H.R. *Biomechanics of Human Motion.* Philadelphia: Saunders, 1962.

RESPIRATION AND OXYGEN THERAPY

Adler, R., and Brodie, S. Postoperative Rebreathing Aid. *Am. J. Nurs.* 68:1287, 1968.

Awad, J.A., et al. Prolonged pulmonary assistance with the membrane gas exchanger: A case report. *J. Pediatr. Surg.* Dec. 1973.

Cotes, J.E. *Lung Function—Assessment and Application in Medicine,* 4th ed. St. Louis: Mosby, 1981.

Council for National Cooperation in Aquatics. *The New Science of Skin and Scuba Diving.* New York: Association Press, 1967.

Cox, P.M., Jr., et al. Respiratory therapy. Pressure measurements in endotracheal cuffs: A common error. *Chest* 65:84, 1974.

Egan, D.F. *Fundamentals of Inhalation Therapy,* 3rd ed. St. Louis: Mosby, 1977.

Flatter, P. Hazards of oxygen therapy. *Am. J. Nurs.* 68:81, 1968.

Johnson, L.R., et al. Use of arterial P_{O_2} to study convective and diffusive gas mixing in the lungs. *J Appl. Physiol.* 36:91, Jan. 1974.

McPherson, S.P. *Respiratory Therapy Equipment.* St. Louis: Mosby, 1977.

Mushin, W.W., Rendell-Baker, L., Thompson, P.W., and Hillard, E.K. *Automatic Ventilation of the Lungs,* 3rd ed. St. Louis: Mosby, 1980.

Nett, L., and Petty, T. Acute respiratory failure: Principles of care. *Am J. Nurs.* 67:1847, 1967.

Sovie, M., and Israel, J. Use of cuffed tracheostomy tube. *Am. J. Nurs.* 67:1854, 1967.

Tinker, J.H., and Wehner, R. The nurse and the ventilator. *Am. J. Nurs.* 74:1276, 1974.

Winter, T.M., and Lowensteen, E. Acute respiratory failure. *Sci. Am.* 221:23, 1969.

SOUND AND LIGHT

Ashkin, A. The pressure of laser light. *Sci. Am.* 226:63, Feb. 1972.

Backus, J. *The Acoustical Foundations of Music.* New York: Norton, 1969.

Chaney, R.B., et al. Relation of noise measurements to temporary threshold shift in snow-mobile users. *J. Acoust. Soc. Am.* 54:1219, 1973.

David, E., Jr. The reproduction of sound. *Sci. Am.* 265:72, Aug. 1961.

Dobelle, W.H. et al. Artificial vision for the blind: Electrical stimulation of visual cortex offers hope for a functional prosthesis. *Science* 183:440, 1974.

Donald, I. Proceedings: Placental localization by sonar—A safe procedure. *Br. J. Radiol.* 47:72, 1974.

Elenaga, J. The synthesis of speech. *Sci. Am.* 226:48, Feb. 1972.

Gombrich, E.H. The visual image. *Sci. Am.* 227:82, Sept. 1972.

Josephs, J.J. *The Physics of Musical Sound.* New York: Van Nostrand, 1967.

Michael, C. Retinal processing of visual images. *Sci. Am.* 220:104, May 1969.

Morrison, R.J. Comparative studies: Visual acuity with spectacles and flexible lenses, ophthalmometer readings with and without flexible lenses. *Am. J. Optom. Physiol. Opt.* 50:807, 1973.

Nagafuchi, M. Filtered speech audiometry in normal children and in the mentally retarded. *Audiology* 13:66, 1974.

Nassau, K. The causes of color. *Sci. Am.* 243:124, Oct. 1980.

Nauton, R. *An Introduction to Audiometry,* 3rd ed. Minneapolis: Maico Hearing Instrument Co., 1972.

Neisser, U. The process of vision. *Sci. Am.* 219:204, Sept. 1968.

New Orleans Academy of Ophthalmology. *Symposium on Glaucoma.* St. Louis: Mosby, 1981.

Olsen, H.F. *Music, Physics, and Engineering.* New York: Dover, 1966.

Pettigrew, J. The neurophysiology of binocular vision. *Sci. Am.* 227:84, Aug. 1972.

Rushton, W.A.H. Visual pigments and color blindness. *Sci. Am.* 232:64, March 1975.

Silverman, W.A. The lesson of retrolental fibroplasia. *Sci. Am.* 236:100, June 1977.

Staab, W.J. *Hearing Aid Handbook.* Blue Ridge Summit, Pa.: Tab Books, 1978.

Stein, H.A. and Slatt, B.J. *Fitting Guide for Hard and Soft Contact Lenses.* St. Louis: Mosby, 1977.

Zweng, H.C., Little, H.L., and Vassiliadis, A. *Argon Laser Photocoagulation.* St. Louis: Mosby, 1977.

ELECTRICITY

Adolph, E.F. The heart pacemaker. *Sci. Am* 216:32, March 1967.

Andeoli, K. The cardiac monitor. *Am. J. Nurs.* 69:1238, 1969.

Arfwidsson, L., et al. Chlorpromazine and the antidepressant efficacy of electroconvulsive therapy. *Acta Psychiatr. Scand.* 49:580, 1973.

Bauman, G. Steps toward building a living cell. *Chemistry.* 51:16, Nov. 1978; 51:11, Dec. 1978; 52:12, Jan. 1979.

Bergveld, P. *Electromedical Instrumentation: A Guide for Medical Personnel.* New York: Cambridge University Press, 1980.

Fitzwater, J. Planning an intensive care unit. *Am. J. Nurs.* 67:310, 1967.

Goldberger, A.L., and Goldberger, E. *Clinical Electrocardiography.* St. Louis: Mosby, 1977.

Grais, I.M., et al. A 12-lead patient cable for electrocardiographic exercise testing. *Am. Heart J.* 87:203, 1974.

Hilsenrath, J., et al. Pitfalls in the prediction of coronary artery disease from the electrocardiogram or vectorcardiogram. *J. Electrocardiol.* 6:291, 1973.

Hodgkin, B.C. Some consequences of electric shock. *J. Main Med. Assoc.* 65:1, 1974.

Kolff, W. An artificial heart inside the body. *Sci. Am.* 213:38, Nov. 1965.

Lester, H.A. The response to acetylcholine. *Sci. Am.* 236:74, Feb. 1977.

Lown, B. Intensive heart care. *Sci. Am.* 219:19, July 1968.

Mackay, R.S. *Biomedical Telemetry,* 2nd ed. New York: Wiley, 1970.

Margaria, R. The sources of muscular energy. *Sci. Am.* 226:84, March 1972.

Nyboer, J. *Electrical Impedance Plethysmography,* 2nd ed. Springfield, Ill.: Thomas, 1970.

Regan, D. Electrical responses evoked from the human brain. *Sci. Am.* 241:134, December 1979.

Robinson, J.L., et al. A new bi-polar coagulator. *Biomed. Eng.* 8:516, 1973.

Rudin, S.G., Foldvari, T.L., and Levy, C.K. *Bioinstrumentation—Experiments in Physiology.* Millis, Mass.: Harvard Apparatus Foundation, 1971.

Russin, A., O'Gureck, J., and Jacques, R. Electronic monitoring of the fetus. *Am. J. Nurs.* 74:1294, 1974.

NUCLEAR MEDICINE

Bernier, D.R., Langan, J.K., and Wells, L.D. *Nuclear Medicine Technology and Techniques.* St. Louis: Mosby, 1981.

Das, K.R. Dosimetry of gold-198 seed implants. *Australas Radiol.* 17:321, 1973.

Duffy, G.J. et al. New radioisotope test for detection of deep venous thrombosis in legs. *Br. Med. J.* 1:712, 1973.

Early, P.J., Razzak, M.A., and Sodee, D.B. *Textbook of Nuclear Medicine Technology,* 3rd ed. St. Louis: Mosby, 1979.

Gore, W.G., et al. Iodine 125-labeled thrombosis monitor. *Biomed. Eng.* 8:306, 1973.

Jahns, M.F., et al. A computerized scintillation data system for the gamma camera. *Comput. Biol. Med.* 2:251, 1972.

McCready, V.R., et al. Proceedings: Radioisotope localization of the placenta. *Br. J. Radiol.* 47:72, 1974.

Meyers, W. The Anger scintillation camera becomes of age. *J. Nucl. Med.* 20:565, 1979.

Rollo, F.D. (ed.). *Nuclear Medicine Physics, Instrumentation, and Agents.* St. Louis: Mosby, 1977.

Shtasel, P. *Speak to Me in Nuclear Medicine.* New York: Harper & Row, 1976.

Sorenson, J.A., and Phelps, M.E. *Physics in Nuclear Medicine.* New York: Grune & Stratton, 1980.

Wagner, H.N. (ed.). *Principles of Nuclear Medicine.* Philadelphia: Saunders 1968.

COMPUTED TOMOGRAPHY, ULTRASONOGRAPHY, AND RADIOIMMUNOASSAY PROCEDURES

Bennett, M.J., and Campbell, S. *Real Time Ultrasound in Obstetrics*. St. Louis: Mosby, 1980.

Bowie, L.G. *Automated Instrumentation for Radioimmunoassay*. Cleveland: Chemical Rubber Company Press, 1980.

Brown, R.E. *Ultrasonography: Basic Principles and Clinical Applications*, 2nd ed. St. Louis: Green, 1980.

Devey, G.B., and Wells, P.N. Ultrasound in medical diagnosis. *Sci. Am.* 238:98, May 1978.

DeVlieger, M. et al. *Handbook of Clinical Ultrasound*. New York: Wiley, 1978.

Hagen-Ansert, S., and Teichman, E.P. *Textbook of Diagnostic Ultrasonography*. St. Louis: Mosby, 1978.

Metreweli, C. *Practical Abdominal Ultrasound*. London: Heineman Medical Books, 1978.

Moss, A.J., Dalrymple, G.V., and Boyd, C.M. *Practical Radioimmunoassay*. St. Louis:Mosby, 1976.

Nanda, N.C., and Gramiak, R.G. *Clinical Echocardiography*. St. Louis: Mosby, 1978.

Rose, J.R., and Goldberg, B.B. *Basic Physics in Diagnostic Ultrasound*. New York: Wiley Medical, 1979.

Sabbagha, R.E. *Diagnostic Ultrasound Applied to Obstetrics and Gynecology*. New York: Harper & Row, 1980.

Phillips, B.J. *Manual of Echocardiographic Techniques*. Philadelphia: Saunders, 1980.

Raskin, M.M. *Comparative Abdominal and Pelvic Anatomy by Computed Tomography and Ultrasound*. Cleveland: Chemical Rubber Company Press, 1977.

Ter-Pogessian, M, Raichle, M., and Sobel, B. Positron-emission tomography. *Sci. Am.* 170, Oct. 1980.

Thorell, J.I., and Larson, S.M. *Radioimmunoassay and Related Techniques*. St. Louis: Mosby, 1978.

INDEX

aaron's way

aaron's way

the journey of a strong-willed child

KENDRA SMILEY
with Aaron Smiley

MOODY PUBLISHERS
CHICAGO

All Scripture quotations, unless otherwise indicated, are taken from the *Holy Bible, New International Version*®. NIV®. Copyright © 1973, 1978, 1984 by International Bible Society. Used by permission of Zondervan Publishing House. All rights reserved.

Cover photo: © 2003 Andrew Kolb/Masterfile

Library of Congress Cataloging-in-Publication Data

Smiley, Kendra, 1952-
    Aaron's way : the journey of a strong-willed child / by Kendra Smiley,
  with Aaron Smiley.
       p. cm.
    Includes bibliographical references.
    ISBN 0-8024-4349-4
    1. Child rearing—Religious aspects—Christianity. 2. Control (Psychology)
in children. 3. Smiley, Aaron, 1981- I. Smiley, Aaron, 1981- II. Title.

BV4529.S477 2004
248.8'45--dc22

                                                                        2003024377

3 5 7 9 10 8 6 4 2

Printed in the United States of America

To John,
a husband and dad
who has devoted his adult life
to loving Jesus and bringing out the best in his family

Contents

ACKNOWLEDGMENTS

There are many people who deserve a thank-you when it comes to the writing of this book. Because I am sharing this responsibility with my son, Aaron, I will try not to be redundant.

Thank you to:
- My dear friend and office manager, Pam Rush, who makes every project I undertake so much easier.
- My extended family (near and far), who have encouraged me as they have encouraged my children (strong-willed or otherwise).
- My Moody Publishers editors, Elsa Mazon and Ali Childers, whose enthusiasm and prayers were and are a wonderful blessing.
- My husband and #1 supporter, John, who never gets billing on the front of a book but whose input is fundamental and essential.

- My son Aaron, for being so transparent and so honest.
- The remainder of my children and children-in-law: Matthew and Marissa, Kristin, and Jonathan for being the supportive, creative, and delightful people that they are. I am blessed.
- And neither last nor least, my very best friend, Jesus.

Your turn, Aaron.

Thanks to:
- My first grade teacher, Grandma, who taught me how to read and expected the best out of me every day. Thanks, too, for always finding something for me to eat and for sharing memories with me.
- Mrs. Janet Crouch for caring about me as a person first and a student second.
- Mrs. Ada Murray for her discipline that made our class feel so safe.
- Mrs. Marty Lindvahl for her forgiveness and for never keeping score.
- Mr. Mark Lindvahl for modeling excellence day in and day out.
- Mr. Bill Meddler for challenging me to learn how to think.
- Mr. Quentin Ryder for allowing me the freedom to express myself and to share my ideas.
- Ed Lockhart for trusting me with adult responsibility at such a young age.
- Lawrence and Loretta Alt for talking and dreaming with me when I was a boy, but treating me like an adult. Thanks for encouraging me when it seemed like no one but my family believed in me.
- Uncle Wynn for all the effort he put into being an uncle. That effort provided *all of us* with many great experiences.

- Grandpa for giving me the opportunity to be a part of agriculture, an interest that means so much to me, and for being generous in every way.
- Mom for putting my ideas and memories of childhood into this book so that they can help others. Thanks for crying and laughing with me through the ups and downs.
- Dad for never giving up hope that I would finish my journey and for the rules that were reasonable and never changed on a whim. Thanks for the fair punishment. You had the hardest job because you put limits on my life that I hated but didn't want to live without.
- Matthew, my older brother, for sharing his friends with me and including me. Thanks for always being a rational voice in my sometimes irrational world.
- Jonathan, my younger brother, for being so flexible and being willing to play with me when we were younger and later help me with all the different work we've done together.
- Kristin, my wife, to whom I do not deserve but who is a blessing from God. Thanks for caring for me and loving me through my ups and downs. Your love helps to make my life complete.
- Jesus Christ for saving me from my sins. I will never fully understand why He sacrificed so much for us, but I gratefully accept His free gift of salvation.

Introduction:

Aaron and Others

The LORD your God has blessed you
in all the work of your hands.
He has watched over your journey
through this vast desert.

✧ DEUTERONOMY 2:7 ✧

"**Aaron wants** two cookies," said Matthew, his older brother. "He wants a glass of cold milk too." Once again, Matthew was correct. Aaron was content to allow his older brother to speak for him for the first eighteen months or so of his life. And why not? He was doing a great job of getting Aaron just what he wanted. Who would have ever suspected that this quiet little guy would be our strong-willed child? Yet, we had an inkling from early on.

"No! Again!" Aaron said in loud, defiant protest as I calmly explained that our time on the amusement ride had come to a close. "NO!" he said even louder, with fast and furious tears. "AGAIN!"

As I pried his little fingers from the safety bar, another rendition of "It's a Small World" played in our ears. I won the battle, but I had a very unhappy young man on my hands. He wanted to go again, and it was obvious that he felt certain that he should have what he wanted. Generally speaking, what Aaron wanted was to be in control of himself. That's what your strong-willed child wants too! It's not just Aaron, it's Aaron and others.

It has been said that a journey begins with a single step. Many times the journey of a strong-willed child starts even before the first step is taken. A strong-willed child, it would seem, is born strong-willed, with intense opinions and a demanding nature. He's born with the desire to control his own destiny and the uncanny ability to actually secure that control in many instances. Strong-willed children come into the world strong-willed and, for that matter, probably exit as strong-willed adults. By that time they have potentially tempered the negative aspects of their strong-willed nature. They have learned to cooperate, acquiesce, and respond to the leadership of others when it is in their best interest. In all hopes, they also capitalize on the pluses of their strong-willed nature. That is the goal we have for our high-maintenance children. Our desire is to contribute in a positive way to their development as responsible adults. Our job is not to break their spirit, but to shape their will.

This is not an easy job. Actually, parenting *any* child is challenging, but a strong-willed child adds a dimension not found in raising a more compliant child. I know that my husband, John, and I were blessed with a 100 percent, genuine, certifiable strong-willed child. And this child was recognized as such, almost immediately, by his father—a strong-willed child turned responsible, loving, cooperative adult. This father-son combination provided me with valuable insight and ultimately gave each one of us— mom, dad, and son—the motivation and understanding we needed to write this book. You'll find encouragement and instruction through this description of one strong-willed child's journey.

The journey of your strong-willed child will not be precisely like Aaron's, but undoubtedly it will be a journey with bumps and curves and perhaps even an occasional detour. It is our hope that Aaron's journey will help you realize that (1) you are not alone, (2) the strong-willed child has thought and behavior patterns that can be understood and anticipated, and (3) there are insights and strategies that can help you and your child on the journey. It can sometimes seem like a passage through a long, dark tunnel, but there is hope, not just for survival but also for success, as measured by the emergence of a responsible adult with godly character.

Before we begin our journey, it is essential to establish the definition we will use for a strong-willed child. Dr. James Dobson, largely respected as the expert on this topic, wrote that a strong-willed child "seems to be born with a clear idea of how he (or she) wants the world to be operated and an intolerance for those who disagree."[1]

That generally summarizes the strong-willed child's black-and-white thinking and intolerance. We can add a few more qualifiers to that definition. A strong-willed child *appears* confident. A strong-willed child can be both charming and defiant. He is *very* persistent and is even willing to take punishment to win. A strong-willed child is generally gifted in manipulation. In fact, he is able and willing to cause emotional upset if it is a means to gain control of his life. A strong-willed child does not necessarily want to control everyone else; he simply does not want to be controlled. Yet, he will strive to control someone threatening to control him. Who is that typically? That's right! It's you, his loving parent. His parent who might be wondering why this child is so difficult and so opinionated—so "right" all the time.

"He that complies against his will,
is of his own opinion still."[2]

BUTLER

If a strong-willed child is one who knows how he wants the world to operate and is intolerant of those who disagree, what is he NOT? Are there behaviors that are sometimes misinterpreted as strong-willed? Absolutely! My mother used to say that her children were not stubborn; they merely had "exceptional resolve." Yes, we did, but we were not strong-willed. We were stubborn—oops, sorry, Mom—I mean we had exceptional resolve, but there were very few issues that I was willing to go to the mat for. That is in contrast to a strong-willed child, who many times chooses fighting and punishment over acquiescence. Being strong-willed goes well beyond being stubborn. A strong-willed child resolutely defends his position and questions any and all authority over him to determine his or her "right" to retain command. That is more than stubborn.

The rod of correction imparts wisdom,
but a child left to himself disgraces his mother.

❖ PROVERBS 29:15 ❖

Sometimes undisciplined children are mistakenly labeled as strong-willed. A child who has managed to gain control over the adults in his life may or may not be strong-willed. Remember, a strong-willed child's goal is not to control everyone else. Instead it is to maintain control over himself (even though some adult is clamoring for that control). A bright child who has determined how to outwit his loving (though unprepared) parent might repeatedly "get his way" and make others miserable when he does not get it. But this lack of parental discipline does not confirm a strong-willed child. It merely encourages inappropriate behavior. **Warning: Lack of discipline can, however, encourage, reinforce, and empower a true strong-willed child to even greater heights (or should I say depths?) of poor behavior.

A strong-willed child is not bad or stupid or mean, although he may be classified as such by others who do not understand or appreciate his positive attributes.

Even a child is known by his actions,
by whether his conduct is pure and right.

✢ PROVERBS 20:11 ✢

Do you have a strong-willed child? For practical purposes, I will refer to the strong-willed child in the male gender. This is not to imply that only boys are strong-willed. In my findings, I discovered almost an equal number of boys and girls classified by their parents as strong-willed. Furthermore, the strong-willed child is not of any particular birth order. Our strong-willed son is our second child. Yours may be your youngest child or maybe a first-born or an only child. I have also discovered that strong-willed children do not have only one personality type. (See Tim LaHaye's *Spirit-Controlled Temperament*[3] or Florence Littauer's *Personality Plus*[4] for a discussion of personalities.) Persons with the choleric personality type are described as thinking they are "always right," and I have definitely known of strong-willed children in that category. However, there are girls and boys who are choleric in their nature and are not strong-willed. My husband, the former strong-willed child, is predominantly phlegmatic in personality (having a quiet will of iron). Therefore, personality type does not guarantee or disqualify someone as a strong-willed child. A strong-willed child can be male or female, oldest, middle, or youngest, and have any personality profile. There are similarities but no carbon copies. Do you have a strong-willed child?

And do not forget to do good and to share with others,
for with such sacrifices God is pleased.

✢ HEBREWS 13:16 ✢

The phone rang, and it was a good friend on the line. "Do you think we might be able to drive down to visit you on our way to Indianapolis for Easter break?" the caller inquired. "There are two reasons for our visit. Number one, you once told me that your son Aaron was a strong-willed child, and it looks as though we've been blessed with one, too. It's hard to believe that your respectful college-age son was once a strong-willed child. And it's equally hard to believe that our iron-willed twelve-year-old will ever turn out to be a respectful college student! And, number two, this same son recently announced that he wanted to grow up to be a farmer and a pilot, and the *only* person we know with that career combination is John. If we can stop by, maybe John can convince Nate that to farm and fly he must pass seventh grade. Then perhaps Nate can go and play, and we can pick your brains about your experience as parents of a strong-willed child."

I welcomed our friends to our home. After John spent time talking with Nathan about the rewards and requirements of his chosen career fields, the adults settled down for a discussion about parenting a strong-willed child. Our conversation had just begun when, to our surprise, our strong-willed child turned young-adult, Aaron, arrived at home from college a day early for Easter break. As soon as he realized the topic of our conversation, he entered into it with enthusiasm. I sat in awe as I witnessed his ministry to our guests.

"I don't know what Nate was thinking when he pushed the issue with his science teacher," lamented his father.

"I know *just* what he was thinking," Aaron replied. And he went on to describe the thought process of a strong-willed child who has declared war on an unsuspecting adult perceived as either vulnerable or deserving of the treatment.

Nathan's dad sat in wonder. Could there be another human on this earth who not only understood his son's behavior but also declared it predictable? Was it possible that his son was not an anomaly, a quirk of nature? Did this child respond and react in a certain way because of how God created him and not just to frustrate and annoy his parents?

As Aaron recounted story after story (some familiar, some shockingly new to me), I saw a look of hope in our friends' eyes.

Hearing the struggles and successes in Aaron's journey as a strong-willed child gave these parents hope. Hearing the insights and actions that helped us navigate his journey gave them tools. Hearing about the people whose behavior positively influenced Aaron gave them new vision, suggestions, and solutions to help Nathan grow into confident manhood.

A Closer Look with Aaron

I remember the day when I came into the family room and found the adults gathered there discussing strong-willed children. Ah, I thought, now this is one topic that I truly understand. My biggest surprise was to realize that our visitors thought their son, their strong-willed child, had some sort of problem—that he was somehow diabolical or targeting them out of meanness. Meanness? The only difficulty I could imagine was that Nathan was a very strong-willed guy. I started to tell some of my own horror stories of wanting to have control, and Nathan's parents, especially his dad, were amazed that I had one or two that outdid their son's. Nate started looking better and better to his folks thanks to some of my antics as a strong-willed child. And when I could actually finish a story that Nathan's dad started, "guessing" with great accuracy what had happened next in the particular battle, I was sure that the visiting parents were going to declare me a genius! Was I? Am I? No. (Both of my brothers got higher ACT scores than I did, although my mom thinks I'm pretty smart!) What I am is an adult who was a strong-willed child not too long ago. I realized that day that I could help parents like Nathan's learn how to appreciate and handle their strong-willed child.

※

And thus the idea was born. Aaron traveled the bumpy road of a strong-willed child, and we, his parents, were with him along the way. The trip was not always an easy one for him or us, but the destination made the journey worthwhile. Luke opened his written account of Jesus' life with these words, "Therefore, since I myself have carefully investigated everything from the beginning, it seemed good also to me to write an orderly account for you . . ." And that is precisely what we have tried to do. Throughout this book you'll hear our stories and those of others. Join us as we take you down the road we've traveled. Join us on *Aaron's Way: The Journey of a Strong-Willed Child.*

"When the train goes through a tunnel and the world gets dark, do you jump out? Of course not, you sit still and trust the engineer to get you through."[5]

CORRIE TEN BOOM

❶
The Journey Begins:
Birth to Pre-Kindergarten

For you created my inmost being;
you knit me together in my mother's womb.
I praise you because I am
fearfully and wonderfully made;
your works are wonderful, I know that full well.

❖ PSALM 139:13–14 ❖

Our strong-willed child, Aaron Joseph Smiley, arrived in February of 1981. I still remember the moment of his birth. Dr. Tanner (our family physician and the father of five sons) announced, "It's a boy!"

Naively, my enthusiastic response was, "Oh good! I already know how to *do* boys!" I don't remember seeing Dr. Tanner's eyes roll to the back of his head, but undoubtedly he wondered how I could make such a ridiculous statement!

Our first son, Matthew, was two years old at the time. I would classify him as a compliant child. It was not that he always obeyed us perfectly, but he did "aim to please." As a former teacher, I had perfected the "schoolteacher look." You know how it goes: lips drawn tightly in a pseudo-pucker, eyebrows knit together, and a very stern countenance. That "look" was completely effective with my oldest son. A stern look or a gentle scolding generally brought about conviction, legitimate repentance, and a heartfelt vow to "do better." (Can you see why I had such confidence when son number two was born? Just look at how well I had been doing with son number one!)

But the truth was that I did not know how to "do boys" any more than I had conquered the art of parenting. And that was a truth that I was soon to discover. Forget "the look" when it came to Aaron.

"Liam, my four-year-old and I were driving home from preschool, and he asked if we could have lunch at McDonald's. I explained that it wasn't a possibility. Because Liam is strong-willed (and not to be deterred by such a flimsy statement), he pursued the idea with great determination. When I finally convinced him that I was NOT going to stop at the fast-food restaurant for lunch, Liam folded his arms across his chest and humphed, 'Well, Mommy, you are making a bad choice.' My own words, frequently spoken, were repeated in an effort to manipulate me with guilt and gain control."

While Matthew was compliant and aimed to please, Aaron had different ideas. I used to explain his strong-willed nature this way: "If we draw a line in the sand and tell Aaron not to cross it and why, *and* we tell him the penalty for disobedience, he will immediately step up to the line, as close as he can possibly get, and inquire, 'What did you say you were going to do to me if I step over this line?' Then he reviews the consequences and determines whether or not to cross the line. And many times, over the line he goes." Ah, a strong-willed child.

Aaron did not always use defiance to try to get his way. This sweet little boy came into the world looking just like the Gerber baby, complete with wispy blond hair, big blue eyes, and a ready smile. One of my earliest recollections of his manipulation skills involved the use of charm, not defiance. When he was just a little over two years old, I remember scolding Aaron. I don't recall the issue, but I do remember his actions. When I finally paused in my reprimand and took a breath, Aaron smiled his deep-dimpled smile, reached out with his chubby little hands and patted me gently on both cheeks. "Dat be alwight, Mommy," he cooed in an effort to comfort me, his overwrought mother. Ah, what a sweet, caring child. Wait a minute! I wasn't the one in need of comfort. I wasn't the one in trouble, he was! I'm sure Aaron thought, "If this works, why not go for it?"

"Christine did not want to go to the first day of preschool. I talked her into getting into the truck, and then she was all excited and really wanted to go. We got there, and all of a sudden, she was mad—screaming, crying, absolutely mad—and she could not believe that I was going to make her go into this lady's room. So she would not go, would not go, would not go. 'Mom, I hate that lady, don't make me go, don't make me go. I can't believe you're doing this to me. I don't want to go, she's mean, she's mean, she's mean! I hate that lady. I hate that lady 'cause she hates me. Take me home right now!'"

Strong emotion can definitely sway a parent. "I can't believe you're doing this to me" can make any parent step back and think. Hopefully, the parent filters this sentiment through the mind to realize that strong emotion and words like "mean" and "hate" are words used to manipulate and gain control. Also, strong emotion can translate into a tantrum, which can add the term "embarrassment" to your list of sentiments.

Very few tricks that Aaron tried (charm, guilt, strong emotion, or otherwise) worked with his dad. Remember, I told you that John is a strong-willed child turned responsible adult. He knew the tricks and the importance of wise parenting.

"It is a wise father that knows his own child."[1]

SHAKESPEARE

At one point, John and two-year-old Aaron were literally eyeball-to-eyeball on the stairs, and the words from John's mouth were as follows: "Aaron, you will not win. When I tell you to do something, you must do it." If only that was the last time he had to make that statement! Even at an early age, Aaron desired control of his world.

"On another occasion, Emily was sitting at the table doing a craft project with her dad, and she told him out of the blue that he was a genius. Her dad asked her why, and she said, 'because you do everything I say.' Emily was three-and-a-half when this happened."

Aaron accepted Christ as his Savior at around four years old. It was actually the result of the guilt and remorse he felt about his own out-of-control, strong-willed behavior. He was having a very bad day and was in trouble with everyone in the family—Dad, Mom, and his older brother.

Here is a little background. Beginning when he was a toddler, Aaron was interested in agriculture and animals. I remember pulling into a cornfield on one of the family farms and hearing little Aaron pipe up from the backseat, "Dat torn looks dood!" (Translation: That corn looks good!) Our older son did not notice the status of the corn and had no opinion about its potential yield. Aaron's paternal grandfather is a farmer. This common love of agriculture made these two fast friends from the very beginning.

Now, back to the story of Aaron's personal encounter with Christ. As I said previously, that day he was behaving quite poorly (gross understatement). Bedtime finally came, and with it, the hope for a better tomorrow with less confrontation. Finally, there was peace and quiet. The next morning Aaron was up quite early. He waddled down the stairs in his footie pajamas, dragging one of his favorite blankets. When he arrived at the threshold of the kitchen, he stopped abruptly, waited for my attention, and then proceeded with his announcement.

"I asked Jesus into my heart last night," he declared. I was thrilled about this and immediately began to ask him about the details.

"That is just great!! Tell me all about it," I pried. "What happened to help you make this decision?"

"Well," he began, "I was sooooooo bad yesterday that everyone was mad at me. I figured that even *Grandpa* would have been mad."

(Remember, as far as Aaron was concerned, he and Grandpa were as tight as you could get. So the thought of Grandpa being mad was a very serious thing!)

He continued, "But I knew that even if everyone else was mad at me, Jesus loved me, so I asked Him into my heart."

By the way, that conversion experience was real and is often referred to by Aaron as "the most boring testimony possible." Personally I call it "the testimony every mother wants her child to have." Understanding God's love was important and would temper Aaron's behavior somewhat, but it definitely did not turn him into a compliant child.

You have made known to me the path of life;
you will fill me with joy in your presence,
with eternal pleasures at your right hand.

❖ PSALM 16:11 ❖

The same year, Aaron became a big brother. This event was exciting for everyone in the family. And Aaron was no exception. I can still picture his little face tightening up with excitement and hear him say, "I love Jonathan so much—I just want to squeeze his guts out!" The fact that everyone else in the family thought that he just might do that very thing was a little scary. But we kept a cautious eye on the baby and Aaron and watched as the little strong-willed child assumed the role of nurturing big brother.

And he has given us this command:
Whoever loves God must also love his brother.

❖ 1 JOHN 4:21 ❖

A Closer Look with Aaron

My earliest memory of being a strong-willed child and having an intense desire to be in charge of my own life was when I was four years old. We bought a small house in town and proceeded to tear down our old farmhouse in order to build a new home on that location. Even though Matthew and I were little, there were things we did to help my dad with his project. Because there were raw materials in the old house that could be utilized in our new one, he was literally tearing the house down rather than burning it. One of our jobs was to sort various building supplies, like hardwood, from the useless things, like shingles.

One day we were carrying materials from one pile to another. It goes without saying that this was a silly job. As a four-year-old, I could see little importance in simply reorganizing the junk! And if such a stupid job really was legitimate, for goodness sakes, let's get a tractor going to at least make the task easier and more fun. I made that suggestion, and it fell on deaf ears. Dad, for some reason—probably because a tractor was really not necessary—said that our work assignment was NOT going to change.

If there is one thing a strong-willed child dislikes, it is doing any task or assignment that he deems useless—especially if he suggested a "better way" to do the meaningless job, and it was rejected. And on that particular day, that is precisely what happened! I wanted my idea to be honestly considered. Using a tractor made complete sense to me. I wanted to defend my position, but I wasn't given that opportunity.

My older brother might have thought the sorting job was a bad idea too, and he may even have liked my idea to use a tractor; but he didn't choose to cause a problem. I did. I simply decided that I would not do what Dad had ordered and expected. Dad would have to pay the price for not considering my great idea. I wouldn't work as hard as he wanted me to, and he would have to shift some of his attention to me and away from his agenda. I remember Matthew telling me that my slowdown strike was a bad idea. When Dad noticed my manipulation of the situation, slowing down but not completely disobeying (a gentle way to say defiance under control), he told me precisely what I was supposed to do, and he also told me the consequences for disobedience. I would be paddled. I pondered my options, much to my brother's discomfort. "You better do it, Aaron," he said. "Dad's serious!" I knew that he was serious, but I had to decide if my work slowdown, impeding Dad's progress, was more important than the pain I'd receive. And, guess what? I decided to go for the paddling.

As I cried, Dad announced that he expected me to do as I was told. As you may guess, I weighed the pros and cons of another confrontation. My brother (the compliant one) by now determined that I needed my head examined. "Come on, Aaron, do what you're sup-posed to do." He could not fathom the thought that winning my case was so important that I would pay a price. That was the first time, but not the last, that I realized we were wired differently. He couldn't under-stand my strong-willed nature, and I couldn't under-stand why he couldn't understand. (But I did appreciate his sympathy when I decided to go for a second round before my Dad was able to make his point that defiance would not win.)

Your word is a lamp to my feet
and a light for my path.

❖ PSALM 119:105 ❖

By the time Aaron reached school age, there was no doubt in my mind that Aaron was a high-maintenance child. In my think-ing, that title "high-maintenance" put most of the responsibility on us as parents. We didn't have a label or an excuse for behav-ior that needed to be corrected. Just as a fine-tuned race car de-mands more sophisticated and time-consuming maintenance, I realized that our potential "top performer" demanded more so-phisticated (read: frequent and intense) effort. As he prepared to go to the adventure called school, I prayed that the adults who would have his attention for the majority of the day would ap-preciate his attributes, keep him under control, and help to mold and nurture his development.

Because John and I are both teachers by training, we instilled in Aaron a respect for education and the teaching profession. He also knew that we would reinforce any discipline administered in school. It was our hope that the teachers would care about Aaron enough to control and encourage him. Some did, some did not.

Avoiding the Discipline Detour

My son, do not despise the LORD's
discipline and do not resent his rebuke,
because the LORD disciplines those he loves,
as a father the son he delights in.

❖ PROVERBS 3:11–12 ❖

The responsibility of disciplining a strong-willed child is by far one of the largest potential detours in the journey. Over and over again the question arises, "Does a two-year-old really need discipline?" The question in return is, "Does your two-year-old choose to defy you?" And if the answer is "yes," that is also the answer to question number one. The time to begin to discipline or train your child is not based on a chronological date or stage, but it is based on the individual development and needs of your child.

I knew a young couple that announced from the beginning of their daughter's life that they were not going to discipline her until she was able to talk. To them, words would indicate understanding on the part of their child, and that was the guideline they established. Although their daughter was still not talking at eighteen months old, they continued with their strategy: "no discipline until Mandi is able to talk to us." Mandi was no dummy, and I watched her get by with some pretty defiant behavior while keeping her mouth shut. In fact, it is my theory (never to be proven or disproved) that Mandi was capable of talking long before she finally uttered her first words. She *knew* that sooner or later she would talk, and the jig would be up. Then her parents would go to "part two" of the program and begin to discipline her. She held out as long as she could, until around the age of two as I recall, and then she gave in. If Mandi was a strong-willed child like Aaron, she might still be silent today! A certain age or stage is not necessarily the perfect indicator of when discipline should begin. You need to discipline your child when he chooses to defy you. And there is usually little doubt when a strong-willed child chooses defiance.

> O LORD, *do not rebuke me in your anger*
> *or discipline me in your wrath.*
>
> ❖ PSALM 6:1 ❖

Disciplining a strong-willed child can be frustrating. The ability of a young strong-willed child to keep arguing when most adults are exhausted often leads to the incorrect conclusion that discipline does not work. Again, that is an *incorrect conclusion*. But in order to see the benefits of loving discipline, there are some very important "Rules of the Road" for disciplining children. First of all, discipline should NEVER be done in anger. When you do this, you have lost control. The reason you are disciplining your child is to control his wayward behavior and to teach him correct behavior. An out-of-control adult is not effective and is actually counterproductive.

There were times when I found myself angry because of the behavior of Aaron. At that point, I left the room to regain my composure and returned to administer the appropriate discipline. Did I always do that? No, and my older sons developed a stand-up comedy routine based on me swinging my arm wildly into the backseat of our car, hoping to connect with one of them and stop their fighting, arguing, complaining—any or all of the above. In those instances, I was attempting to control my children's behavior when I wasn't even successful at controlling my own. That goofiness did NOTHING to gain control of the poor behavior (except maybe distract them as they dodged and giggled at my ridiculous actions). That was not discipline—it was slapstick comedy with the potential to cause an automobile accident.

There are two more important points to be considered in the discipline of a strong-willed child of any age. Number one, you must pick your battles wisely. If you chose to, you could fight all day with a strong-willed child and exasperate your child. What is really important? Not a bad question to ask yourself.

I was doing a book signing in a major metropolitan area, and a young woman came into the store with her toddler in a stroller. The two-year-old was wearing a biking helmet and my first thought was, "My goodness, this mom must be a very poor stroller driver!" As she made it to the front of the line, she pointed down at her little darling and said, "He wanted to wear it, and I decided it wasn't worth a fight." How true! (And how wise.)

> "The main thing is to keep the main thing the main thing."
>
> UNKNOWN

When Aaron and Matthew were just twenty-one months and four years old, respectively, the family made a very long trip by car to California. My husband, John, an Air Force Reserve pilot, was scheduled for some additional flight training, and the reporting time was not negotiable. We loaded the kids into our sedan and started on a fifty-two-hour road trip. In order to maintain peace and harmony, we planned to do as much traveling around-the-clock as we could, and I organized many diversions for the boys. I hung men's shirts on the hooks by each of their seats. The bottom hems were sewn shut. Inside the shirts-turned-bags were all sorts of goodies, including metal cookie sheets that were used for desktops and magnetic playing boards. I wrapped up little toys and markers (each individually packaged) as well as magnetic letters and shapes. Every waking hour each boy was given a new package to unwrap. Believe it or not, this kept the boys happy during the trip.

After we were in the car for about thirty-six hours, we decided to stop at a restaurant rather than eat another of the lunches I packed. The boys enjoyed their meal, and then we allowed them to run around the area in which we were seated. There were no other patrons close enough to be bothered by their footloose behavior, although I'm sure some wondered why any parent in his right mind would allow such commotion. We decided that corralling them was not a battle we wanted to fight. After all, they would soon be corralled for the remainder of the trip, still some sixteen hours. What battles do you want to fight?

No discipline seems pleasant at the time, but painful.
Later on, however, it produces a harvest of righteousness and peace
for those who have been trained by it.

❖ HEBREWS 12:11 ❖

The next important point is this: the battles that you choose to fight, you must win. Remember what John said to Aaron during their eyeball-to-eyeball meeting on the stairs? "You will not win." Did Aaron believe him? Maybe not that day. Did he eventually believe him? Yes. When you pick a battle, you must win. That is why you don't want to engage in every possible skirmish.

A man reaps what he sows.

❖ GALATIANS 6:7 ❖

As soon as he can reason, a strong-willed child wants to know the consequences that will result from his misbehavior. Aaron got as close to "the line" as he could to weigh the joy of control against the pain of the consequences before choosing whether or not to cross the line. That is knowledge every strong-willed child desires. When you are determining the consequences, NEVER make a threat you cannot keep. The mother who shrieks, "If you do that you'll be grounded for life," is simply showing her lack of control. The strong-willed child knows immediately upon hearing the idle threat that he has the upper hand. No mother is capable of enforcing the threat to ground her child for life (and maybe not even for a month), and indeed, the parent also would be grounded. Believe it or not, I hear fathers threaten, "I'll break your arm if you do that." Or, "I'll knock you to the other side of tomorrow if you don't stop." These unattractive, highly offensive statements are not credible and result in the strong-willed child winning the battle. The unreasonable threats made by a loving (though unthinking) parent will never be carried out.

"We were at my sister's house having dinner, and she was trying to get her little boy to stay in his seat. He wanted to get down and play with my two-year-old. He did not want to finish his dinner. Finally, she let him know that he was not trapped at the table forever. She simply stated, 'You can get down and play once all of that food is off your plate.' At that point, she continued eating until Brian announced, 'Mom, I can get down now.' All the food was off his plate and placed on the table.

Being clear about your instructions is as important as being clear about the consequences. In the early stages of the journey it is not too early to discipline your child. Undoubtedly the form that discipline takes will vary as the child matures.

"For three days I tried to get my three-year-old to clean the toy room. Each time, she would very calmly tell me 'no' and walk away. Finally, Halloween rolled around, and we were talking about trick-or-treating. I told her, 'If you want to go trick-or-treating, you have to clean the toy room.' And she said she didn't want to. I said, 'Sarah, if you don't do your job, you are not going trick-or-treating.' She didn't want to cooperate, so I let twenty minutes go by, and then I sat her down. I explained very calmly: 'Your sisters are going to get their costumes on, and Daddy's going to take them outside, and they are going to run around and yell "trick-or-treat" and get lots of candy.'

She nonchalantly said, 'I already have candy,' and walked out of the room. A couple of hours later it was almost time to go, and the older kids decided to put on their costumes. I was getting their makeup done, and Sarah came running in reporting, 'I want to pick up the toys. I want to go, too.'

'No, it's too late,' I said. 'It's time to get the costumes on.'

So she walked out of the room, and fifteen minutes later she walked back in. 'I cleaned the whole toy room by myself.'

That is what I wanted her to do for three days, so I helped her get into her costume and let her trick-or-treat. Mistakenly, I thought I won the battle. And then I realized that my three-year-old won, because she did *exactly* what she wanted to do *when* she wanted to do it and not when I asked her to do it. I actually lost the battle.

Blessed is the man you discipline, O LORD,
the man you teach from your law;
you grant him relief from days of trouble.

❖ PSALM 94:12–13 ❖

So how do you discipline a strong-willed child? There are many different ways. A parent can give the child an explanation or a "time-out." These are sometimes quite effective. So is spanking. We did not use only one form of discipline with our strong-willed child or with our other children. The most important thing to remember is that discipline must NEVER be administered in anger. Spanking is a form of discipline that does not require a long time frame to accomplish. Someone asked me once if I typically used a time-out with my strong-willed child. No, I didn't. As I reflected on Aaron's strong-willed nature, I realized that if I had, he would probably *still* be in time out and would not have been able to attend college. No, we had a paddle, "Mr. Sore Butt." I know, it's not a very elegant title, and I'm sure my mother would have preferred the name "Mr. Sore Bottom." Nevertheless, it was effective.

Punishment is meant to deter negative behavior, that is, to dissuade and discourage inappropriate or potentially dangerous behavior and defiance. Your punishment must have impact. A light tap on the bottom of a diapered strong-willed toddler is not accomplishing your goal of punishment.

It has been over two decades since I taught school, and many things have changed in the classroom and the nation since that time. But I still remember the day that a student named Joseph stepped over the line in my class, and I escorted him to the principal's office. We decided that Joseph's behavior warranted a spanking, and with the principal's approval and watchful eye, I administered one swift swat to Joseph's bottom. He went back to the room, and I followed at a much closer distance than he realized. I entered the room just as Joseph was bragging that it "didn't hurt at all!"

"No problem," I said, "we can take care of that." And back we went to the office. I swatted harder the second time. My initial punishment was not sufficient to deter poor behavior. If you do not see a change in behavior when your punishment is administered, it could be because you did not use adequate force.

"My niece, Melissa, was in kindergarten for two months when her parents got a call from the teacher. 'I don't know what to do with Melissa,' lamented the instructor. 'Every day when it is time to pick up the toys, she refuses to help.'

Melissa's dad promised to try and get to the bottom of the problem. That day when she came home from school, he confronted her about not picking up the toys at school. Her reply was simple, honest, and straightforward. 'Look, Dad,' she said, 'all the teacher does is make you sit in the corner in time-out if you don't help pick up the toys. I'd rather sit in a chair than have to pick up all those toys. So, I just make sure that she sees me not picking up the toys, and pretty soon she sits me on the chair. I get to watch all the other kids pick up the toys and it's over.'"

The previous story is an obvious example of the attempted deterrent failing miserably. Melissa preferred punishment to losing control and having to perform an unpleasant task—in this case, it was picking up toys.

Simply let your "Yes" be "Yes," and your "No," "No";
anything beyond this comes from the evil one.

❖ MATTHEW 5:37 ❖

What about the parent who sets the boundaries, witnesses her child cross them, follows through on the predetermined discipline, and then begins to feel sorry for her child, and cancels the punishment in midstream, before the entire "sentence" is served? Woe to that parent. The fancy term for what transpired is intermittent reinforcement. Modern science has proved that sporadically reinforced behavior is very difficult to extinguish.

Parents exercise intermittent reinforcement for two main reasons. They feel sorry for their child. (After all, Little Johnny has been inside for two whole days now, and all the neighborhood kids are frolicking right outside his window.) Or, they are too exhausted to administer the punishment. Strong-willed children are persistent, and as the parent of one, it is important for you to be *more* persistent. If your strong-willed child can wear you down or convince you that you were overboard with your discipline, he will. If you are inconsistent with your discipline, your strong-willed child will battle longer, imagining that this is another time that you will give in. If you are consistent, the chances of your strong-willed child eventually giving up the fight are increased.

"I have a very strong-willed child. She is so difficult that it's been putting a wedge between my husband and me, especially at bedtime. We're having difficulty finding time alone together, because she doesn't want to go to sleep at night. She says she is scared. She starts cracking her knuckles and licking her lips. She doesn't want to sleep alone. We are asked why mommy and daddy get to sleep in the same bed, and she can't sleep in the bed with us. A lot of times we end up giving in because we're so tired, and we just want to go to sleep. All of our alone time is gone, and it's really causing friction in our relationship. HELP!!!"

Giving in is the opposite of winning the battle. It is losing the battle. Remember that winning the battle is one of the "Rules of the Road" for a successful journey. I was most guilty in this department when it came to potty training. Again, there was a stark contrast between my first, compliant child, and my second, strong-willed child.

I can still hear my mother saying, "All you have to do is feed him oatmeal for breakfast, every day at the same time, and then have him sit on the potty." It sounds so simple, doesn't it? And that is exactly what I did with Matthew, and within weeks he was done having accidents. (Do I need to remind you again of my amazing skills as a mother? Just wait, my big head will deflate soon enough.) So what did I do with Aaron? Why, of course, I did the same thing. And when he didn't respond as rapidly as his older brother, I immediately tried plan B. That was the reward plan, as I recall. And in a few days when that had no obvious result, I went to plan C. I think that one was some kind of a point system. By plan F, I was thoroughly frustrated with Aaron and with potty training. Talk about something that he alone was able to control. In retrospect, I realize that I was not consistent or persistent. I intermittently reinforced his noncompliance by switching strategies. In short, I didn't do a very good job. Fortunately, he finally decided that it was in his best interest to join the ranks of the potty trained.

Bowel and bladder control (how glamorous) are very difficult for the parent of a strong-willed child to control. The best plan is to convince your child that he wants to acquiesce and control these functions as the adult population does. And whatever your method, stick to it.

Parents, every day you have to decide where you'll draw the line—what behavior is permissible and what will not be tolerated. You'll have to be ready for battle every day until your child makes his own decision to stop battling. The hope is that his determination for control will lessen each day. But until he decides to acquiesce, I guarantee you that you will be pushed and tested. The more often you give in, the worse it will get, and the longer the process will take. If you hold firm, your strong-willed child will eventually give up engaging in many of the fights. Don't be short-sighted. Raising a strong-willed child is not a sprint; it's a marathon.

A Closer Look with Aaron

Mom told you the story of Dad and me "eyeball-to-eyeball" on the stairs. And she wondered out loud if I actually believed Dad when he said that I had to obey him. It wasn't really a question of whether or not Dad was telling the truth. That wasn't the issue. The issue was whether or not I wanted to suffer the consequences of disobeying him. My Dad was always fair and never administered punishment in anger, but when you were spanked for disobeying, you knew you were spanked.

Some parents find that discussing an issue with their children is an effective way to discipline them. My guess is that if that system works, the child is not actually strong-willed. As an adult, I recently saw a television show about disciplining children. One mother told the story of her three-year-old not wanting to stay in his bed at night unless she was in his room. Her solution? She pulled a rocker into the room and told her son that she would sit and read while he fell asleep, but if he got out of bed, she would have to leave the room and shut the door. Interestingly, her plan worked after a week. At first, she had to stand up from her chair when he hopped out of bed. But after she threatened to leave the room, he went back to his bed. She reported that soon he was staying in bed and falling asleep right away.

As I listened to this mother giving her discipline advice, I realized that gaining control over that situation would have been no problem for me (or any strong-willed child). I would simply get out of bed and force the mother to leave the room. Then, after she shut the door, I would scream bloody murder. At that point, the mother would have an obvious dilemma. Because no one was in the room to stop me, no one to physically put a hand over my mouth, the odds are she would feel compelled to check to be sure I was alright and try to calm and silence me. And as soon as she reentered the room, I would immediately jump into bed. Then if she had decided to spank me, she'd feel guilty because I was back in bed, right where I was supposed to be. If I got out of bed again and she left, I'd simply repeat the scenario. The probability of this woman's plan working with a strong-willed child was, in my estimation, not very high. A strong-willed child is capable of almost immeasurable persistence, unless he is ultimately convinced that the discipline is too uncomfortable.

What is difficult for most people to understand is the motivation of a strong-willed child. The strong-willed child will choose the stated punishment in order to be in control, especially if the child considers the punishment moderate. Too many times the parent is shortsighted. A strong-willed child may choose to miss a party, not get dessert, or be sent to his room, because those things are really no big deal. A guilt-ridden parent (this is false guilt induced by the strong-willed child) is not likely to choose to fight that particular battle again. The child is now running the show. The strong-willed child has chosen a small sacrifice for a big payoff. And usually, the parent has no clue. "Why didn't he want to go outside? All I asked him to do was pick up his toys. Doesn't this child ever give in? I hate seeing him missing out on all the fun." If a strong-willed child can make his parent feel guilty for carrying out the punishment, he may have lost a skirmish, but he has advanced in winning the war. What is the key for the parent? Keep things in perspective. Always look at the big picture.

And, if and when you leave the room to gain your composure, go ahead and get your blood pressure down and calm yourself, but don't ever change the edict you made prior to your exit. Don't change the rules. Think before you make them. Be fair and stick to your decision.

You probably also need to know that, even after the punishment has been administered, it's really not over. I remember getting paddled and screaming like crazy. The punishment definitely hurt, but the screaming was totally to get Mom's sympathy. I would respond to reasonable and adequate spanking like someone was ripping my leg out of the socket. I tried to convince the folks that I was going to be permanently injured. And Dad (the guy who'd "been there and done that") would say, "You have three minutes to cry until I spank you again." That was smart and effective. If he hadn't said that, I would have continued the manipulation indefinitely.

Mom's suggestion that discipline ought to never be administered in anger is a good one. But I want you to think about a more radical possibility. We've all heard the cliché "This is going to hurt me more than it hurts you." Some of you have probably said it. That statement implies that the parent enforcing the discipline will be experiencing some degree of pain, which is probably true to some extent. Now, here's my idea. What if an attempt was made to discipline with no emotion whatsoever? We all know that it is wrong to discipline in anger. I'm suggesting that neither disciplining in sympathy nor frustration nor pity nor commiseration is the optimal condition. My dad was able to accomplish nonemotional punishment the great majority of the time (as opposed to my mother). He dispensed the predetermined punishment as a sales clerk might hand you a product you purchased. You selected the item, paid for it, and now it was yours. The sales clerk has little emotion (positive or negative). She is giving you what you have chosen and deserve. In the same way, a parent giving a strong-willed child what he has chosen and deserves should try to do it without emotion. The truthful message to your child should be "This is going to hurt *you* more than it hurts me." That doesn't indicate lack of love or caring; it is merely healthy detachment in delivering punishment.

The Journey Continues:
Kindergarten to Grade Six

Give, and it will be given to you.
A good measure, pressed down,
shaken together and running over,
will be poured into your lap.
For with the measure you use,
it will be measured to you.

✧ LUKE 6:38 ✧

The verse above is one example of a paradox found in God's Word. The Bible contains numerous paradoxes. When we give, we receive. We die to live (Matthew 10:39, Philippians 1:21). Many who are first will be last, and the last first (Mark 10:31). These seemingly contradictory statements can be confusing, like the paradoxes in the life of a strong-willed child. For example, the strong-willed child wants to control himself and yet wants to be controlled by his loving parents and other responsible adults. That apparently inconsistent attitude is manifest in the life of a strong-willed child. He wants to live in the presence and structure of just and fair rules.

As Aaron's parents, it was our hope that his desire to control his world would be lovingly overruled by his teachers. We looked forward to his teachers being caring enough to make him mind authority. We hoped that they would respect and appreciate his creative, out-of-the-box thinking and his enthusiasm for life. In the grade school years, Aaron intersected with many adults who encouraged him and some adults who were very discouraging.

In grade school, Aaron was often disappointed by lack of fairness in the classroom. The reasoning of a strong-willed child is more black-and-white than that of other children. They trust that when rules are made, they will not be changed. They are intolerant of behavior that is not perceived as just.

One Monday, a primary school teacher of Aaron's declared that "Button Day" was on Friday, and instructed the children to wear as many pins and buttons as they could find. There was to be a prize awarded for the most buttons. Aaron took this challenge seriously. He raided my desk and found campaign and advertising buttons, and he decked himself out from head to toe. Upon arriving at school that day, proudly displaying what was undoubtedly the winning button collection, his teacher told him that he had to give some of his buttons away to the students who had forgotten about the assignment. What? Before the judging? Certainly that couldn't be fair! Tears flowed as his teacher forced him to give up pieces of his blue-ribbon button collection. She expected him to unselfishly share his buttons and to do it in a good-natured manner, even though this was a total change in the rules for the day. Aaron would not have minded sharing his buttons if that was the plan from the beginning. If the explanation of "Button Day" was to collect and distribute buttons, he would have been more than happy to be the chief button provider.

Aaron couldn't understand why his teacher was changing the rules in the middle of the game. He expected her to understand that he had searched long and hard for this collection of buttons. She never indicated that he was in charge of bringing extra buttons for the less motivated or more forgetful students in the class.

"Life is not fair. God is good. And this too shall pass."[1]

SHIRLEY DAHLQUIST

She forced him to give up his buttons, and he did as he was told but not before his feelings were hurt. That didn't seem to matter to his teacher.

The unfair order of Aaron's teacher wounded him. And, unfortunately, that set the stage for the rest of the year with this particular teacher. She did not like Aaron. Perhaps she didn't before the infamous button incident but definitely not after it. She perceived him as selfish and didn't like his questioning her rule change. At one parent-teacher meeting, she informed us that Aaron was merely "an average child with no great potential." We should not expect anything more than that. As someone who expects the best in people and in situations, this thought was not only foreign to me but also distasteful. An adult who changes the rules and makes the strong-willed child pay for the change is, if not a roadblock on the journey, at least a great big speed bump!

> "There is something very powerful about ... Someone believing in you, someone giving you another chance."[2]
>
> SHEILA WALSH

Another of Aaron's elementary school teachers had a totally different approach. Rather than declare him to be something less than he was, she specifically asked us how we thought she could best help Aaron develop his potential. We encouraged this teacher to hold a tough line with him, not let him dictate, and make him abide by the rules she established. Delightfully, she followed our suggestions, and Aaron's experience with her was very positive. This teacher was wise and mature, and Aaron reaped the benefit of her wisdom. As the parent of a strong-willed child, you can offer understanding and insights that can assist a teacher in helping your child.

> *Therefore I do not run like a man running aimlessly;*
> *I do not fight like a man beating the air.*

❖ 1 CORINTHIANS 9:26 ❖

In second grade Aaron and I had a battle that is permanently etched in both of our memories. He was in the top reading group at school, and the students were given an assignment to write a creative poem for an area-wide competition. The problem with it, as Aaron saw it, was that it was not for a grade in class. They were writing this poem to be entered in a contest. Aaron didn't see any use in that! What a silly waste of time. He didn't want to enter a contest.

Thankfully, he didn't express these feelings to his reading teacher. Instead he announced to me on the evening before the due date, that he was NOT going to do the "stupid" assignment. Of course, I told him differently. It was a battle that I was going to fight and win. Aaron was fired up for the conflict, and it was one of the most intense battles of the will we ever fought. At one point, I literally pinned him in the chair, almost sat on his lap and informed him, "I will not create this poem for you. *You* are going to do it. I have a piece of paper, and I will take down dictation as you compose the poem." (How many times is a stenographer that forceful?) We sat there as he protested loudly. I did not give an inch, literally, and finally he started shouting out ideas for the poem.

Ultimately his creative work evolved into this: "Busy bear, busy bear, don't bother me! I have work of my own—like gathering honey and other nutty things. So, DON'T BOTHER ME!" Years later, we laughed at his finished product. There was no doubt that he was trying to tell me not to bother him. Fortunately, I didn't read between the lines that evening. I was just so glad he had actually produced something to hand in to his teacher for the contest. I was exhausted from the battle, but it was one worth fighting.

By the way, "Busy Bear" won second place in the three-county contest. Aaron received a certificate and savings bond at a ceremony held at the junior college. And he thoroughly enjoyed the accolades. "Busy Bear" was an important victory, and it was one battle that I fought without the aid of my husband, John. Usually he was there to lead the charge, but he was gone on assignment with the Air Force Reserve, and this was one of those times when *I* was flying solo.

Years later I discovered a biographical form that Aaron completed in the third grade. Among the many sentences that the students were asked to complete were these:

"Something that I am afraid of is _____." Aaron drew scribbles through that line. In his mind there was NOTHING he was afraid of.

Then came the next line:

"Something that I am NOT afraid of is _____." In that blank he had boldly written "Mom." I shall be forever thankful that I was not alone raising my strong-willed child.

"Every experience God gives us,

Every person he puts in our lives,

Is the perfect preparation for the future

That only he can see."[3]

BETH MOORE

49

There was one year in grade school that was especially unpleasant for Aaron. Amazingly, John and I didn't even realize what was transpiring until the day we got *the phone call* with only six weeks left in the school year. "Could you please come in for an appointment?" was the request. "I've been having a little trouble with Aaron."

"A little trouble" did not prepare us for what we heard later that day. As John and I sat down with the teacher, she told us about Aaron's behavior in the classroom. She gave the class an assignment that Aaron decided not to do. The class was to write in their journals every day. "That's silly," he declared. So what did he do? He organized a coup d'état, and most of the class boycotted the assignment. (Remember, I told you that a strong-willed child could be charming and persuasive!) We didn't know anything about this incident or his rebellion until we met with the teacher that day.

Was I embarrassed? Yes! Was I disappointed in Aaron's poor choice? Yes! Was I amazed that this teacher had let the conflict get to this point? Yes! She was unable to control this elementary school boy. He was allowed to take control, and now in tears, she was meeting with us, his parents. "What should I do?" she wanted to know.

Even though part of me wanted to scold the teacher for relinquishing her power to Aaron, we knew that it was in Aaron's best interest for us to take control of the situation. We told her that she would have no more trouble with Aaron and that he would write in his journal as instructed. (This guarantee was a little easier to make because there were only a few weeks left in the school year.)

That evening we made it very clear to Aaron that he could no longer behave poorly in class, and if he did, he would pay the price at home. He believed us, but he was not happy. My recollection is that every single day of school, for the rest of the year, he cried before the bus arrived. (And I cried after the bus drove away.) The teacher wanted to win the battle, but her authority had been so undermined throughout the year that she could not. We came to her rescue, though Aaron felt that he needed rescuing, too. Yet, forcing him to give up control was definitely in his best interest.

A Closer Look with Aaron

I can't believe that the teacher *only* told Mom and Dad about the journal boycott. That was not the first incident. I guess it was the straw that broke the camel's back. And I know why it did. The majority of the class decided to "get on board" with my refusal to complete the daily assignment. And because of that, our teacher couldn't use the fear of poor grades as motivation. If she did, most of the kids in the class would have received an F on the assignment. Can you imagine explaining the failing grade to someone's parents? "Your child flunked this assignment, because I couldn't convince her to do it." Now *that* sounds professional. So she finally had to call in the big guns. She told my parents about the problem at hand—but believe me—she lost control long before that.

The problems began about four or five weeks into the school year. We were in a science unit studying solids, liquids, and gases. She gave us the following definitions for each: A solid is something that holds its own shape and has a constant volume. A liquid takes the shape of the container and has a constant volume. A gas takes the shape of the container and has a variable volume. Simple enough, right? (Can you believe that I can still remember these?)

Well, I got to thinking, what is ice cream? Is it a

solid or a liquid? So I raised my hand and asked, "What is ice cream? When you scoop it out, it takes the shape of the ice cream scoop like a liquid. But when you put it from the scoop into the bowl, it holds that shape and doesn't take the shape of the bowl."

I was not trying to be a smart aleck when I asked that question. It just fascinated me. My teacher, however, was not fascinated. She did not know the answer, did not care about the answer, and was completely annoyed that I asked the question. Her reply was something like, "It doesn't make any difference." What kind of answer is that? I raised my hand again, and when she acknowledged me I simply said, "You don't know, do you?" That completely flustered her, and she ignored me and moved rapidly through the lesson.

The ice cream question was the beginning of the end for that teacher. I didn't care at all that she didn't know the answer. I didn't know the answer either. If I did know, I wouldn't have asked. But it was a good question, an interesting one, and she wouldn't admit that it was one we'd have to research. She was not equipped to deal with a grade school kid who could ask a question that she couldn't answer. From that point forward, she was a target for me. I looked for opportunities to frustrate her, wondering if she would ever take charge.

I could tell you story after story. (Sorry, Mom.) For example, we had to write sentences with our spelling words, and we were told to put the punctuation at the end of the sentence. I did—*way* at the end—as close to the right-hand margin as I could get, regardless of the length of the sentence. She'd mark all my sentences wrong, but she could not give me a poor grade, since I (technically) followed her directions.

We had a rule—no gum, water, soda, etc., during the day—only milk during milk break. That was the rule, although she chose to drink coffee and soda all day long. So one day I saved my milk and brought it out much later to sip as I was reading my history lesson. "Aaron, milk time is over. There are no drinks in class," was her reaction. And my reply? "You have your coffee, and I have my milk. If there are no drinks in class, then why do you get to have one?" She informed me that the rule only applied to students, not to teachers. Since I decided that the rule was unfair, and I knew that she was vulnerable, I chose to make an issue of it. My battleground wasn't the classroom. Instead I took my protest to the playground. While we waited in line to play four-square or to swing, I'd say to one of my classmates, "Can you believe that our teacher gets to drink in class? That's not fair." It didn't take much to get the majority of the class inflamed over the injustice of the rule. Before long, every drink of coffee or pop she took was met with one of my contemporaries saying, "No drinks in class."

The pressure became too great for her. Rather than control the students in her class and insist they stop their new aggravating habit, she put a little table out in the hall. And that was where she left her coffee and soda when she entered the room. Then when she wanted a sip, she stepped outside. Evidently, she felt guilty about drinking in the classroom. Her giving in to our pressure diminished our respect even more. And it liberated the kids in the class. Slowly but surely, she was disintegrating in front of us. She lost the respect of the majority of the students in the class. And she seemed to be helpless when it came to reestablishing her rightful control. After all, what was she going to do? Call in the principal or phone our parents and tell them that we had forced her to drink her coffee in the hall? She was obviously embarrassed and wanted to keep the uprising confidential.

I didn't enter school that fall with the idea of making her life miserable. And she probably didn't imagine that she could be challenged so severely. The hole in the dam occurred when I realized her unwillingness to admit that she didn't know something. I didn't expect her to know everything, but I did expect her to be truthful and honest about what she did not know. Her immaturity triggered my strong-willed nature, and I took control.

After the meeting with my parents, she knew that I was going to have to tow the mark. She finally had leverage. Unfortunately, by this time she had an understandable personal vendetta against me. She was so angry at this point that she wanted revenge. And she was able to achieve it to some degree.

There was a student in our classroom who had many serious problems. Years later, in fact, he was imprisoned, and he has not, up to this point, been his contributing member of society. The possibility of a bleak future was evident even in the grade school years. One of his antisocial behaviors was spitting on his classmates. Whenever he behaved in this manner, the teacher disciplined him until one day when he haphazardly spit on me. "Aaron spit on me first," was the defense of this young man. "No, I didn't," was my reply, backed by several unsolicited testimonies from my classmates. "Well," responded the teacher. "I wasn't there to see it, so I can't be sure who was at fault." Touché! The spitter walked away with no punishment, noted who was the "free target," and I spent the remainder of the school year dodging the vengeance of the teacher in the form of an out-of-control classmate. I knew I couldn't strike back, because she would never come to my defense. It didn't matter if I was innocent or not. In her mind I was guilty. It was a painful last few weeks of school.

"We should be in the business of building people up. There are too many people in the demolition business today."[4]

NORMAN VINCENT PEALE

That was an awful year for Aaron! In our school district, the policy used to be that parents could request a specific teacher for the coming year. I went into the principal's office weeks after our meeting with Aaron's teacher and simply said, "I don't know who would be the best teacher for Aaron. I need your help. Please suggest someone who will love him." And she did.

The next year was a time of healing for our strong-willed son. His subsequent teacher did love him. She appreciated his out-of-the-box thinking and encouraged him to solve problems that were not necessarily in her lesson plans. If Aaron asked a question in class about something only vaguely related to the discussion, she would reply, "That's a great question, Aaron. I'm not sure what the answer is, but maybe you can look it up in the encyclopedia during break and find something." Her reaction could have been, "Aaron, we are NOT discussing that right now. Please do not waste my time with your unrelated questions." To a strong-willed child, an answer like that is internally translated: "I don't know the answer, and I'm not going to admit it, because I'm trying to convince everyone that I know everything." And with that answer, a strong-willed child has discovered a chink in the armor—an adult who is not willing to admit that he or she doesn't know everything—an insecure, immature, and vulnerable adult.

His new teacher was not threatened by his questions. She was not a pushover; she was simply respectful of him. And he returned the respect. At one point in the year, the class bully lashed out at the teacher, and she had to physically restrain him—at least she tried. As she literally wrestled with his arms and upper body, Aaron came to her rescue and grabbed the bully's legs, preventing the child from continuing to kick their teacher. Aaron could not bear the thought of someone hurting this kind person. He risked his well-being (and becoming a later target of this bully) as an act of loyalty to someone he loved and respected. What a change in behavior, a logical response to his new teacher's behavior.

> "The man incapable of making a mistake is incapable of anything."
>
> ABRAHAM LINCOLN

Everyone must submit himself to the governing authorities,
for there is no authority except that which God has established.

❖ ROMANS 13:1 ❖

In the years of kindergarten through grade six, Aaron's life intersected with more positive, encouraging adults than discouraging ones. He encountered more than one wonderful teacher who demanded and deserved his respect. One of those came in the later years of elementary school. The defining battle with this teacher came almost immediately upon introduction. Aaron was testing to see what *this* school year would bring. His teacher pronounced a word differently than Aaron had heard it pronounced, so he chose to correct her. When she disagreed, he jumped up from his seat, got a dictionary, and started to look up the word. His teacher marched over to Aaron, shut the dictionary with force, and loudly slammed it on his desk. Then she told him to go outside the room. In a few minutes, she joined him in the hall, handed him a piece of paper and a pencil and told him that he would not speak to her that way. Furthermore, he was to write her an apology.

Stunned and a little surprised, he wrote that apology and took it in to her. She read it and responded to Aaron, "I forgive you," and that was the end of it. She immediately buried the hatchet. She was mature. She was secure. She demanded and deserved Aaron's respect. Not only did he give it to her, he loved being in her classroom that year. He felt safe, comfortable, and respected. He knew where the boundaries were placed and that they would be enforced in no uncertain terms.

Aaron treated both of these excellent teachers with the utmost respect. Their personalities and methods of teaching and interacting with their students were not identical. However, these women were mature and confident and appreciated the finer points of a strong-willed child. They were very positive influences in Aaron's life. Remember, one of the paradoxes of a strong-willed child is that this child who desires control of his world also desires to be lovingly controlled. More paradoxes to come. Next we will examine what people assume about the strong-willed child versus what is actually true.

Not many of you should presume to be teachers, my brothers, because you know that we who teach will be judged more strictly.

❖ JAMES 3:1 ❖

4

Assumptions vs. Actualities

Now we see but a poor reflection as in a
mirror; then we shall see face to face.
Now I know in part; then I shall know
fully, even as I am fully known.

✦ 1 CORINTHIANS 13:12 ✦

People make many assumptions in regard to strong-willed children. Once accepted, assumptions that are false can impede the progress of the strong-willed child on his way to competent, confident adulthood. If you are operating under any of these kinds of assumptions, you and your strong-willed child are bound to run into roadblocks on the journey.

There is little research available on this topic of strong-willed children. Perhaps it is because it is difficult to qualify and quantify their behavior. In the absence of extensive scientific study, I have conducted my own informal research. Obviously, I have personal experiences to draw from, and I have spoken to and heard from thousands of parents who identify one or more of their children as strong-willed. This personal analysis has led to the debunking of several assumptions that can impede the progress of a strong-willed child on his lifetime journey.

Assumption vs. Actuality #1
Tough vs. Tender

Consider the ravens: They do not sow or reap,
they have no storeroom or barn;
yet God feeds them. And how much more valuable
you are than birds!

❖ LUKE 12:24 ❖

It is often assumed that the strong-willed child is as tough as nails and that he has no problem with being mistreated or ridiculed. This is an untrue assumption that often results from the strong opinions and in-your-face behavior of a strong-willed child. It is a misconception that a strong-willed child is insensitive. Because they consistently question authority and are willing to choose punishment over compliance, it is assumed that they are unfeeling. A strong-willed child appears to push his way through all situations, yet if one looks closely, he wears his heart on his sleeve. The truth of the matter is that strong-willed children are very compassionate. Although they are typically the last to "give in," they are often the first to feel compassion for another and offer comfort.

The strong-willed child, like everyone else, has the desire to be loved, appreciated, and treated with respect. But because their responses tend to be so abrupt, people assume that they want that same sort of abruptness in return. Because of the way these children are wired, others do not perceive them as capable of or interested in tenderness. But a strong-willed child can be hurt just like any other child and is often misread because of his strong responses. An adult who does not encourage the strengths of a strong-willed child, deeming him tough and more welcoming of criticism than another child, is an adult who may create a serious roadblock in the journey of a strong-willed child.

God is our refuge and strength,
an ever-present help in trouble.

❖ PSALM 46:1 ❖

As the parent of a strong-willed child, it is important for you to provide a safe environment where your child is encouraged, corrected, and supported. You may have to be his advocate in situations beyond the home. When I say this, I don't mean that you unequivocally take the position of your strong-willed child and defend behavior that should not be defended. Remember, we did not support Aaron's coup in grade school. We did, however, sympathize with him when an earlier teacher unfairly changed the rules on "Button Day." When you are your child's advocate, you are doing what is in his best interest. It is NEVER in your child's best interest to condone improper behavior or poor choices. Remembering that your strong-willed child has tender emotions that are often vulnerable will keep *your* heart tender toward him and keep him moving forward.

Look for opportunities to encourage your child. God's Word is filled with reminders and instructions to encourage. One of my favorites is "Therefore encourage one another and build each other up, just as in fact you are doing" (1 Thessalonians 5:11). God's Word is encouraging encouragement, and it is applauding the encouragers, too. Catch your strong-willed child in the act of doing something right, and then encourage him to keep it up.

First Thessalonians 2:11–12 is another verse that can apply to the parenting of a strong-willed child: "For you know that we dealt with each of you as a father deals with his own children, encouraging, comforting and urging you to live lives worthy of God, who calls you into his kingdom and glory." It is your privilege and responsibility to provide encouragement to your tenderhearted, strong-willed child. Don't let his rough exterior fool you. As an adult, Aaron has commented on the powerful effect of being raised with a "You can do it!" attitude. That message was so ingrained that it was an unquestioned assurance in his life. It fostered confidence and made pretentiousness and arrogance unnecessary.

Assumption vs. Actuality #2
Discipline Doesn't Work vs.
Increased Demand for Loving Discipline

Train a child in the way he should go,
and when he is old he will not turn from it.

❖ PROVERBS 22:6 ❖

Wait! Didn't we already have a whole chapter on discipline? Yes, and at the risk of repeating some of the same principles, we are going to tackle the topic once again. Discipline is one of the more difficult components of raising a strong-willed child.

Let's begin with a simple premise that we haven't covered yet. *Why* do we discipline our children? We discipline them to teach them to obey. This is important for a number of reasons. Children are unaware of the dangers of the world. Teaching a preschool child to not go out in the street or to hold your hand in the parking lot is an act of love. Teaching your grade school child to not accept a ride with a stranger or to be discreet when supplying information to a caller is important for his well-being. We make our children mind in order to keep them safe.

I was traveling recently and encountered a situation where discipline was definitely lacking. I was at Chicago O'Hare airport, one of the busiest airports in the world, and I was moving toward the baggage carousel to wait for the arrival of my luggage. The previous flight was very full, and there was a large crowd gathering as the bags began to move around the conveyer. In front of me was a young mother with two children in a double stroller. One was a baby and the other was a boy probably about three years old. As I observed this family, I saw the boy squirming in his seat and heard his mother say, "Sit down, please. Daddy will be right back. You can't get out in this crowd."

The little boy completely ignored his mother and actually seemed to pursue his escape with even more fervor. In a matter of seconds, he was entirely reversed in his seat, standing up, and preparing to bail out of the stroller. "SIT DOWN! You cannot get out of there!!" said his mother sternly.

I wanted to say, "Oh yes he can. Look at him." And before I could even finish my thought, he was out and racing toward the baggage carousel. I lost sight of him almost immediately and imagine that his mother did, too. I'm assuming that somewhere between his point of escape and the Tri-State Tollway, his father intercepted him. His mother could not control him enough to keep him from potential harm. I'm sure I was openmouthed for more than a few seconds as I pondered the possible negative outcome of this young man's breakout. This independent, potentially danger-prone youngster was in charge, and I am willing to guess that his mother was in for big trouble.

Obey your leaders and submit to their authority.
They keep watch over you as men who must give an account.

❖ HEBREWS 13:17 ❖

Disciplining our children is not only for their safety. We also discipline them to teach them respect for others and for property. Jumping on Grandma's couch may not harm the child, but it is not respectful to Grandma or to her property. Teaching our children to respect others helps them to develop into responsible adults, able to interact in our society. I read an issue of the *Ladies' Home Journal* that contained an article entitled "The Perils of the Pushover Parent." Not only was this article soundly supporting the necessity of discipline, but it said that "parents who chronically cave in to their kids' whims are actually doing them harm."[1] Continuing on, it read:

Kids may relish their grasp on power, but child-rearing experts from across the political spectrum agree that it can be hazardous to their long-term emotional health. "Kids absolutely need structure and limits," says Laurence Steinberg, Ph.D., a psychology professor at Temple University, in Philadelphia, and author of *You and Your Adolescent.* "Children learn to control their impulses by having rules imposed, then gradually learn to internalize those rules. Kids whose parents have never set limits often have difficulty controlling aggressive impulses, or even mustering the self-control to sit still in school." And a child who has never been allowed to feel frustration or pushed to do something he doesn't like is getting a dangerously lopsided view of the world . . .

Adversity and frustration are an inevitable part of life, and to survive in the real world, you must know how to cope with them," says psychologist Diane Ehrensaft, Ph.D., author of *Spoiling Childhood.* "A parent who doesn't teach that skill isn't preparing her child for adulthood, and may be creating a self-centered, unpleasant person who will be unable to make the compromises necessary to establish solid relationships or get along with colleagues."

A child who is disciplined will develop better self-discipline, an attribute that is vital to mature adult behavior. It is largely the responsibility of you, the parent, to see that discipline is carried out. Don't be a pushover parent.

> *Teach me to do your will, for you are my God;*
> *may your good Spirit lead me on level ground.*
>
> ❖ PSALM 143:10 ❖

Finally, and perhaps most important, a child who has learned to obey his parents is more likely to choose to obey God.

"The wise in heart accept commands, but a chattering fool comes to ruin" (Proverbs 10:8).

"Hear, O Israel, and be careful to obey so that it may go well with you and that you may increase greatly in a land flowing with milk and honey, just as the LORD, the God of your fathers, promised you" (Deuteronomy 6:3).

"But Samuel replied: 'Does the LORD delight in burnt offerings and sacrifices as much as in obeying the voice of the LORD? To obey is better than sacrifice, and to heed is better than the fat of rams'" (1 Samuel 15:22).

"This is love for God: to obey his commands" (1 John 5:3).

Undoubtedly, choosing to obey God is a wise and wonderful choice. To reinforce that to your strong-willed child may take extensive time and energy. But that time and energy *is* well spent. Remember, the assumption that discipline does not work is untrue. The actuality is that the strong-willed child has an increased need for loving discipline. However, don't expect your strong-willed child to appreciate the discipline you administer, at least not when he is young and currently in the situation. Hopefully when he is an adult he will see the benefit of discipline. He may even choose to express his thanks to you. But in the early years don't look for it, expect it, or be disappointed if appreciation is not one of your child's responses to discipline. Remember, he is only a child.

Assumption vs. Actuality #3
Anomaly vs. Enormity in Number

The LORD made his people very fruitful;
he made them too numerous for their foes.

❖ PSALM 105:24 ❖

Rather than an anomaly, a freak of nature, I discovered that the number of strong-willed children is quite prolific. I wonder how many are the oldest child in a family, presuming that their demanding nature might deter their parents from further family additions.

A multitude of parents tell me that they are relieved to discover that they are not alone. Their child is not the only one having a strong will. And consequently, there is help and hope.

It is a relief to realize that *your* strong-willed child is not the only one who acts the way he acts or thinks the way he thinks.

> "I attended your seminar on strong-willed children at the Hearts at Home conference in November 2002. I listened to you with tears in my eyes because I can relate so well to everything you described."

I could map out a continuum to display the various perspectives on the role that discipline plays in a strong-willed child's life. One group of parents' view (and excuse for the lack of self-control a child exhibits) is summed up by the declaration "Of course he misbehaves. He's a strong-willed child." There is a great hazard in adopting this strategy. Being strong-willed is no more a justification for poor behavior than being a man or a woman or a blond or a redhead. Children do not need excuses for disrespectful conduct, they need parental discipline and encouragement of good behavior.

Others do not see the uniqueness of a strong-willed child and want to use the same methods with all children. I have heard, "None of *my* children are strong-willed." (That's possible.) And I've heard, "Don't you think that *all* children are strong-willed?" No, I don't. All children are children, with childish ways. They are not all strong-willed.

People with opinions on another point of the continuum view the pronouncement of a child as strong-willed equivalent to saying that he is bad. We all have strengths and weaknesses among our personality traits that are seemingly inherent. So does a strong-willed child. Some people might say that these children have more weaknesses than strengths, but I disagree.

> "I am a strong-willed adult raising a strong-willed child. At a retreat, my husband was asked to name my best and worst qualities. 'Becky's best quality is that she is determined. Her worst is that she is pig-headed.' My son is teaching me how those can be the best and worst qualities in a person."

My favorite definition of a weakness is "A strength carried to extremes." I believe I first heard it from the lips of Florence Littauer, the author of *Personality Plus.*

"Face your deficiencies and acknowledge them; but do not let them master you. Let them teach you patience, sweetness, insight. When we do the best we can, we never know what miracle is wrought in our life, or in the life of another."[2]

HELEN KELLER

Think about the explanation that a weakness is a strength carried to extremes. On any given day, it is possible for any of us to take some wonderful, positive, God-given attribute and allow it to be carried to extremes, and it becomes a weakness.

Perhaps our youngest son, Jonathan, expressed it best. One

evening Aaron, the strong-willed child turned adult, was describing a situation, which, to him, even as an adult, was very black-and-white. His passion and excitement were escalating as he gave his account. Finally, as he paused to take a breath, his younger brother sighed and said of that passionate, one-track thinking, "I know that's one of his better points, but sometimes I can't stand it."

Assumption vs. Actuality #4
Targeting You vs. Testing You

Don't be afraid to tell God exactly how you feel
(He's already read your thoughts anyway).[3]

❖ ELISABETH ELLIOT ❖

The parents of a strong-willed child are prone to lose sleep trying to answer the question, "Why does this child hate me?" Relax and fall asleep—he doesn't hate you. He loves you. Your strong-willed child is not targeting you with his unrelenting challenges of your authority—he is testing you. He is curious to discover if you are *now* ready to relinquish control of him. Do you love him enough to hold your ground? After all, he is eager and willing to be in charge if you abdicate the throne. The stark difference between a strong-willed child and one who is not is that the strong-willed child will continually retest you. Winning one battle does not mean a victor is declared. Strong-willed children have more fortitude than most adults. And, unfortunately, too many parents give up before their strong-willed children do. They throw in the towel, and, in fact, no one is the winner.

Because a strong-willed child requires such high maintenance, and because he is more than aware of the extra energy his parents must exert, the possibility exists (at least in his mind) that his parents might give up. He thinks they might surrender and no longer be willing to go the extra mile (or two or three or four . . .) that is demanded in parenting a strong-willed child.

Furthermore, a strong-willed child often receives a smaller amount of positive strokes than a more compliant child. The strong-willed child is periodically irritating the adults in his life rather than pleasing them. He is hence more likely to test adults to see if they are still willing to show him love and be consistent by setting and reinforcing the boundaries.

> "When it comes to relating to the strong-willed child, these [basic parenting] principles take on even more significance, since almost everything in the relationship—both positive and negative—tends to be more extreme."[4]
>
> CYNTHIA TOBIAS

Because the strong-willed child is challenging, being his parent can be very tiring. It is easy to be exhausted and angry in the process of parenting. On more than one occasion, I found myself upset with Aaron and angry with myself for being upset with him. The continual testing and the numerous battles took a toll on my patience and my good attitude.

Interestingly, the majority of the notes that I receive from parents of strong-willed children either began or ended with a declaration of the parents' love for their difficult child. The rest of the note typically depicts an incredible strong-willed antic or adventure. Why do we (especially mothers) feel the need to announce our love for our strong-willed child? Because sometimes, even when we know we love him, we're not so sure we like him.

Your strong-willed child can make you very angry. In fact, the chances are great that he *will* make you angry. Why? It is because the antics of the strong-willed child are capable of embarrassing you as a parent. They confirm in the minds of others your inability to be the perfect parent.

"I love my son so much. He has so many good qualities. He is funny, and extremely energetic. He's athletic, loving, and confident in his abilities. I am a stay-at-home mom and have an absurdly difficult time having the energy to endure him every day. I do not think he will be starting kindergarten next year. His birthday is in May, and I want to see if one more year of maturity will help him to sit still in school. On the upside, I never get a complaint about him at preschool or when he goes to a friend's house. People think I make these stories up about him."

Still, a strong-willed child is exasperating and unashamedly irritating until control is gained. Those are the kinds of things that trigger anger. It's okay. What is not okay is to discipline in anger. Remember, your strong-willed child is not targeting you; he is testing you. Be sure to get an A on the test. Don't surrender to the pressure to give up and give in. Show him you love him by being the authority figure. You are not a target.

Therefore, since we are surrounded by such a great cloud of witnesses, let us throw off everything that hinders and the sin that so easily entangles, and let us run with perseverance the race marked out for us. Let us fix our eyes on Jesus, the author and perfecter of our faith, who for the joy set before him endured the cross, scorning its shame, and sat down at the right hand of the throne

*of God. Consider him who endured such opposition from sinful men,
so that you will not grow weary and lose heart.*

✦ HEBREWS 12:1–3 ✦

Assumption vs. Actuality #5
Prison vs. Presidency

*Let me tell you the secret that has led me to my goal.
My strength lies solely in my tenacity.*[5]

✦ LOUIS PASTEUR ✦

There is no doubt in the mind of the parents that their strong-willed child has determination. In fact, that word seems much too mild. A strong-willed child has determination combined with resolve and fortitude and stamina and good old-fashioned grit. The key for every parent is to focus that determination and to point it in a positive direction and not to let it be a roadblock on the journey. As James Dobson observes:

> "It would appear that the strong-willed child may possess more character and have greater potential for a productive life than his compliant counterpart. However, the realization of that potential may depend on a firm but loving early home environment."[6]

Debunking the assumption that the strong-willed child is en route to disaster has to do with how you, a significant adult in your child's life, handle your assignment of parenting. You are dealing with a persistent personality, and you must be able to stand firm. Strong-willed children are born leaders. That is not the issue. The question is, which group will they lead?

Will your strong-willed child find the cure for the suffering of cancer, or will his behavior create more pain and suffering in the world? Will he lead the youth group or the local gang? The research group or the chain gang?

"Leaders are ordinary people with extraordinary determination."[7]

ALIVE AND WELL IN THE FAST LANE

Adolf Hitler announced to the German Army upon the assumption of its command, "After fifteen years of work I have achieved, as a common German soldier and merely with my fanatical willpower, the unity of the German nation and have freed it from the death sentence of Versailles. My soldiers! You will understand, therefore, that my heart belongs entirely to you, that my will and my work unswervingly are serving the greatness of my and your nation, and that my mind and determination know nothing but annihilation of the enemy—that is to say, victorious termination of the [second world] war."[8]

Could there be a more disgusting representative of a will of iron being used for evil and not for good? The possibilities are endless, on both ends of the spectrum. However, the determination and leadership skills of a strong-willed child can most definitely work for his benefit.

"Our pediatrician said that these [strong-willed] children may one day grow up to find a cure for cancer, but you have to channel their strong will in the right direction."

"Our ten-year-old son, Ricky, had been friends with a boy in the neighborhood for years. Then one day he severed the relationship completely. Why? He discovered that the other boy was smoking marijuana. Ricky, a very strong-willed child, drew the line and would have nothing to do with something that he knew was wrong."

Wise men store up knowledge.

❖ PROVERBS 10:14 ❖

One of the major pursuits of a strong-willed child is to gain or maintain control of his life. That is even more important to him than leading others. He does not want someone else to control him. In order to be successful in this pursuit, it is my contention that many strong-willed children possess above-average intelligence. I am not saying that every strong-willed child is a genius, and I am also not saying that every compliant child is dull. I do suggest the possibility that a strong-willed child, in order to create such chaos at times, must at least be ingenious. Trust me— the supposition that a strong-willed child is bright will NOT necessarily manifest itself in the academic world (at least not right away). Still, anyone who can get the best of an adult who is potentially twenty-five or more years older must be bright.

It would seem that this child was thinking, "What can I say to motivate my parents to give in to my wishes? Ah, I know! I'll use guilt, a highly effective weapon in the battle for control." This child was bright enough to choose an appropriate weapon from his arsenal.

Let's return, for a minute, to the scene of the coup that Aaron organized in elementary school. John and I did not know about

> "When Brian was little, he could be quite a challenge. When he signed up for high school wrestling, God gave me a peek into *why* He'd created Brian that way. All of that strong-willed energy went into his left shoulder, which *would not* hit the mat. I was *so* grateful for that insight and for God's keeping me from thinking it was necessary to *unstrung* my son. Today he is an engineer. I'm thankful for strong-willed engineers for they are the ones who build our homes, bridges, and world. May they never cave in to anyone's negative pressure, no matter who it is from."

it until the fateful meeting with his teacher. I discovered much later that at least one of my good friends was not in the dark about the goings-on in this particular classroom. Jackie, one of Aaron's school friends, was one of the classmates who boycotted the journaling assignment at Aaron's recommendation. Probably a year after this incident, her mother and I were chatting. (I'm guessing it was at least a year, because that is how long it took me to cope with the shock and embarrassment of his takeover.) When I asked Kay if she knew about the incident, she gently admitted that she did. And then she added her daughter Jackie's analysis of the situation: "Jackie told me that Aaron really *should* be in charge because, and I quote, 'He is much smarter than our teacher!'"

"Peter was making the transition into first grade. We pulled up to school when he declared that he did not want to go to school, and he was NOT going to go to school. As we walked toward the front door of the school he looked at me and said, 'I thought you were going to home-school me.'"

So, is Jackie's evaluation the basis for my idea that the strong-willed child is typically a bright and creative child, an out-of-the-box thinker? No, more than the opinion of one loyal friend is the realization that many strong-willed children have managed to gain control of adults, one or even two generations older.

"The class was told that they had forty minutes to complete the true/false test. This test was not for a grade. The results would simply be recorded in each student's permanent record. Ian decided that this was a silly waste of time. Being a creative, strong-willed child, he decided to challenge himself by purposefully answering each question *wrong.*

When the test was graded and scored, he had earned a 2%. Wow! Actually he had gotten 98% correct—well, sort of. Try explaining *that* one to his teacher. The permanent record reported a score of 2%."

Because of their ability to solve (as well as create) problems, it is wise to give a strong-willed child responsibility that is age appropriate. By giving him a worthwhile task to do, it builds up his sense of worth and allows him to develop credibility in his own mind. Putting trust in your strong-willed child to do the job assigned to him validates his abilities and illustrates your confidence in him. It is also important to assist him to realize, even at an early age, how important it is to align himself with positive people of good character. Encourage your strong-willed child to accept responsibility for himself in other arenas also. This can help him feel significant and keep him on the path most destined for success.

A Closer Look with Aaron

As a grown man, looking back to my days as a strong-willed child, I am certain that I was most impacted by the assumption that I was tough and not tender. I realized that this assumption was incorrect when I realized that I could replay in detail so many of the hurtful experiences that resulted from my strong-willed nature. I know that my older and younger brothers had unpleasant things happen to them in their growing-up years, things that seemed unfair or cruel, and the chance exists that their episodes were no less painful than mine. The difference is that in most cases they cannot vividly recall the incidents, issues, or the intensity of the episodes. I can.

There were definitely repercussions from when my strong-willed nature got out of control. At those times, I knew that what I was doing was wrong and that I was misbehaving. In spite of that admission, it is hard to imagine that some of the treatment that I received from one or two noteworthy adults was justifiable. The fact that I can remember dramatic details, like an out-of-control classmate being allowed (and in a sense, encouraged) to spit on me, is an illustration of the pain it caused.

Parents, your strong-willed child is hoping, no, begging, for you to take and maintain control and to comfort him when the world lashes back in what they think is self-defense. This does not come in the form of justifying his inappropriate behavior. It is in acknowledging his mistake and granting him grace to grow and mature; the grace he requires to move forward and learn more about how to control his strong-willed nature. He needs you to show your love for him by being his parent, not his friend or his oppressor.

Being strong-willed may be programmed from birth, but each time a "strength is carried to extremes," it is a conscious decision. It is not a reflex. We are waging war, partly with our own strong emotions. Please come to our rescue by loving us enough to help us learn proper control—godly self-control of our actions and emotions. We will test you and almost everyone. We are looking for adults who are patient and persistent in their love. We want you to be willing to invest your time, energy, emotion, and love in us.

The Junior
High Journey

*Adolescence is an awesome
and sacred time of life—
it is not a disease to be cured.*[1]

❖ WAYNE RICE ❖

Junior high is a time of change for every young
man and woman. For our strong-willed child, it was time for a
fresh start. Aaron was changing buildings, administrators, and
class structure—and hopefully changing and maturing in the ability to control his strong-willed ways. Junior high provided him
with increased opportunities for responsibility. Now he could join
student council, various sports teams, drama, and special music
groups. There was more diversity in the classes that were offered,
with different students making up each one.

For us as the parents of a strong-willed child, the junior high
years were still a time of guiding, encouraging, correcting, and
molding. The job was not finished just because Aaron reached his
teen years. While there were no more classroom coups, there were
still incorrect decisions resulting from his strong-willed nature.

Again, Aaron intersected with some wonderful, encouraging
adults and, again, occasional roadblocks. Perhaps one of the most

encouraging situations actually occurred at church. Aaron was the only student in the junior high Sunday school class. In fact, his individualized instruction went even further. He had not one, but two teachers. A husband and wife team, Lawrence and Loretta, taught the class.

> *We always thank God for all of you,*
> *mentioning you in our prayers.*
> *We continually remember before our God and Father*
> *your work produced by faith,*
> *your labor prompted by love,*
> *and your endurance inspired by hope*
> *in our Lord Jesus Christ.*

❖ 1 THESSALONIANS 1:2–3 ❖

Do you find it unusual that these two adults would *both* commit to a single junior high school boy? It might seem more reasonable if one of them had opted out of the instructor's role. (Or maybe even both of them.) After all, Aaron could always go up to the high school class or down to the class for fifth and sixth graders. Everyone would understand and be supportive. But that wasn't the case. These two wonderful Christian people gave Aaron their undivided attention Sunday after Sunday.

Because he was the solo student, his questions never disrupted the rest of the class members. On any particular Sunday, Aaron, Lawrence, and Loretta might begin in Genesis and end up somewhere in the New Testament, pursuing one of the mysteries that caught Aaron's attention.

Lawrence and Loretta encouraged Aaron's questions and spiritual growth. They were not put off by his desire to set his own course. Instead they were willing to navigate that course with him.

As the parent of a strong-willed child, it is important that you search for adults who are capable of and willing to affirm your son or daughter. I have discovered that being kind and respectful to the children at church is an important calling for every adult, but some are more enthusiastic than others about that proposition. If you discover that your child is in an environment that is hostile (because of the adult's natural attitude or the adult's reaction to your child), I would recommend that you find different surroundings.

As a frequent speaker at *Hearts at Home* conferences and a contributor to their magazine, I am asked many questions about family relationships. One day I received an e-mail from a mother about a problem that revolved around her daughter. It seemed that this girl "was picked on" by the students, and ultimately the teacher, in the public school that she attended. The parent withdrew her daughter from this school and entered her in a private, Christian school. Before long, the same pattern of behavior began. Her daughter was once again "being picked on," and the teacher was again a negative part of the mix. The mother's question to me was, "Do you think I should home school my daughter? Everywhere we go, she is singled out for criticism."

"Lord, Change me."

EVELYN CHRISTENSON

My reply was probably not what this mother was expecting. I assured her that in spite of the rough times her daughter was experiencing, there was some possible good to be found in the situations. Both schools, both groups of students, both teachers responded to her daughter in the same way. The variables were the schools, the students, and the teachers. The constant was her daughter. That was, in a very real sense, the good news. Why? Because this mother still had the chance to positively influence the behavior of her daughter. Her opportunity to control any of the other factors was slim. I suggested that perhaps it was her daughter's behavior that was the problem, behavior that could hopefully be modified by the mother.

In this case, I did NOT recommend finding an alternative

environment. If your child is creating and re-creating a hostile setting, it is time to work with your child, not with others.

For nothing is impossible with God.

❖ LUKE 1:37 ❖

The junior high years brought another positive encourager into Aaron's life. In third grade, Aaron was finally given a horse. The lobbying for this creature began near the time Aaron started talking. For years we held him off by explaining that there was already one horse on his grandfather's farm down the road, and one horse was enough. The fact that this horse was old and decrepit and seemed to have a death wish for anyone who dared to saddle it, didn't influence our decision. We were NOT going to buy another horse while that one was alive. Speaking of a death wish, I would not be surprised if Aaron's prayers included a petition for God to "get rid of Fury" as soon as possible. Eventually, Fury did die, and that gave Aaron the green light to push for a "decent horse."

Aaron's grandfather purchased Lady, the new horse, later that year. As a six-year-old mare she was big, but not too big for Aaron. He spent hour after hour with Lady, one day literally investing three hours in coaxing her to take a bit. He was inordinately patient with her (something he seldom was with human beings). When Lady reached her two-year anniversary of living on the farm, Aaron announced that she was lonely and that "being a social animal" she needed another horse for companionship. My recollection is that the answer was a firm, "No one is buying another horse." When you live on a working farm, you may have the space for multiple horses, but most animals are expected to be productive and not just for entertainment.

Aaron, our strong-willed, out-of-the-box thinker, came up with a plan so that he could have his way and not ask anyone to invest money in the purchase of another horse. By this time, we were friends with Ed, the owner of several horses. Ed and his family lived about ten miles away. His middle son and our eldest son, Matthew, were very good friends. They played together on sports teams from junior high on, and because of that, the lives of our two families intersected.

One day Aaron approached Ed with a plan. "If I can earn enough money, can I pay you to have your stallion breed my mare?" And Ed's reply? "I'll let our stallion stay with your mare if you will break him before he comes back home. I won't charge you a stud fee if you won't charge me to break the stallion." What a *terrific* deal! As far as Aaron was concerned, it was a huge win-win—for him! If everything went as planned, he would have two horses, his mare and a foal, *and* he would have the fun of breaking Ed's horse! Dad agreed and everything went as planned. Duchess, the foal, was born in about a year, and the stallion, named SW, was broken and trained in great fashion.

That first partnership began a long relationship. By the time Aaron was in junior high, he was breaking horses for Ed on a regular basis and also learning how to sell them. I once heard him say, "I try to treat my horses the way I want adults to treat me. I care about their well-being and am patient with them, but they don't get away with any disobedience!" Aaron's horses were therapeutic for him. And so was his relationship with Ed—an adult who respected Aaron's abilities, gave him responsibility, and taught him how to be a "horse trader" in the very best sense of the phrase.

In today's society, those of us who live on a farm are in the minority. In fact, we now represent less than 3 percent of the population. So what take-home value is there from this positive scenario for those of you who do not have a bale of hay within twenty miles? The key here is not necessarily the horse itself. It is the *responsibility* that Aaron was allowed, encouraged actually, to accept. How can that be replicated in the city? One option is smaller animals. If your son or daughter desires to have a pet, this can be a great opportunity for responsibility —so can a paper route or any other part-time job. Be sure that *you* aren't assuming the responsibility, however. That defeats the purpose.

> *Don't let anyone look down on you*
> *because you are young,*
> *but set an example for the believers*
> *in speech, in life, in love, in faith and in purity.*

❖ 1 TIMOTHY 4:12 ❖

Another positive influence in Aaron's junior high journey was his principal. One of Aaron's passions is politics. He is interested in it today and was interested in it in junior high school. This interest was fueled by discussions with willing adults like his junior high principal. He and Aaron came from opposite ends of the political spectrum. When the daily news contained a controversial political issue, Mr. Burkey would seek out Aaron and ask his spin on the current event. One day when Aaron was in eighth grade, we received a letter from the principal's office. (Now *that* can bring terror into the heart of any strong-willed child's parent.) In the letter, Mr. Burkey complimented Aaron on his sound grasp of national issues and his ability to debate those issues confidently and from an informed position. The principal also commented on the maturity with which Aaron listened to a differing point of view. (I do believe that I saved that note! In fact, I might have even mounted it in a scrapbook!)

What was Mr. Burkey doing? He was allowing Aaron to share his views in a controlled environment and encouraging him to think. Never did he say, "I'm much too busy and important to discuss things with *you,* a junior high boy." No, instead, he enjoyed the banter and debate with a young man who was learning to express himself and, more important, learning to listen to others' opinions.

Do you ever take the time to encourage your child to express himself? Are you interested in his thoughts and opinions? All too often, we as the overtired, disheartened parents of strong-willed children do not encourage our children to think and to express themselves, especially if they do not walk in lockstep with our thinking. When was the last time you engaged your adolescent strong-willed child in a conversation about something *he* was interested in? Why are our "conversations" more like reading an instruction manual? "Do this and then do this. And whatever you do, don't do that!" Or we talk simply to discover facts. "What time does the bus leave? Did you tell Chad we would pick him up?" Your strong-willed child's journey will be enhanced if you look for ways to converse with him about *his* interests.

> *Fathers, do not exasperate your children;*
> *instead, bring them up in the*
> *training and instruction of the Lord.*
>
> ❖ EPHESIANS 6:4 ❖

By junior high school, it is necessary for parents to develop alternative ways to discipline their strong-willed child. Obviously as children mature, it becomes inappropriate to spank. The hope is that, by this time in their lives, it is no longer necessary. But discipline will undoubtedly still be needed on occasion. There are several disciplinary measures that are effective with adolescents. The loss of independence or the suspensions of recreational activities (i.e., limiting or prohibiting computer play, television, or time on the phone with friends) are, when necessary, very effective means of control as a child gets older. How do you determine the appropriate discipline? You go back to the basics. Number one, you must be able to enforce any discipline you threaten. Number two, you must be certain that the discipline will be effective. Ask yourself, what occupies my child's free time? What does he enjoy doing? If you choose to limit his access to something he is genuinely interested in, your discipline will be effective.

Usually, by junior high school, the way a strong-willed child maintains control has also matured and developed. A friend who is one of my favorite examples of a strong-willed child turned responsible adult, told me that his most successful weapon of control in the junior and senior high years was his use of humor. After all, how could someone punish a person who brought such great laughter and delight to the scene? Although this was not a ploy used extensively by our strong-willed child, I can understand how it could be effective.

I already told you that it was not completely smooth sailing by junior high school. There were still some roadblocks along the way. Like another teacher who decided that Aaron's "creative alternative" to her instruction was *rebellion* and that it must be squelched with force as soon as possible. (Can you guess that we did not agree with her reaction?) In this instance, a trip to school was made to get the precise facts. During our visit, the teacher calmed down and actually admitted that even if Aaron did have a legitimate suggestion, she was not willing to examine it. (As my own father used to say, "Don't confuse me with the facts. My mind is made up!") How's that for being flexible and adapting to circumstances? We explained to Aaron that she was the teacher, he was the student, and she had more power. We also let him know that, in our opinion, she was wrong and he was right. Although we did not agree with her stance, we told Aaron that he would have to comply. This woman was an adult and his teacher, and he had to treat her with respect whether or not he respected all of her judgments.

This was a teachable moment for Aaron. Life is not fair. We were able to persuade Aaron to let go of some battles—something that he would have to do throughout life, even as an adult. Part of his willingness to comply was fueled by the fact that we talked honestly with him. The truth is very powerful. Just because this woman was the teacher and an adult did not make her correct. She merely held a position that was to be respected. If we had failed to examine the situation correctly and instead made a blanket statement that she was right and Aaron was wrong, he would have had a more difficult time giving up control.

Jesus said, "If you hold to my teaching,
you are really my disciples.
Then you will know the truth,
and the truth will set you free."

❖ JOHN 8:31–32 ❖

In this case, the "truth" was twofold. The teacher had made a poor decision, *and* it was in Aaron's best interest to comply. That truth set Aaron free to acquiesce and move forward. Learning that "the better part of valor is discretion," is an important lesson from Shakespeare for a strong-willed child.[2] As the parent of a strong-willed child, it is imperative that you are honest and straight-forward with him. Try to avoid assuming a position on either end of the continuum. It is incorrect to *always* conclude that your strong-willed child is right. That is one radical end of the spectrum. The other is to determine that you must protect any and all unsuspecting adults from your strong-willed child, *always* making him the bad guy, the one at fault. Neither extreme is valuable. Each situation must be scrutinized and evaluated. Search for the truth and do not be reluctant to share it with your child. If you are willing to handle each conflict one by one, with wisdom and fairness, your strong-willed child is more likely to acknowledge your appraisal and act accordingly.

A Closer Look with Aaron

I like absolutes. I believe that is true for the majority of us who are strong-willed. In junior high, life begins to become less and less "absolute," less black and white. As teachers introduce more topics and issues that can be interpreted as gray, they also give opinions. It is quite possible that the opinions presented (sometimes even presented as fact) will not reinforce the standards you have for your family.

It is necessary for the parents of a junior high strong-willed child to be honest with him at all times and in every instance. At this stage, topics are more thought-provoking and more controversial. We strong-willed children really do love a world that is black-and-white. We thrive on absolutes. In junior high school and beyond, the world is shouting that there are many, many gray areas.

God's Word has absolute truths that need to be re-inforced with even more enthusiasm as the world strives to make His truths subjective. I remember dis-cussions about creation, gender differences, morality, and tolerance of others' beliefs (not necessarily the Christian's) being prolific in the junior high years. I am not saying that we should relinquish thinking and evaluation of situations that are truly gray in nature. I am saying that junior high is the time to live by the straightforward truth of the Bible.

We had a ritual in our home, one that always ex-isted as far as I know. When Mom or Dad said prayers with us at night they always concluded by making a statement and then asking a question. "Mom and Dad love you lots and lots. And who loves you best?" That was our cue to holler, "Jesus!" We did it night after night by memory. We didn't have to contemplate that question. The answer was always the same. It was a ritual—a habit—our nighttime routine. And the truth! By junior high we were on our own with our evening prayers, but that truth was firmly planted in our minds and hearts.

What would have happened if my parents' words and actions had not been in sync? Obviously, I would have believed their actions. I'm not trying to suggest that they never messed up or made mistakes. Of course they did. But their desire was to live the truth and to encourage all of their children to do the same.

In junior high school, an adolescent becomes more aware of varying life-styles, thoughts, and standards of behavior. More and more viewpoints are introduced, with the majority of them contrasting the basics of Christianity. I suppose that if your child is enrolled in a Christian school, this conflict may be slightly postponed. But for those who attend a public school, junior high is a place where values are questioned.

When I was growing up, there was a quote on the wall of one of the Sunday school rooms that read: "If you don't stand for something, you'll fall for anything." What are the basic beliefs and standards of your family? Do you live by them, or do you just espouse them? Your strong-willed child is watching you. Do you encourage your strong-willed child to tell the truth, and then you tell the person on the other end of the phone that your spouse, sitting in the family room, is not home? Every junior high child needs the assurance of your constancy, your black-and-white behavior. For example, my dad never swore. There was no question in my mind about whether something could push him over the edge and cause him to change. He was constant. And his example was something I could count on and a pattern that I could replicate.

Parents, it is also important to know *why* you have taken a stand. The strong-willed child wants to know why. My compliant brothers probably had no problem giving in when the explanation to the order was "because I said so." That was never enough for me. Be ready to substantiate your conviction.

If your strong-willed child wants an explanation of why "We're all going to church," the correct answer is NOT "because I said so," nor is it "because if we don't go to church, we'll go to hell." That's not the truth. Why DO you want your junior high child to go to church? Can you answer his question about this or any other conviction you may have? Be prepared! Now is the time to "walk your talk" more than ever before! That will help your child trust you and accept your appraisal of situations and recommendations of how to handle those situations.

Value his intelligence. Have legitimate conversations with him. Just because he is an adolescent does not mean that he does not need your input or that he will disregard it. He has a strong desire to know the absolutes found in God's Word and have them reinforced by his loving parent. This will help him as he comes to his own conclusions about some of the legitimately gray areas. And it will help him articulate his own convictions.

I met a young man whose parents were of two different races. Their child was completely Caucasian in his looks, with blond hair and blue eyes. When he studied genetics in junior high science and learned about recessive and dominant genes, the question arose in his mind, "How could I have inherited such Aryan looks when the other race is dominant?" It was then that he learned that one parent was not his biological parent. The lie that had been accepted as truth for years led to extreme pain in this family.

I've often wondered if the children who are told that Santa Claus is real question other things that their parents have said. Is Jesus real? Be honest with your children, especially with your strong-willed one. Allow your strong-willed child to ask questions. In fact, encourage legitimate questions. Yes, there are gray areas in life. Help your child understand which areas those are. Let him ask questions, and then try your best to answer them. It is important to him.

"As your kids get older, LOOSEN your grip on things that have no lasting moral significance, and TIGHTEN your grip on things that do."[3]

WAYNE RICE

The fact that Aaron accepted our decision about how to handle an unwavering, inflexible teacher marked a great move toward maturity. It also indicated to us that some of the goals we had as parents were being met. It is important that the parents of strong-willed children set goals to mark the journey, determine the desired destination, and plot progress toward that end.

Determining the Desired Destination

*I press on toward the goal to win the prize
for which God has called me
heavenward in Christ Jesus.*

✣ PHILIPPIANS 3:14 ✣

The strong-willed child is not without a goal. From the beginning, the goal of a strong-willed child is to be in control of his life. And he pursues that goal with determination and gusto, needing little outside motivation or stimulus. Unfortunately, being in control is not always in his best interest. That is why you, as the parent of a strong-willed child, need to develop goals for his journey.

Setting goals for your strong-willed child can be tricky business. Remember, the strong-willed child desires to be in control of his own life. If you set goals for him, you are, in a very real sense, seeking to control him. The strong-willed child who gets an inkling of the idea that his parents are directing him this way or that way, is likely to turn around and go in the opposite direction.

So do you still set goals? Absolutely! But these are definitely not shared with your strong-willed child. In a sense, these are actually goals for the parents. They are set to guide them as they shape the journey of their strong-willed child.

First of all, as parents, we wanted Aaron to know in his heart, not just his head, that Jesus loved him and that He wanted the very best for him. That was the most important goal. Our goals had to be specific, attainable, and measurable. A specific goal was that Aaron would accept Christ as his Savior at an early age, and the sooner the better. This was both specific and attainable. It was measurable by his proclamation of the decision and subsequent behavior. In order to facilitate this goal, we attended Sunday school and church each Sunday. We read the Bible to the kids each day and prayed as a family at breakfast. We prayed with Aaron each night before bedtime. And we accepted the challenge to do what was by far the most difficult and the most powerful thing— to live our faith before him each day (to walk our talk), not perfectly, but with conviction and with love. There is no formula to guarantee that a child will accept the love of God and become a Christian, but it is a goal worthy of every attempt made to encourage it.

With the help of God's Holy Spirit, the strong-willed child begins to understand that using his strengths negatively and in destructive ways isn't the best plan. Winning at the expense of another is not what Christ supports or applauds.

"The man who walks with God always gets to his destination."[1]

HENRIETTA MEARS

When I was a child, I talked like a child,
I thought like a child, I reasoned like a child.
When I became a man, I put childish ways behind me.

❖ 1 CORINTHIANS 13:1 ❖

We also determined that it would be attainable and very positive if, by the end of junior high school, Aaron had the skills to make consistently good decisions without our direct coaching, reactive intervention, or punishment. After all, when a child is only nine years old he is halfway to adulthood and being able to live on his own. If we missed our self-imposed deadline of Aaron's exit from junior high, we still had a little time (high school) to be a direct, daily influence in his life. Specifically, we hoped to observe Aaron choosing not to battle for his own way with authority figures, but to be mature even when dealing with immature adults. We would measure this by his behavior and reports from him and his teachers.

Obviously, we desired for Aaron to ultimately use his gifts and talents to be a productive member of society. We wanted to help him evaluate other people's input (positive and negative), teaching him to apply it to his life when it was relevant and reject it if it was not godly input. It is important to take criticism or comments of any kind "straight to the top" to the Lord, the Creator of strengths and the Corrector of weaknesses. It is important to teach your child to be discerning. If someone's purpose is merely to discourage your strong-willed child and convince him that he is less than who he really is in Christ Jesus, that person's opinion is not to be valued. If, on the other hand, the suggestions or criticisms have validity, it is time to prayerfully work for change.

> **"Inch** by inch, it's a **cinch."**
>
> UNKNOWN

Aaron's achievement of the majority of these goals was *almost* attained by the end of junior high. The important thing to remember about most goals is that even if you don't hit the bull's eye each time, you are much closer than you would have been without the establishment of goals in the first place.

Reaching the goals we set was not an overnight accomplishment. I will be the first to admit that in this age of instant mashed potatoes and e-mail, I am not very good at delayed gratification. So, how do you continue the process when it seems to take so long?

First of all, it is important to assess your starting point and, if at all possible, begin the process of moving forward immediately. By noting your starting point, all progress can be celebrated. As soon as we recognized that Aaron was a high-maintenance child, we began to look for ways to help him use his strong-willed nature in positive ways. The fact that Aaron took control of his grade school classroom and organized a coup was certainly an indicator that we still had ground to cover, but it was not a step backward. Helping a strong-willed child learn to temper his strong-willed nature and make good choices is a process—a long and sometimes arduous process—but a process necessary to facilitate his journey. It is like building a retaining wall of brick, a ten-story retaining wall, brick by brick. Aaron's coup did not pull bricks from under the structure, it merely indicated that the wall was still under construction. As much as we could, we would help Aaron become a good decision maker by the end of junior high.

"Have patience. Have patience.

Don't be in such a hurry.

When you get impatient,

you only start to worry.

Remember. Remember.

That God is patient, too.

And think of all the times

when others had to wait for you!"[2]

THE MUSIC MACHINE

We would help him construct that retaining wall, the wall that would hold back his strong emotion and the overwhelming desire to control his circumstances so that his journey could continue in a positive, forward manner. In order to reach that goal, we monitored his classroom behavior (with even more zeal after his successful grade school takeover). I didn't want to be surprised by a teary phone call from a teacher again! Undoubtedly, in the past I ignored some obvious indicators and just hoped for the best, which will not suffice for a strong-willed child. By keeping in closer communication with his teachers, we could intervene and help modulate his behavior when necessary, before the situation was out of control (and he was completely in control).

Whenever it was suitable, we dialoged with Aaron about what behavior was appropriate and what was not. We did not automatically agree or disagree with his behavior, but we tried to help him come to mature conclusions about what action would be most beneficial.

"The most important single ingredient in the formula of success is knowing how to get along with people."[3]

THEODORE ROOSEVELT

I remember talking with Aaron about a particular interaction with a group of his peers in junior high school. He attempted to control not only himself but the entire situation, when he deemed the behavior of his peers to be in error. Was he correct in his estimation of his friends' behavior? Yes, I believe he was. Was he positively influencing his friends with his abrupt, immovable response to their behavior? No, he was alienating them. As I recall, my exact words to him were, "It doesn't matter if you are right and everyone hates you. You haven't made a difference." Diplomacy is something a strong-willed child must cultivate. I am not suggesting that the strong-willed child give in to destructive, life-threatening choices or allow them in others' lives. But that was not the case in Aaron's particular situation. I am suggesting that the strong-willed child think through his motivation for (and method of) confronting and desiring to control others.

Little by little, brick by brick, we witnessed that retaining wall of discernment growing taller. We applauded Aaron's progress. Parents should always look for opportunities to catch their children doing something right. When his choices were not as good, we punished him if necessary and helped him determine what might have been a better route to take.

How did we know that we did not reach our parenting goal by the end of junior high? Because he still failed to comply with a teacher whose request was, in his way of thinking, silly. And his decision to maintain his own control in that instance resulted in a trip to the principal's office. Was the trip worth it? The strong-willed boy thought it was. After all, the teacher's demand was obviously frivolous. But was the trip worth it? We, the parents of that strong-willed boy, did NOT think it was. Aaron could have made the choice to follow the rule, silly or not, and avoid the punishment. He had not "arrived" at the desired destination by then, but progress was still being made. The retaining wall was growing taller, and that was facilitating forward progress.

The first semester in high school, Aaron chose to get a B in a class in which he could have easily earned an A. Why? Again, his ultimate desire was for control. He decided that he could get an adequate grade without doing the work assigned. (It's rather interesting to me that he managed to receive no worse than a B.) The B grade was not as much the issue as was his choice not to do the work. We were disappointed in his decision, but the retaining wall was still getting closer to completion.

By the sophomore year, Aaron was making good choices with great consistency. And we were not involved in direct coaching, reactive intervention, or punishment. That is not to say that he never again made a poor choice. The difference was that he was now evaluating his own behavior and making necessary adjustments. We saw definite signs that he was tempering his wonderful gift of a strong-willed nature. The retaining wall was now complete to our specifications, and it was doing its job of holding back Aaron's desire for control at all costs.

> "When we set exciting worthwhile goals for ourselves, they work in two ways: We work on them, and they WORK on us."[4]
>
> EDGE LEARNING

Even though it is not a good idea to push your goals on another person, especially your strong-willed child, it is not a bad idea to help your child establish his own vision. The well-known proverb "Where there is no vision, the people perish" is derived from Proverbs 29:18. I have always broadly applied this verse to encourage young people, in my own home and in the youth group at our church, to establish a vision—a goal—a target on the wall. I challenged them to picture where they would like to be in ten years or what they would like to be doing. It is my theory that when this is done, when a vision is established, youth are much less likely to engage in activities which will hamper the achievement of those goals. Remember our Easter weekend visitor, Nathan, who told his parents about the dream he had to be a pilot and a farmer? That vision demanded certain things (like the successful completion of seventh grade). Defining his vision could help give him the motivation to accomplish the necessary task at hand, a task that was an essential step to seeing that vision accomplished.

Recently I discovered an exercise Aaron completed in grade school, a fill-in-the-blank. "When I grow up, I want to be a _____." And scribbled in the blank was the word "vet" (a veterinarian). Today he is en route to that dream. It was his vision, his goal, and his self-determined target on the wall. It kept him from myriad poor choices that could have made his dream "perish." Help your strong-willed child determine *his* vision. Your ability to do that will be based on your relationship with your child. It is difficult to help someone achieve his goal or even *define* his goal if you do not truly know the individual. Be sure your strong-willed child knows of your great interest in him as a unique human being.

The strong-willed child is not interested in *your* vision for him. Your vision is for you and you alone. Don't be discouraged if you do not hit the bull's-eye right away or even if you do not hit it at all. The benefit of establishing a goal, of drawing a target on the wall, is that regardless of whether you hit the exact center of the target, at least you are facing the right wall!

"In absence of
clearly defined goals,
we become strangely loyal
to performing daily
acts of trivia."

UNKNOWN

A Closer Look with Aaron

Reflecting on the content of this particular chapter is by far the most difficult for me. Why? It is because I had no idea that my parents had a plan or a goal that they were trying to accomplish. And that is very good. I can't stress enough the importance of the parents of a strong-willed child functioning as a team, seriously evaluating their starting point and their finish line, and NOT sharing that information with their child. My folks were both good communicators, but they did not communicate their goals with me.

The idea that my parents had set goals for guiding my journey was a complete shock to me. I discovered that the day of Easter vacation, when I walked in on the family room discussion about raising a strong-willed child. Mom told you that I surprised her with some of my stories, stories of my strong-willed nature out of control. Well, she and Dad surprised me when I heard them tell our friends about setting specific goals. They shared with Nathan's parents their hopes that, by the end of junior high school, I would be consistently exercising control of my strong-willed disposition. I was stunned! They had a game plan. They had an undisclosed objective! Why was that necessary? Weren't we on the same team?

The realization that they had a strategy was astounding to me. In fact, when I became aware of that, I also became aware that I must have demanded a great deal of their energy. Believe it or not, that never occurred to me before. Later that weekend, I had one of my last strong-willed child "implosions" as I realized the truth of how difficult I must have been to parent. I got very emotional and, in a sense, felt detached from my two biggest supporters. My parents and I must have been on opposite teams! They had a plan, a strategy, a plot, and I was not privy to it. On Easter Sunday, late in the day, we had a discussion about what I recently discovered.

The truth was that we were NOT on opposing teams. We were on the same team. It was just that they were the co-captains of the team, setting the game plan, and I was a player who didn't need that much information. They knew that the possibility existed that if I knew about their goals, I would go in exactly the opposite direction—just for the sake of having control. And, if I decided that I was failing to meet their goals, I might become disappointed and possibly even defeated.

That Easter day when I imploded, I remember my parents' reaction. They were sorry that I was emotional and upset, but it gave them an opportunity to share with me what I have filed in my brain as a parenting tip for future use. Yes, I was difficult at times (I added the "at times" so that I'd feel better). "But," my dad explained, "when you are a parent your job is to 'do what it takes' for each one of your kids." And that is what they did. It took more energy and effort in the areas of disciplining me and teaching me godly self-control. "Do what it takes" meant other things for my brothers. That is part of what it takes to work together as husband and wife, as a team, to guide the journey of your strong-willed child. Set goals, determine your desired destination, and celebrate with one another (not with your child) as you and he reach those goals.

"Maturity is when the child knows himself, accepts himself, controls himself, and is able to use what he is and has creatively and constructively."[5]

WARREN WIERSBE

103

7

The Journey Goes On:
High School

*I'm not all that I ought to be, but I thank
God I'm not what I used to be. If I keep
praying and asking God to make me to be
what He wants me to be, some day I will
be what I need to be . . .in my walk with
the Lord, I'm not saying I'm better than
others—I'm just better than I was.*[1]

❖ FRANCES KELLEY ❖

As you already know, by high school great progress was made in the area of helping Aaron bring out the best in his strong-willed nature, but the journey was not complete. Aaron was assuming more and more responsibility for his decisions. And more and more of those decisions were very good ones.

His errant B grade in an easy course first semester of his freshman year was not repeated. He began to understand the personalities, quirks, and demands of his various teachers and of others with authority over him. He actually tried to function in their established parameters rather than his own. Helping your child realize that people have different personalities with varying strengths and weaknesses will help him to work with those individuals rather than be combative. It is also important to acknowledge the fact that aging does not equal maturing. Too often the strong-willed child cannot understand why an adult would act with immaturity. The answer is simple—some are not mature. By dialoging about this truth, you have freed your strong-willed child to accept the possibility of immaturity and to rise above the occasion.

"While great brilliance and intellect are to be admired, they cannot dry one tear or mend a broken spirit. Only kindness can accomplish this."[2]

JOHN M. DRESCHER

Aaron found his academic niche in the areas of science and social science, two interests of his since childhood. He loved chemistry. Was this child really mine? (When I sat in an orientation session for the University a few years later, I heard a college chemistry professor explain that chemistry is really problem solving. Interesting. No wonder Aaron liked it so much.) Civics and political science were fascinating to him and fed his love of politics. Those teachers too appreciated his ability and propensity to think outside the box, making their classes even more enjoyable for Aaron. They enjoyed his questions and took time to answer them, encouraging Aaron to problem solve and dig deeper into the areas that interested him.

What kind of questions does your strong-willed child ask? The answer to that question might give you a clue about his interests, passions, and intellectual strengths. Then you can find ways to encourage and develop those interests and passions.

Of course, not everyone appreciated Aaron's questions. He ran into at least one coach who didn't want any questions at all and made that known by ridiculing Aaron when he asked one. Aaron's interest in that sport was definitely squelched, which, by the way, was NOT football. Football and Aaron were a perfect match! I always claimed that it was a wonderful channel for controlled aggression. Being aggressive was a necessary part of success in the game, which provided an outlet for him. I've known of other strong-willed children who participated in the martial arts. And sometimes simply providing a place where your strong-willed child can freely run will do the trick. Physical activities serve as a release for some of the strong emotions possessed by your strong-willed child.

Music was also an attraction and a talent that Aaron was able to cultivate with the loving attention of the directors of both the band and chorus. His high school was not very large, a little more than four hundred students, and the athletic and music departments discovered how to work together and share those teenagers interested in both endeavors. During his senior year, Aaron was one of the co-captains of the football team and also one of the drum majors for the marching band. How did he do it? At halftime he removed his shoulder pads and climbed onto the director's stand. When the show was complete, he headed for the locker room. It was the best of both worlds! (And, as I said, it was a very small high school.)

Football and Aaron were a good match. Basketball and Aaron were not. Basketball, in fact, became an activity that Aaron chose to give up in order to have his way. He knew that we thought that there were positive things to be gained from participating in the sport. He also knew that we did not think it was an absolute necessity. Aaron was not a basketball star, but he had good fundamental skills that he had potential to build upon and have fun with. But Aaron wasn't willing to work; he wasn't willing to give it a try.

Sometimes the negative results of a decision don't manifest themselves immediately. That's what happened with basketball. After two years of participation in high school, he wanted to stop playing. We allowed him to make the choice not to be involved in it.

And we knew that it probably was not a good choice and one he would regret. (See Aaron's comments on this decision at the end of the chapter.) But it is important to allow your child to make choices, especially if they are not a matter of life and death. Bad choices can teach lessons. There did not have to be any consequences imposed by us for Aaron's decision; the decision itself created the logical consequences. We knew that we were not in the business of raising basketball players but in the business of helping our strong-willed child reach maturity as a man of good character.

"A life of obedience is not a life of following a list of do's and don'ts, but it is allowing God to be original in our lives."[3]

VONETTE Z. BRIGHT

One of the biggest conflicts Aaron encountered in high school was definitely a result of his strong-willed nature, but it was his strong willed nature under control and used for good! Aaron used his strong will as a wonderful asset by directing it in a positive way.

During his senior year, Aaron was the president of the local chapter of the Fellowship of Christian Athletes (FCA). They met every other Wednesday evening in the basement of a local business. When Aaron's senior year began, he requested that a notice of the meeting be put into the school announcements. His request was denied, stating that the school did not highlight any activities that were not specifically associated with academics. Was that response accurate? Not really. At the same time that the FCA announcements were NOT being read, announcements about teen dances at the American Legion and applications available for babysitting opportunities were read. "Hmmm," Aaron wondered, "how were local teen dances and babysitting specifically associated with academics?"

After consulting us, he contacted the Center for Law and Justice and discussed his situation with a lawyer that was on staff. This gentleman assured him that the school could not deny his announcement if it was including others that were outside the realm of academics. Aaron and his vice president, Jackie (the same young lady who had declared him smarter than the teacher during the coup incident), began to collect copies of the daily announcements, gathering evidence for their claim.

The lawyer sent Aaron booklets containing the pertinent laws and various legal interpretations. Enough booklets were received to potentially distribute to each administrator and every member of the board of education, which included his very own father. John was serving on the school board at that time. As we perused the information, it was very clear that the school was in the wrong for denying the inclusion of FCA announcements. There was truly no gray area because of the other announcements they allowed.

Interestingly enough, we already knew that the school did not include the FCA announcements before Aaron brought it to our attention. Matthew was FCA president two years before. When Matthew asked for their announcements to be included, his request was also denied. The difference was that he did not think it was a battle worth fighting. He just found an alternate way to get the word out to the students. When John began his first term on the school board a year later, he determined that the refusal to make FCA announcements was not the most pressing problem to address. So the issue was on hold until Aaron brought it up. "It is not fair. It is not right. In fact, it is not legal for the administration to deny the rights of the FCA organization," he declared. Aaron was right, and John could not deny it. This became an issue that would be brought by John to the school board.

Sometimes I lose sleep because of my vivid imagination. In this case, I could just imagine the possible headlines in the local paper—School Board Member's Son Sues School! Yikes! I wished that Aaron and the administration had not come to loggerheads, but Aaron was right to stand up and demand that the law be followed. I recall that I had a few fitful nights' sleep until the school board, without Aaron's even attending a meeting, ruled to allow the FCA to participate in the school announcements. It was a battle that Aaron fought (with the help of his father) and won. And he did it with maturity and integrity. Very few people ever knew that it was a battle, only the two or three student leaders of FCA, their parents, the administration, school board, and school lawyer. When it was said and done, it was a victory, not just Aaron or for the local FCA chapter, but also for the rights of Christian organizations—rights that are often ignored by others.

"Some men march to the beat of a different drummer; and some polka."[4]

ANONYMOUS

Have you ever wondered if standing up for what is right, even in the face of strong and powerful opposition, is the right thing to do? Does it make a difference? At the end of the year, at senior awards night, Aaron received the Principal's Award for leadership. (She obviously observed something she appreciated.)

One of the characteristics of a strong-willed child is an intense degree of loyalty. This manifested itself in varying ways throughout the years. In high school, when one of his female friends who was insecure in a particular area came under unfair attack from a teacher, Aaron was quick to defend her by attacking her attacker—verbally, of course. It wasn't the best decision, and, several apologies later, Aaron seemed to be back on track. The attribute of loyalty was manifest in an extreme way, which ended up being the wrong way. But that was happening with less frequency.

Another strong-willed characteristic is the courage to take chances. Aaron was not afraid of the unfamiliar and had a degree of confidence to try new things that exceeded many of his peers. The most obvious example occurred when he determined that he would try a new summer job, rather than repeat his former one in an office cubicle, trapped with an adding machine and a ledger sheet at the local CPA. He purchased an ad in the *Thrifty Nickel* and launched his own business, A & L Horses. The "A" was for Aaron, and the "L" was for Lady, the horse his grandfather purchased for him during the summer before fourth grade. He had no guarantee that anyone would be interested in having him "break and train horses" or "give beginning riding lessons," but he wanted to give it a try.

With college right around the corner, Aaron was aware of his need to generate income that summer, and he was sure he could do it in the horse business. A & L Horses was an extension and expansion of what he had begun years earlier with his encourager Ed. But now he had the advertisement of a sign at the end of our lane, business cards, and stationery. Not bad for an eighteen-year-old risk-taking entrepreneur. I still remember Matthew going to work that summer and wondering out loud why Aaron didn't have to work. Aaron discovered the key. Find something you like to do. Find someone who will pay you to do it, and you'll never "work" another day in your life.

How about your strong-willed child? Where can you allow him to take chances that are not a matter of life and death? What encouragement can you give your strong-willed child to try something intriguing though unknown? Confidence is a terrific attribute that you can applaud.

A Closer Look with Aaron

My dad loves basketball. He was a very successful high school coach when he was teaching school. And he could see that I had the potential to do well and enjoy participating in the sport. But I didn't want to do it, because it was an activity that I didn't originally choose. There was no doubt in anyone's mind that I decided that I didn't like basketball. It must have been hard on him, but to his credit, he let me make the decision whether or not to stick with it. Actually, I was shocked! I chose to stop playing after my sophomore year. It was a bad choice, and I even knew it at the time. It didn't matter though; I wanted control. I quit for the satisfaction of being in control. I know this is difficult for anyone who is not strong-willed to understand, but, trust me, it is how we strong-willed people think.

My dad could have made me do it, but he didn't. With the freedom to choose came the responsibility for my actions and, in this case, the disappointment. Up to that point, I was seldom disappointed in my own behavior, even though my parents were at times. Now I was given the opportunity to make the decision, and I was responsible for my choice. Before, it always seemed that the result (in many cases the punishment) was the consequence of someone else's decision. I didn't completely make the association between my actions and the ultimate outcome. Because of that disassociation, I could be angry with someone else, typically the person delivering the consequences. With this particular decision, I could only be angry with myself.

What actually happened was that my parents allowed me to make a poor choice. I was allowed to fail. It was all my responsibility. I realize that this is a delicate dance for most parents. Obviously, no parent wants to see his child fail, but allowing failure is unbelievably important in the journey toward maturity. I'm not advocating that control of each and every choice should be given to your strong-willed child. That would be irresponsible.

Instead it is more like this: A young child sees a shiny knife on the counter, and he begs and cries for control of that "pretty thing." Of course, the responsible parent does not give in to those tears. As the child grows older, the parent teaches the child about the potential hazards of a knife and also gives him instructions about how to use it safely. Ultimately, the child is allowed to use a knife. The odds are that the trained young person will not do permanent damage to himself or something else. He might, however, be responsible for an occasional nick or cut. Do you see the analogy? You want to protect your child from harm until he is adequately instructed and able to understand the consequences of a mistake. Then you give him the responsibility to suffer the consequences or to enjoy the rewards of his own decisions. You see, it is twofold. The strong-willed child who is never allowed to fail is not only protected from disappointment, he is also sheltered from the positive results of his decisions.

As far as the incident with the Fellowship of Christian Athletes, that was a real adventure! Again my parents made the choice to allow me to pursue this matter. I am sure that being on the board of education made the possible confrontation very awkward for my dad. In a sense, he chose whether to remain comfortable or to face possible attack. In my black-and-white way of thinking, his choices were to (1) do what was right or (2) ignore the issue. So obviously, in my opinion, there was really no choice at all. There was only one correct answer. I am grateful that Dad made the "right" choice. I really think I'd still be mad today if he had not. After all, it was a legal issue and not just my opinion. But what about a different issue that your strong-willed child determines is important but is not as obvious and not a legal matter? When that is the case, you must make the decision about whether or not to allow your strong-willed child to pursue the concern. You must use clear-cut reasoning. It is important to be logical, honest, and compelling. Your own comfort will not be deemed an appropriate or persuasive reason to avoid the potential conflict. Remember, your strong-willed child is constantly on the lookout for people of character who stand for what is right. Be sure that you fall into that category.

"The two most important things we give to our children are roots and wings."

UNKNOWN

It is true that high school was a critical time for me as I learned to respond to situations rather than react to them. I was coming to grips with the truth that I had the option of choosing or NOT choosing to try and control a situation. Furthermore, I was aware of the fact that there were things I couldn't control. And I was learning that I could always choose to control myself.

Encouragers
and Discouragers

Where seldom is heard,
a discouraging word.
And the skies are not cloudy all day.[1]

✧ BREWSTER HIGLEY, "HOME ON THE RANGE" ✧

Wouldn't it be wonderful if those words from the song "Home on the Range" were completely accurate? Such a blessing if we seldom heard or said a discouraging word? Yes, it would be, but it is not reality. All of us have encountered, and been responsible for, discouraging words. And it is quite possible that strong-willed children have heard more discouraging words than their compliant counterparts. That possibility goes back to one of the assumptions vs. actualities—the incorrect assumption that these children are tough not tender. Because we have already discussed and debunked that assumption, we are aware that the discouragement faced by a strong-willed child is often taken to heart. Likewise, the encouragement that he receives will endear a strong-willed child to the encourager, potentially for a lifetime.

Aaron has a long list of people whom he would classify as encouragers and, thankfully, a shorter list of discouragers in his life. You have already met many of the people who would appear on these lists. You have been introduced to encouragers and discouragers. Rather than repeat those introductions, let's take a look at the common denominators. What traits and characteristics are shared, first of all, by the list of encouragers?

The most obvious shared quality is that these people did not embrace the assumptions typically made about a strong-willed child. For whatever reason, these folks saw the strong-willed nature for what it really is—a positive attribute. And they also realized that, like every other strength, it had to be controlled to be most effective. As I gazed at the list of encouragers, I was able to notice another generalization. Although the ages, professions, genders, political affiliations, and religious beliefs varied, all of the people who made the list of encouragers were "real." They did not pretend to be something they were not—they did not live a lie. Instead, they were mature and secure in themselves. They had nothing to prove, especially by demeaning or discouraging a child.

"Influence often isn't noticed until it blossoms later in the garden of someone else's life. Our words and actions may land close to home, or they may be carried far and wide."2

PAM FARREL

Are there people like that who can intersect with your strong-willed child? Yes, there are. Look for adults who are ready, willing, and able to let your strong-willed child know that he is loved. Find individuals who realize the potential of your child, recognize his talent, and appreciate his out-of-the-box thinking. Discover the people who will show respect for him and take him seriously—those who deserve and demand his respect.

What were the common denominators in the list of discouragers? In general, the discouragers believed the lies. They accepted one or more of the incorrect assumptions about a strong-willed child. For one reason or another, they did not take the time to search for the truth and apply it.

> "As a nonconformist throughout high school, I was left with a deflated sense of self-worth and ended up in a relationship that further eroded my self-esteem. It has been a long, tough journey, but with the help of God and certain individuals that could see my gifts, I now have a loving marital relationship, three great kids (one of which is a strong-willed child—the middle one), and an appropriate sense of self. I am college educated, have a good job, am a leader in my church, and am becoming a mentor to other women. I champion my strong-willed child's cause, having recently taken on her school principal in her defense. I don't want her to suffer the way I did."

"God does not dispense strength and encouragement like a druggist fills your prescription. The Lord doesn't promise to give us something to take so we can handle our weary moments. He promises us Himself. That is all. And that is enough."[3]

CHARLES SWINDOLL

In general, the list of discouragers contained people who were irritated and bothered by the unconventional thinking of a strong-willed child. They manifested this annoyance by frustration and disrespect. In some cases, discouraging adults actually teased and provoked Aaron. These discouragers tended to underestimate his skills and abilities and viewed his strong-willed nature as a liability rather than an asset. In general, those who made this list were insecure and lacked the maturity that one hopes comes with adulthood. Remember, just getting older does not make a person more mature.

For we are to God the aroma of
Christ among those who are being saved
and those who are perishing.
To the one we are the smell of death;
to the other, the fragrance of life.

❖ 2 CORINTHIANS 2:15–16 ❖

Peers have an influence on children and adults alike. Again, Aaron could list encouraging peers and discouraging ones. But more than by reaction to his strong-willed nature, these two groups were polarized because of Aaron's faith and their faith (or lack of it). The encouragers usually shared Aaron's love for Jesus, and the discouragers found it distasteful.

For the most part, family members were encouragers to Aaron. What about his brothers? In Aaron's case, Matthew and Jonathan were and are encouragers. In order to understand that phenomenon and to make an attempt to foster those dynamics in your own home, several concepts need to be presented and understood.

First of all, sibling rivalry is a possibility in any home. However, there are things that you can do to reduce the risk of its development. Rivalry implies competition. One of your missions should be to eliminate competition between your children as much as possible.

I know of a set of twins who were continually pitted against one another in competitive ways. Before they could even walk, their father would put them at one end of the living room, stand at the other end and have them crawl and race to him. And, destructively, he would then declare a "winner" and a "loser." By the time the twins were college graduates, they were, for all practical purposes, enemies. How sad!

In addition to eliminating competition whenever possible, it is also important to acknowledge that everyone has faults. Realizing that is a great help in developing and maintaining any human relationship. Shortcomings are not an excuse for misbehavior, but they are an opportunity to extend grace. I have always told the boys to allow at least three glaring faults in any individual. That practice greatly aids their ability to get along with others, even in their own family.

> "Brotherly love is still the distinguishing badge of every true Christian."[4]
>
> MATTHEW HENRY

Encouraging siblings to applaud each other's achievements is also an important deterrent to sibling rivalry. To this day, I hear Aaron bragging about the accomplishments of his brothers and vice versa. And in a family, there is a definite connection and correlation between accomplishments and family dynamics. The others in that family influenced one member's victory.

This leads me to a question that I am frequently asked: "Do you think my other children are negatively affected by having a sibling who is strong-willed?" Time after time, my answer to that question is emphatically, "No. Simply having a strong-willed family member is not a negative factor." Every single family has their own dynamics, and each member contributes to the whole package. Does a strong-willed child "ruin" the family dynamics? No, not any more than any other family member does. Let me give you an example by introducing you to a family I know.

Greg and Joan have two children. They are two years apart. Phillip is the older child and Elizabeth his younger sister. The dynamics of their particular family were not affected by one of the children being strong-willed. Instead it was something much more dramatic. Within hours of Elizabeth's birth, the doctor announced that she was profoundly mentally retarded. Today, at twenty-two years of age, Elizabeth is unable to walk, talk, or personally meet her basic daily needs. Yet I have never once heard her parents ask, "Do you think Phillip is negatively affected by having a sibling who is profoundly mentally retarded?"

Elizabeth needed, and actually still needs, more time and energy than Phillip. She is definitely more of a daily challenge. But she contributes to making their family what it is. Elizabeth influences the very nature of their family unit. And her parents would be the first to tell you that her influence is positive.

I specifically asked Matthew about how he interpreted the impact of having a brother who was strong-willed. And I quote some of his responses:

"I only felt deprived or negatively affected by it when we practiced basketball. Aaron took more than his share of time and attention, because he didn't want to practice. I *did* want to, and he distracted Dad. That affected my fun . . .

"As far as discipline, I never thought he was disciplined unfairly (although sometimes I wanted to be involved in the judiciary process). He deserved to be spanked! I enjoyed making sure he got punished, but, ironically, I felt bad when it actually happened . . .

"Sometimes I was embarrassed for him, wondering why having control was so important . . . important enough to alienate other people. I liked to be liked, so it was hard to understand . . .

"The summer after my senior year in high school we took a big family vacation to Alaska, and during that time it dawned on me that it was great to have Aaron around. We had actually become friends . . .

"It was obvious to me that Aaron and I didn't think the same way, but by the time I was ending high school, and definitely when we were both in college, I realized that it was good to be strong-willed. He was confident and could accomplish things the average person could never do!"

So what about your family? Does the fact that your strong-willed child is more high maintenance than his siblings negatively affect your family? It certainly doesn't have to if you are realizing the blessing you have been given in this particular child, this gift from God. There *may* be a negative influence on your family, however, if you have failed to avoid some of the roadblocks and detours along the way. There are several potential hazards that can turn your family dynamics upside down.

One of those obstructions is believing that you must deal with each child in your family equally. What? Isn't that the right thing to do, to make everything equal? No, equality is not the key; it is equity. Equity implies justice, fair play, and impartiality. Equality is uniformity and sameness. I have yet to meet two children whom I can treat with complete uniformity. Why? It is because no two children, strong-willed or not, are exactly the same. Equity in a household does not mean cookie-cutter responses from the parents. The compliant child probably does not require the consistent parental monitoring that a younger or older strong-willed child requires. Likewise, a strong-willed child may not be reprimanded in one arena or another because of the desire of the parent to choose the battles wisely.

Another potential rut in the journey of your strong-willed child is to allow his siblings to make him the scapegoat in all circumstances. Jonathan, Aaron's younger brother, learned from his oldest brother at a very early age to declare, "Aaron did it," whenever it might seem applicable. Did Aaron "do it"? Well, if I were a gambler, I would go with "yes." Did Aaron *always* "do it" when he was accused by his brothers? Absolutely not! Don't allow your strong-willed child to take the rap unfairly. This will alienate him from you and his siblings. In the same way, avoid the peril of disciplining every one of your children when the discipline should only be directed to the strong-willed child. Many times the convoluted thinking is that if the siblings share the responsibility of the others' good behavior and share the punishment with the guilty strong-willed child, then they will apply positive "peer/sibling pressure." I suppose that could happen, but the parent who gives a blanket punishment to all involved runs the risk of polarizing the children. It makes them not only frustrated with the discipliner, but also very angry with their strong-willed sibling. The chances of constructive adult relationships between siblings later in life diminish.

"Life affords no greater responsibility, no greater privilege, than the raising of the next generation."[5]

C. EVERETT KOOP

And finally, here is a roadblock that I encountered and needed to overcome. Since our other two sons were not strong-willed, I would ask them to do a certain task, because I knew it would not be a battle. I avoided asking Aaron to do his share of the family chores, ones he should have been expected to do, because I didn't want to engage in combat. This kind of behavior by a parent, if it goes uncorrected, can definitely cause the siblings of a strong-willed child to be negatively affected.

The key to fostering sibling love and encouragement is to be aware of the possible deterrents and to adjust and make corrections when they are needed. Remember, it is not a liability to your other children to be the brother or sister of a strong-willed child, unless you are failing to avoid roadblocks in the journey.

Never underestimate the power of your encouragement in the life of your strong-willed child. It is vital to your strong-willed child's well-being and successful journey. So, in a practical sense, what form should your encouragement take? Every parent can potentially apply the strengths of our original list of encouragers at the beginning of this chapter. The difference is that your intersection in the life of your child has more longevity, demands more patience and consistency, and is more important than any other person's contribution. Now is the time to remind you of the magnitude of the positive influence of your loving discipline. Never doubt the extent of your example—what you are modeling with your life. Always keep in mind that your strong-willed child is a gift from God, a gift whose possibilities are limitless. That is the mind-set of an encouraging parent.

Look for opportunities to encourage responsibility, to help your strong-willed child realize his significance in your family and ultimately in the world. For Aaron, we were able to encourage him in his love of animals. Lady, the horse, was a noteworthy encourager in her own right. Likewise, we surrounded Aaron with caring adults and included him in meaningful conversations with them as soon as he was able. We were very fortunate to have a supportive extended family. Aaron's grandparents, who live less than two miles away, were hands-on encouragers to him as were his other grandmother and relatives from Iowa to Indiana to Tennessee. They supported Aaron and applauded his abilities and talents. We believed that God had a special plan for Aaron. And we believed that He would "carry it on to completion until the day of Christ Jesus" (Philippians 1:6). We believed in Aaron, and he *knew* we believed in him. Do you believe in your strong-willed child? Does he *know* you believe in him?

Inevitably, we also made mistakes. That is an unavoidable part of the human experience. Rather than allowing that potential roadblock to paralyze you in your parenting, the key is to make corrections for your mistakes whenever possible, as soon as possible.

By far the most important thing we did as the parents of a strong-willed child (or any child for that matter) was to pray. We prayed for Aaron and about Aaron. We asked for patience, wisdom, guidance, and blessings. And we listened and read the instruction book for parenting and life in general, the Bible.

"When we pray, it is far more important to pray with a sense of the greatness of God than with a sense of the greatness of the problem."[6]

EVANGELINE BLOOD, WYCLIFFE BIBLE TRANSLATOR

In a very real sense, the potential encouragers and discouragers of a strong-willed child have varying degrees of influence based on their relationship to that child. We have been climbing the ladder of impact. Less significant are those people outside of your own home who intersect the life of your child. More influential are his extended family members, then his siblings, and finally, his parents. By far, the most powerful and positive encouraging influence can be found in your strong-willed child's relationship with the Lord, something you can influence through your relationship with God.

I've said it before and I'll say it again, parenting any child is not easy. And a strong-willed child is not just any child. Being the parent of a strong-willed child requires more patience, more persistence, more consistency, more wisdom, more knowledge, and more understanding. There is no formula, biblical or otherwise, to guarantee that your strong-willed child's journey will be a successful one to adulthood. But realizing and communicating to your child his value in God's eyes, helping your strong-willed child internalize the overflowing love that God has for him, and acknowledging the fact that God has a purpose and plan for his life can contribute to the journey toward responsible adulthood.

"A demanding spirit, with self-will as its rudder, blocks prayer ... Prayer is men cooperating with God in bringing from heaven to earth His wondrously good plans for us."[7]

CATHERINE MARSHALL

A Closer Look with Aaron

It was interesting for me to reflect on the various individuals who positively and negatively affected my life. In some ways it was difficult because the discouragers stirred up some painful memories. It never ceases to amaze me that I can bring to the surface feelings from incidents that occurred so many years ago. One of the realities of being strong-willed is that your emotions are more intense—both positive and negative ones. When my older brother and I compare "war wounds" from our high school days, I realize that it took a lot more effort to inflict pain on him than it did for me. He tended to analyze the facts of

an interaction while I reacted to the emotion behind
it. These intense emotions that a strong-willed child
feels can be rough on his parents.

I remember one particular example from my early
high school days. It was summer, and I had gotten a
new horse to break and train. He was a paint horse,
and I named him Pablo Picasso, in honor of his mod-
ernistic coat. He was also the most stubborn horse I
ever encountered. He absolutely refused to move in a
forward direction. After I saddled him and mounted
the saddle, he would back up until he finally backed
into something (usually our machine shed), and then
he would kick like crazy. His behavior baffled me. I'd
never known a horse whose only gear was reverse. I
tried this tactic and that tactic. And finally one night
when I was completely out of strategies, I'd had it!!

I remember my mom coming out of our house, just
when I had reached the end of my rope. I was completely deflated and discouraged by that silly horse. I
was done with horse training, done with horseback
riding, done with anything and everything that had to
do with horses. It was obvious to me that I was a complete and utter failure in this department. With uncon-
trolled emotion, I poured out my frustration and pain
to Mom. The next memory I have is the two of us sitting on top of the trampoline in our backyard. Mom
was calmly and rationally discussing what things I
might be able to do with Pablo Picasso. She reminded
me of many other successes working with animals.
And she generally soothed and comforted my wounded
spirit. Somehow, in her words of kindness, she was
able to encourage me not to quit but to try yet one
more idea. She convinced me that I was NOT a failure,
but I had merely determined the various training

methods that would not work with this horse. (I think that's when she used the analogy of how many ways Edison had discovered NOT to invent the lightbulb.)

The result? My emotions were finally brought back under control, and I was encouraged that there was hope for me, for my horse skills, and for my small business. Encouragement is powerful. Encouragement from your parents is incredibly powerful. My mom thought I could be successful in my endeavors and genuinely believed that now was not the time to panic or give up. Did she have this great insight because of her equine knowledge and ability? Hardly! My mom knows horses have four legs and, well, actually, that might be about it! What she did know was that, with a little encouragement to combat the discouraging behavior of Pablo Picasso, her son could be successful.

Mom suggested to you earlier that because of the tender nature of a strong-willed child (and I would add because of the intensity of their emotions) they are easily hurt by discouragers and extremely loyal to their encouragers. That is true. Even today, if one of the teachers who made my list of encouragers called and needed for me to drop everything in order to do something for him or her, I would make every effort to do so. My appreciation and loyalty probably exceed the normal range, and I think that is true for all of us who have made the journey of a strong-willed child. I am very tuned in to the emotions and needs of those I love. My parents understood the powerful impact of encouragement, and our adult relationship reflects my appreciation of the safe home, the refuge, boundaries, and love they provided. The connection made with a strong-willed child is strong and lasting. It is all the more reason, it would seem, for the parent of such a child to strive to encourage and connect with him in positive ways.

> The blessing of a strong-willed child is that she does *everything* with great passion. She loves her family with great passion and reminds us constantly of the good, both in her life and in ours.

9

It Goes On and On:
College and Beyond

Freedom is not the right
to do as a person pleases,
but the liberty to do as he ought.

❖ CICERO ❖

You are now hearing from me, Aaron. I am officially switching seats with my mom. Now I am in the driver's seat of my journey; Mom and Dad are just passengers. That's how it is when your kids reach eighteen. They are, for all practical purposes, adults. By the end of high school, I was ready to take the controls, and the folks seemed to be pretty comfortable with me at the helm. Comfortable or not, when your strong-willed child (or any other child, for that matter) graduates from high school, he is no longer a child.

At my high school graduation, there were many things to celebrate. I'd been accepted at the University of Illinois, majoring in animal science/pre-vet. I tried out for the Marching Illini and secured a spot in that prestigious marching band. I was going to live on campus in a fraternity with my older brother, Matthew. And the girl of my dreams was still in the picture. Those successes can all be attributed to the effort put into my ongoing strong-willed journey, a journey that never really ends but merely gets smoother and smoother.

College was a challenge that demanded the use of my strong-willed persistence. Needless to say, I was well trained in the area of perseverance, and it paid off. I set a goal of my own choosing. I wanted to get into vet medicine school. With the same tenacity I formerly applied to the less important areas of my life, I attacked this goal. I studied and worked hard to move toward my target of vet school. I remember the first time I thought that maybe I could actually achieve my objective. It was during the first week of classes. I went to college wondering if my academic preparation was adequate. Would I be able to compete with students from bigger schools, schools that had AP (advanced placement) classes and fancy lab equipment? I was nervous until it dawned on me that the professors spoke the same language that I did. I know it sounds silly, but somehow I imagined that I would not even be able to understand the lectures. When I realized that was a myth, I gave it my all. There was nothing to lose by working my hardest. And that is exactly what I did. School was a top priority—not the Marching Illini, not the fraternity, not even my girlfriend, Kristin. I was pursuing my goal with the diligence of a strong-willed individual.

The Marching Illini was, in one sense, a good diversion. I liked to play my horn, and being in the marching band certainly gave me that opportunity. In fact, it was very close to "too much of a good thing." We practiced two hours each day, Monday through Friday and four hours on Tuesday. And then, of course, there were the games all day on Saturday. Plus, I spent a few hours practicing each week so that I would make the Thursday cut. Many times I questioned the wisdom of devoting long hours to something that wasn't directly a part of reaching my primary goal. Because of my strong-willed drive, however, I stayed with it and made the cut each week for the two years I played in the band. I suppose it was a positive break from my obsessive studying. Also, it provided me with another link to my older brother. Matthew was a student coach for the football team, so we intersected even more. For me, the band was sort of a "brother thing."

So was the fraternity. I lived there only because he did. Neither one of us were very committed Greeks. In fact, the fraternity house gave me one last fling at over-the-top strong-willed behavior.

You see, there were some men in our fraternity house who were making very poor decisions. It was voted that our house would be "substance free." That meant no tobacco, liquor, or drugs on the property. Unfortunately, a few guys decided to violate the rule and went as far as smoking pot in the house. That definitely stirred my sense of right and wrong. As far as I was concerned, it was a black-and-white issue (remember how I love those?). Something had to be done. I led the charge to have these men punished for their poor choices. In retrospect, I look back at the battle I fought, and to some degree won, and I question whether or not it was worth the fight. It took a great deal of time and alienated a lot of the guys. But I also developed some genuine, lifelong friendships. The alienation didn't really matter to me, because I saw the goal as more important than their perception of me. By this time in my life, I was used to people jumping to conclusions about who I was and why I did what I did. Those conclusions were usually far from accurate. The two most significant people in my life at that point, my brother and my girlfriend, thought what I did was right, so I walked away from that battle with only a few minor scars.

I told you that I developed some solid friendships during my college years. Interestingly, they were friends that, for the most part, I shared with my older brother. When I reflected on that phenomenon, I realized that because of my rabid study habits, I spent little time exploring friendship possibilities. It was much easier to simply adopt the friends Matthew had already pre-classified and "broken in." That worked very well for me, and I appreciated the camaraderie. Unfortunately, it dawned on me at the beginning of my junior year that most of these friends had graduated and were gone.

My best friend, my girlfriend, Kristin, was still in school, but she attended another university about one hour away. My junior year was a lonely school year for me, but with academics as my primary focus, I had plenty of time to devote to that area. And that is exactly what I did. One of the things I always appreciated about Kristin was her willingness to take a backseat to my immediate goal of getting into vet school. Many dates were spent at the library or studying in the fraternity house. She was patient and supported my sometimes out-of-control preoccupation with my classes. To me that was love in action. I knew that having my classes as my primary focus was what it would take to get accepted to vet school, and she supported my dream.

And that was not the only thing Kristin knew about me. She was well aware of my strong-willed nature and shared the opinion that, when under control, it was a wonderful attribute. She had witnessed that "strength carried to extremes," and I guess (as my mother would say) that was one of the glaring faults she allowed. Interestingly, Kristin was in my grade school class the year that I organized the coup and took over. Even more interesting is the fact that she cannot remember the incident! Now that is love in the form of selective forgetfulness! Kristin's gracious spirit, her love of me and of Jesus endeared her to me more and more each day.

In October of my junior year I applied for early admission to vet school. And then I waited. Before too long, I received notice that I was granted an interview. This was a very significant step. Approximately four hundred students interviewed. Around one hundred students would be admitted. The day I was notified that I was officially a member of next year's class of first year vet students I was ecstatic! My perseverance and persistence paid off. My determination and bulldog-focus led to success. My strong-willed nature was a tremendous asset and helped me reach a goal that I'd looked toward since third grade.

I am a vet student now. It is not easy. But it is the next step in my journey. I am also a husband. Kristin and I married in August before I began school. For a few years I had the encouragement and assistance of many, but I was primarily alone in my responsibility for making forward progress on my journey. Now Kristin and I are traveling together. My parents are and always will be supportive, but they no longer steer the journey nor determine the route. In my childhood they helped navigate. They lovingly disciplined and guided me. They cheered me on at every possible juncture. And they still support me with their prayers. Ultimately, every strong-willed child must travel alone. The key is to equip your child and prepare him for the time when you, his parents, will no longer be in the driver's seat. Your child will take control of the journey, something he has always longed to do.

A Closer Look with Kendra (Mom)

Wow, relegated to the comments on the chapter, things really do change. And I am glad for the demotion. Every parent, one with a strong-willed child or otherwise, has the desire to work himself out of a job. From the very beginning, it was our hope to see Aaron mature into a capable, well-adjusted adult. We knew that once he was able to master control of his strong-willed nature that he would be able to use it for good. I have seen his tenacity in many arenas of life, like in his career pursuit, leadership, quest for what is right, love of others, and faith. I look back on the journey and realize that every battle chosen and won, every discussion of right and wrong, and every identification of his strengths being carried to extremes was worth the time and energy it demanded.

And I also realize that I am looking at these things

in hindsight. The chances are great that you are presently in the infancy of the journey with your strong-willed child. You have no history to guide you, and no benchmarks to celebrate, at least not yet. Take heart. That is the reason that my husband, John, Aaron, and I have spent hours thinking, praying, writing, and rewriting. It is for you—for you and your strong-willed child. My desire is to hear from you many years from now, after your strong-willed child has reached adulthood. I hope to hear that you are no longer the parent of a strong-willed child but instead are the parent of a capable, well-adjusted adult.

Conclusion:

Do You See What I See?

We may run, walk, stumble, drive or fly,
but let us never lose sight
of the reason for the journey
or miss a chance to see
a rainbow on the way.[1]

❖ GLORIA GAITHER ❖

We have come to the conclusion. Not the conclusion of the journey, but the conclusion of the book. I have again taken over the pen as we bring this narrative to a close. The birthing of this book has brought me both pleasure and pain, much like the birthing of my children. And, just as most other mothers feel, the pleasure has overshadowed the pain. Joy remains long after the memory of discomfort has vanished.

This year, Aaron is the teacher of the fourth-and-fifth-grade Sunday school class at the church he and Kristin attend. One day he and another teacher were chatting, and the conversation turned to this book. Aaron shared one or two stories, and the other teacher asked him a question that stymied him, at least momentarily. "How is it that you can be so introspective and so transparent?" Aaron told me about the interchange and said that at the time he had no answer. Then after some thought and prayer, he realized that it was really very simple. "I am able to share about my faults and how I stumbled in my journey because of the grace of God. I made mistakes and will undoubtedly continue to make mistakes, just like everyone else. We are all sinners saved by grace. Without God's grace no one would be forgiven from strong-willed antics or compliant failings."

"Grace is the good pleasure of God that inclines to bestow **benefits** upon the undeserving."[2]

A.W. TOZER

One Easter Sunday I heard a wonderful sermon at the little country church we attend. Pastor Tim Kovalcik preached on John 20:19–30, the story of doubting Thomas. Thomas demanded proof of Jesus' resurrection: "Unless I see the nail marks in his [Christ's] hands and put my finger where the nails were, and put my hand into his side, I will not believe it." Rather than dwelling on the doubt of Thomas, our pastor talked about the fact that Christ met with Thomas after the resurrection and *still* bore the scars of the crucifixion. Why wasn't the risen Lord in a perfect and whole body? It was because He was able to minister through His scars. And so are we. Everyone has scars and wounds from the journey of life. We usually choose to hide those blemishes from others and even try to ignore them ourselves. But God's grace empowers us to persevere through our journey, with all its road-blocks, bumps, setbacks, and detours. We can share the stories behind our scars in hopes that we can minister through them.

It is God's grace that enables us to be authentic and forgiven despite the fact that our lives have been fault-filled. Never underestimate the power of grace in your own life, the life of your strong-willed child, or the lives of those you intersect.

"Nothing we can do will make the Father love us less; nothing we do can make Him love us more. He loves us unconditionally with an everlasting love. All He asks of us is that we respond to Him with the free will that He has given to us."[3]

NANCIE CARMICHAEL

We have taken you through over twenty years of our journey. And, hopefully, you have been encouraged, challenged, and enlightened en route. Now I leave you to travel your own journey. But before this book is closed, and it comes to permanent rest on a shelf, let me share one final thought, beginning with a closer look with Jesus.

"If we love people, we will see them as God intends them to be."[4]

FLORENCE LITTAUER

A Closer Look with Jesus

Matthew 4:18–20: As Jesus was walking beside the Sea of Galilee, he saw two brothers, Simon called Peter and his brother Andrew. They were casting a net into the lake, for they were fishermen. "Come, follow me," Jesus said, "and I will make you fishers of men." At once they left their nets and followed him.

Jesus was walking beside the Sea of Galilee. He saw two brothers, Simon called Peter and his brother Andrew. These two men were casting a net into the lake, a very logical thing for two fishermen to do. Undoubtedly Peter and Andrew looked like fishermen, smelled like fishermen, and their actions supported the notion that they were fishermen. If you and I were walking on the same path as Jesus, walking beside the Sea of Galilee, we would have seen these two brothers. We would have known that they were fishermen. We would have recognized the obvious.

But Jesus saw something more. He saw beyond the obvious. His vision wasn't limited to the apparent. He saw all that you and I would see, but He also saw more. We would just see fishermen. Jesus saw disciples.

Look into the eyes of your strong-willed child. What do you see? Do you only see the obvious—the out-of-the-box thinker who is demanding and difficult? Do you just see an annoying child who is so "right" all the time? Ask God to help you see your child through His eyes. Look through the eyes of Jesus. You just might see a disciple.

"Oh the comfort, the inexpressible comfort of feeling safe with a person; having neither to weigh thoughts nor measure words but to pour them all out, just as it is, chaff and grain together, knowing that a faithful hand will take and sift them, keeping what is worth keeping, and then, with the breath of kindness blow the rest away."[5]

MARIAN EVANS CROSS (GEORGE ELIOT)

Notes

Introduction—Aaron and Others

1. James Dobson, *The Strong-Willed Child* (Wheaton, Ill.: Tyndale House, 1978), 24.

2. Samuel Butler (1612–1680), *Hudibras* Part iii. Canto iii. Line 547, from *Familiar Quotes*, comp. John Bartlett, Familiar Quotations, 10th ed., rev. and enl. by Nathan Haskell Dole (Boston: Little, Brown, & Co., 1919).

3. Tim LaHaye, *Spirit-Controlled Temperament* (Wheaton, Ill.: Living Studies, 1982).

4. Florence Littauer, *Personality Plus* (Grand Rapids, Mich.: F. H. Revell Co., 1992).

5. Corrie ten Boom, *A Woman's Journey with God* (Nashville, Tenn.: Brighton Books by Mary Prince, 2001), 21.

Chapter 1—The Journey Begins: Birth to Pre-Kindergarten

1. William Shakespeare (1564–1616), Launcelot, in *The Merchant of Venice*, act 2, scene 2, lines 72–73 (1600). "Launcelot repeats a proverbial saying as he attempts to make himself known to his blind father, Gobbo." *The Columbia World of Quotations.* (New York: Columbia University Press, 1996).

Chapter 3—The Journey Continues: Kindergarten to Grade Six

1. Shirley Dahlquist, my friend, deceased, former resident of Como, Colorado.

2. Sheila Walsh, from *Calendar—For a Woman's Heart: Thoughts by Women for Women* (Bloomington, Minn.: Garborgs, 1998, 1999).

3. Beth Moore, *A Woman's Journey with God* (Nashville, Tenn.: Brighton Books by Mary Prince, 2001), 110.

4. Norman Vincent Peale, from *Minute Motivators for Teachers*, ed. Stan Toler (Tulsa, Okla.: RiverOaks Publishing, 2002), 10.

Chapter 4—Assumptions vs. Actualities

1. Carol Lynn Mithers, "The Perils of the Pushover Parent," *Ladies' Home Journal* (January 2003) 92, 94–96.

2. Helen Keller, from *Calendar—For a Woman's Heart: Thoughts by Women for Women* (Bloomington, Minn.: Garborgs, 1998, 1999).

3. Elisabeth Elliot quote, ibid.

4. Cynthia Tobias, *You Can't Make Me* (Colorado Springs, Colo.: Waterbrook, 1999), 24.

5. Louis Pasteur, from *To Your Success*, comp. Dan Zadra (Edmonds, Wash.: Compendium, Inc., 1997), 78.

6. James Dobson, *The Strong-Willed Child* (Wheaton, Ill.: Tyndale House, 1978), 21.

7. Pamela Smith and Carolyn Coats, *Alive and Well in the Fast Lane* (Nashville, Tenn.: Thomas Nelson Publishers, 1994), 23.

8. Adolf Hitler, Announcement to the German Army of his assumption of its command, December 21, 1941. Originally published in the *New York Times*, December 22, 1941. Found at www.ibiblio.org/pha/policy/1941/411221a.html.

Chapter 5—The Junior High Journey

1. Wayne Rice, *Understanding Your Teen* (Nashville, Tenn.: Word Publishing, 1999), ix.

2. William Shakespeare (1564–1616), Falstaff, in *Henry IV*, Part 1, act 5, scene 4, lines 119–20. Falstaff's "discretion" meant avoiding danger on the battlefield by pretending to be dead. *The Columbia World of Quotations* (New York: Columbia University Press, 1996).

3. Wayne Rice, *Understanding Your Teen*, gathered from a seminar on this book.

Chapter 6—Determining the Desired Destination

1. *431 Quotes from the Notes of Henrietta C. Mears*, ed. Eleanor L. Doan (Glendale, Calif., G/L Regal Books, 1970).

2. "Patience," *The Music Machine*, 25 min. (Bridgestone Group: Agapeland Recordings, 1990), VHS videotape.

3. Theodore Roosevelt, from *God's Little Instruction Book for Teachers* (Tulsa, Okla.: Honor Books, 1999), 132.

4. EDGE Learning, *To Your Success*, comp. Dan Zadra (Edmonds, Wash.: Compendium, Inc., 1997), 19.

5. Warren Wiersbe, quoted in *Parents and Teenagers*, Jay Kesler with Ronald A. Beers (Wheaton, Ill.: Victor Books, 1984), 409.

Chapter 7—The Journey Goes On: High School

1. Frances Kelley, from "Better Than I Was," in *Quotable Quotes*, comp. Helen Hosier (Uhrichsville, Ohio: Barbour Publishing Co., 1998), 121.

2. John M. Drescher, from "Now Is the Time to Love," in *Quotable Quotes*, 81.

3. Vonette Z. Bright, from "For Such a Time as This," in *Quotable Quotes*, 161.

4. Quote from *To Your Success*, comp. Dan Zadra (Edmonds, Wash.: Compendium, Inc., 1997), 16.

Chapter 8—Encouragers and Discouragers

1. Brewster Higley, "Home on the Range," from *The Western Home* (Originally 1873). Modern version by John Lomax, who, on his first trip west, recorded a black saloon keeper in San Antonio singing "Home on the Range" on an Edison cylinder, and the lyrics were written down and published in the book "Cowboy Songs and Frontier Ballads" by Lomax in 1910. The song became a national favorite and is the state song of Kansas. Steve Schoenherr, University of San Diego Department of History. Found at http://history.sandiego.edu/gen/recording/notes.html. First published April 18, 1995; notes revised March 4, 2003.

2. Pam Farrel, from *Calendar—For a Woman's Heart: Thoughts by Women for Women* (Bloomington, Minn.: Garborgs, 1998, 1999).

3. Charles Swindoll, from "Encourage Me," in *Quotable Quotes*, comp. Helen Hosier (Uhrichsville, Ohio: Barbour Publishing Co., 1998), 58.

4. Matthew Henry, from "Moments of Meditation," in *Quotable Quotes*, 155.

5. C. Everett Koop, from *Minute Motivators for Teachers*, ed. Stan Toler (Tulsa, Okla.: RiverOaks Publishing, 2002), 5.

6. Evangeline Blood, from *Quotable Quotes,* 176.

7. Catherine Marshall, from *Quotable Quotes,* 178. Originally published in *Adventures in Prayer* (Old Tappan, N.J.: distributed by F. H. Revell, 1975).

Conclusion—Do You See What I See?

1. Gloria Gaither, *Calendar—For a Woman's Heart: Thoughts by Women for Women* (Bloomington, Minn.: Garborgs, 1998, 1999).

2. A. W. Tozer, from *Quotable Quotes,* comp. Helen Hosier (Uhrichsville, Ohio: Barbour Publishing Co., 1998), 116. Originally published in A. W. Tozer, *Knowledge of the Holy* (San Francisco: HarperSanFrancisco, 1961).

3. Nancie Carmichael, *Calendar.*

4. Florence Littauer, from *Minute Motivators for Teachers,* ed. Stan Toler (Tulsa, Okla.: RiverOaks Publishing, 2002), 72.

5. Marian Evans Cross (George Eliot, pseud.) from *Quotable Quotes,* 90.

LIVE LIFE INTENTIONALLY!
This phrase describes Kendra's
life and the message that she brings
to her listening and reading audiences.
Kendra would be delighted to hear from you.
Contact her at: www.KendraSmiley.com

Upbeat Living in a Downbeat World

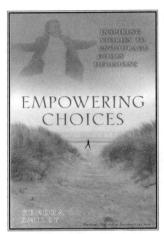

ISBN: 0-8024-4342-7

Empowering Choices
Inspiring Stories to Encourage Godly Decisions

Kendra Smiley introduces us to real women who faced real challenges but who opted to make the empowering choices of forgiveness, prayer, joy, and contentment. You'll laugh and cry when you read about these women who refused to let life's circumstances destroy their positive attitudes. Kendra's philosophy, exuberance and acknowledgement of Jesus Christ as the source of hope will make a lasting impact. Prepare to be encouraged!

High-Wire Mom
Balancing Your Family and a Business @ Home

You will not discover the perfect filing system, or how to evaluate a franchise agreement. You won't get tax tips or information on how to dress for success. Instead, you will see the valuable insights in God's Word as Kendra illustrates His wisdom with lessons she has learned from twenty-plus years as a mom and the owner of a home-based business.

To all moms who hear the call to have a home-based business—here is your tool to equip, encourage, and motivate from Kendra's heart to yours.

ISBN: 0-8024-1361-7

MOODY
PUBLISHERS
THE NAME YOU CAN TRUST.

1-800-678-6928 www.MoodyPublishers.org

AARON'S WAY TEAM

ACQUIRING EDITOR
Elsa Mazon

COPY EDITOR
Ali Childers

BACK COVER COPY
Lisa Cockrel

COVER DESIGN
Ragont Design

INTERIOR DESIGN
Ragont Design

PRINTING AND BINDING
Versa Press Inc.

The typeface for the text of this book is
Giovanni